Advanced Calculus and its Applications in Variational Quantum Mechanics and Relativity Theory

Fabio Silva Botelho

Adjunct Professor, Department of Mathematics
Federal University of Santa Catarina, Florianopolis SC, Brazil

CRC Press
Taylor & Francis Group
Boca Raton London New York

CRC Press is an imprint of the
Taylor & Francis Group, an **informa** business

A SCIENCE PUBLISHERS BOOK

First edition published 2021
by CRC Press
6000 Broken Sound Parkway NW, Suite 300, Boca Raton, FL 33487-2742

and by CRC Press
2 Park Square, Milton Park, Abingdon, Oxon, OX14 4RN

© 2021 Taylor & Francis Group, LLC

CRC Press is an imprint of Taylor & Francis Group, LLC

ISBN: 978-0-367-74645-2 (hbk)
ISBN: 978-0-367-74649-0 (pbk)
ISBN: 978-1-003-15891-2 (ebk)

Typeset in Times New Roman
by Radiant Productions

Preface

The first part of this book develops some topics on multi-variable advanced calculus. The approach presented includes a detailed and rigorous study on surfaces in $\mathbb{R}^n$ which comprises items such as differential forms and an abstract version of the Stokes Theorem in $\mathbb{R}^n$.

In the final of this first part, an introduction to Remannian geometry, which will be used in the subsequent chapters of applications, is developed. In its second part, the book presents applications of the previous chapters, specifically on variational quantum mechanics and relativity theory. Topics covered include a variational formulation for the relativistic Klein-Gordon equation, the derivation of a variational formulation for relativistic mechanics firstly through (semi)-Riemannian geometry, and in a second step, for a more general context, through fundamentals of differential geometry. In the final chapters, the book develops a new interpretation for the Bohr atomic model through a semi-classical approach. The book finishes with a classical description of the radiating cavity model in quantum mechanics.

Finally, it is worth emphasizing that the present book overlaps somewhat in its first two chapters with my previous book "Real Analysis and Applications, published by Springer in 2018, but only on a part related to basic mathematics. The applications, presented in the final section, comprise essentially new developments.

At this point, we start to present a summary of each chapter content.

Chapter 1–The Implicit Function Theorem and Related Results

The first part of this chapter presents the implicit function theorem in its scalar form. Throughout the chapter, various related topics are developed, such as a study on vectorial functions which includes the implicit function theorem in its vectorial form and the inverse function theorem. The final section develops some basic concepts on differential geometry. Topics such as arc length and curvature are presented in detail. The chapter ends with the statement and proof of a well known Gauss theorem concerning the curvature for a surface in $\mathbb{R}^3$.

Chapter 2–Manifolds in $\mathbb{R}^n$

This second chapter develops a study on surfaces in Rn. Among the main topics addressed, it is worth highlighting the issue of the boundary of a surface in Rn and the concerning outer normal field, the concept of differential forms and an abstract version of the Stokes Theorem. The chapter finishes with an introduction to Riemannian geometry.

Chapter 3–A Variational Formulation for the Relativistic Klein-Gordon Equation

This chapter develops a variational formulation for the relativistic Klein-Gordon equation. The main results are obtained through a connection between classical and quantum mechanics. Such a connection is established through the definition of normal field and its relation with the wave function concept.

Chapter 4–Some Numerical Results and Examples

In this chapter, in a first step, we develop an algorithm for solving some eigenvalue problems found in the standard quantum mechanics approach. Several numerical examples are developed in detail, including one about the well known hydrogen atom model. We emphasize that the numerical procedures presented are very robust from a numerical point of view and applicable to a large class of similar problems.

Chapter 5–A Variational Formulation for Relativistic Mechanics Based on Riemannian Geometry and its Application to the Quantum Mechanics Context

This chapter develops a variational formulation of relativistic nature applicable to the quantum mechanics context. The main results are obtained through basic concepts on Riemannian geometry. Standards definitions, such as vector fields and connection, have a fundamental role in the main action establishment. In the last section, as a result of an approximation for the main formulation, we obtain the relativistic Klein-Gordon equation.

Chapter 6–A General Variational Formulation for Relativistic Mechanics Based on Fundamentals of Dierential Geometry

The first part of this chapter develops a variational formulation for relativistic mechanics. The results are established through standard tools of variational analysis and differential geometry. The novelty here is that the main motion manifold has a dimensional range of n + 1. It is worth emphasizing in a first approximation that we have neglected the self-interaction energy part. In its second part, this chapter develops some formalism concerning the causal structure in a general space-time manifold. Finally, the last chapter section presents a result concerning the existence of a generalized solution for the world sheet manifold variational formulation.

Chapter 7–A New Interpretation for the Bohr Atomic Model

This chapter also develops a variational formulation for the relativistic Klein-Gordon equation. The main results are again obtained through an extension of the classical mechanics approach to a more general context, which, in some sense, includes the quantum mechanics one. For the second part of the text, the definition of normal field and its relation with the wave function concept play a fundamental role in the main results establishment. Among the applications, we include a model with the presence of electromagnetic fields and also the modeling of a chemical reaction. Finally, in the last section, we present some results about the Spin operator in a relativistic context.

Chapter 8– Existence and Duality for Superconductivity and Related Models

This chapter develops duality principles applicable to the Ginzburg-Landau system in superconductivity. The main results are obtained through standard tools of convex analysis, functional analysis, calculus of variations and duality theory. In the second section, we present the general result for the case including a magnetic field and the respective magnetic potential in a local extremal context. Finally, in the last section, we develop some global existence results for a model in elasticity.

Chapter 9–A Classical Description of the Radiating Cavity Model in Quantum Mechanics

In this chapter, we address a solution for a problem presented in the first chapter of the David Bohm book on quantum theory (our main reference in this chapter), namely, the problem of obtaining the equilibrium distribution of electromagnetic radiation in a hollow cavity. We present a classical description of such a result, originally established through the Planck hypothesis of radiation quantization in order to fit the theoretical results with the experimental data.

Acknowledgments

I would like to thank some people of Virginia Tech - USA, where I got my Ph.D. degree in Mathematics in 2009. I am especially grateful to Professor Robert C. Rogers for his excellent work as an advisor. I would like to thank the Department of Mathematics for its constant support and for providing this opportunity to study mathematics at an advanced level. Among many other Professors, I particularly thank Martin Day (Calculus of Variations), William Floyd and Peter Haskell (Elementary Real Analysis), James Thomson (Real Analysis), Peter Linnell (Abstract Algebra) and George Hagedorn (Functional Analysis) for the excellent lectured courses.

Finally, I am also pleased to express my gratitude to Professor Simon Crampin of the Physics Department of Bath University - UK, a colleague who, with his valuable suggestions and insights, helped me improve the quality of some important parts of this text, in an academic visit to Bath University between September 2016 and March 2017.

November 2019

Fabio Silva Botelho
Florianopolis - SC, Brazil

Contents

Preface   **iii**

SECTION I: ADVANCED CALCULUS

1. The Implicit Function Theorem and Related Results   **2**

    1.1   Introduction   2
    1.2   Implicit function theorem, scalar case   3
    1.3   Vectorial functions in $\mathbb{R}^n$ 8
          1.3.1   Limit proprieties   10
    1.4   Implicit function theorem for the vectorial case   17
    1.5   Lagrange multipliers   25
    1.6   Lagrange multipliers, the general case   28
    1.7   Inverse function theorem   32
    1.8   Some topics on differential geometry   46
          1.8.1   Arc length   48
          1.8.2   The arc length function   49
    1.9   Some notes about the scalar curvature in a surface in $\mathbb{R}^3$ 52

2. Manifolds in $\mathbb{R}^n$ **61**

    2.1   Introduction   61
    2.2   The local form of sub-immersions   61
    2.3   The local form of immersions   63
    2.4   Parameterizations and surfaces in $\mathbb{R}^n$ 65
          2.4.1   Change of coordinates   65
          2.4.2   Differentiable functions defined on surfaces   66
    2.5   Oriented surfaces   68
    2.6   Surfaces in $\mathbb{R}^n$ with boundary   71
          2.6.1   Parameterizations for surfaces in $\mathbb{R}^n$ with boundary   74
    2.7   The tangent space   80

2.8	Vector fields	85
2.9	The generalized derivative	89
	2.9.1 On the integral curve existence	92
2.10	Differential forms	98
2.11	Integration of differential forms	102
2.12	A simple example to illustrate the integration process	105
2.13	Volume (area) of a surface	107
2.14	Change of variables, the general case	111
2.15	The Stokes Theorem	113
	2.15.1 Recovering the classical results on vector calculus in $\mathbb{R}^3$ from the general Stokes Theorem	114

SECTION II: APPLICATIONS TO VARIATIONAL QUANTUM MECHANICS AND RELATIVITY THEORY

3.	**A Variational Formulation for the Relativistic Klein-Gordon Equation**	**128**
3.1	Introduction	128
3.2	The Newtonian approach	129
3.3	A brief note on the relativistic context, the Klein-Gordon equation	132
	3.3.1 Obtaining the Klein-Gordon equation	135
3.4	About the role of normal field as the wave function in quantum mechanics	139
3.5	A more general proposal for the energy, one more example	140
3.6	Another example, generalizing the standard quantum approach	144
3.7	Recovering the standard quantum approach	146
	3.7.1 About the role of the angular momentum in quantum mechanics	148
	3.7.2 A brief note concerning the relation between classical rotations and the quantum angular momentum	150
	3.7.3 The eigenvalues of L_z	152
3.8	A numerical example	154
3.9	A brief note on the many body quantum approach	156
3.10	A brief note on a formulation similar to those of density functional theory	160
4.	**Some Numerical Results and Examples**	**162**
4.1	Some preliminary results	162
4.2	A second numerical example	167
4.3	A third example, the hydrogen atom	170

4.4	A related control problem	177
	4.4.1 A numerical example	182
	4.4.2 Another numerical example	183
4.5	An approximation for the standard many body problem	185
4.6	Conclusion	188

5. A Variational Formulation for Relativistic Mechanics Based on Riemannian Geometry and its Application to the Quantum Mechanics Context — **189**

5.1	Introduction	189
5.2	Some introductory topics on vector analysis and Riemannian geometry	192
5.3	A relativistic quantum mechanics action	196
	5.3.1 The kinetics energy	198
	5.3.2 The energy part relating the curvature and wave function	199
5.4	Obtaining the relativistic Klein-Gordon equation as an approximation of the previous action	202
5.5	A note on the Einstein field equations in a vacuum	204
5.6	Conclusion	206

6. A General Variational Formulation for Relativistic Mechanics Based on Fundamentals of Differential Geometry — **207**

6.1	Introduction	207
6.2	The system energy	209
6.3	The final energy expression	213
6.4	Causal structure	214
6.5	Dependence domains and hyperbolicity	223
6.6	Existence of minimizer for the previous general functional	226
6.7	Conclusion	233

7. A New Interpretation for the Bohr Atomic Model — **234**

7.1	Introduction	234
7.2	The Newtonian approach and a concerning extension	234
7.3	A brief note on the relativistic context, the Klein-Gordon equation	238
7.4	A second model and the respective energy expression	241
7.5	A brief note on the relativistic context for such a second model	244
7.6	A brief note on the case including electro-magnetic effects	247
	7.6.1 About a specific Lorentz transformation	247
	7.6.2 Describing the self interaction energy and obtaining a final variational formulation	248
7.7	A new interpretation of the Bohr atomic model	252

	7.8 A system with a large number of interacting atoms	257
	7.8.1 A proposal for the case in which N is very large	259
	7.9 About the definition of Temperature	261
	7.10 A note on the Entropy concept	263
	7.11 About modeling a chemical reaction	264
	7.11.1 About the variational formulation modeling such a chemical reaction	265
	7.11.2 The final variational formulation	269
	7.12 A note on the Spin operator	271

8. Existence and Duality for Superconductivity and Related Models **275**

	8.1 Introduction	275
	8.2 A global existence result for the full complex Ginzburg-Landau system	277
	8.3 A brief initial description of our proposal for duality	279
	8.4 The duality principle for a local extremal context	282
	8.5 A second duality principle	288
	8.6 A third duality principle	292
	8.7 A criterion for global optimality	297
	8.7.1 The concerning duality principle	299
	8.8 Numerical results	303
	8.8.1 A numerical example	307

9. A Classical Description of the Radiating Cavity Model in Quantum Mechanics **311**

	9.1 Introduction	311
	9.2 Regarding an approximate classical description of the last section previous result	312
	9.3 Conclusion	316

References	**317**
Index	**321**

ADVANCED CALCULUS

I

Chapter 1

The Implicit Function Theorem and Related Results

1.1 Introduction

In this chapter we develop in detail the statement and proof of the implicit function theorem and some related results. The main references for this chapter are [23] and [39].

Consider first a function $f : D \subset \mathbb{R}^2 \to \mathbb{R}$ of C^1 class and let $(x_0, y_0) \in D^0$.

Assume $f_y(x_0, y_0) \neq 0$. Under such hypotheses, the implicit function theorem, to be stated and proved in the next lines, guarantees the existence of r, $r_1 > 0$ and a function $y : B_r(x_0) = (x_0 - r, x_0 + r) \to B_{r_1}(y_0) = (y_0 - r_1, y_0 + r_1)$, defined implicitly by the equation

$$f(x, y) = f(x_0, y_0) \equiv c,$$

that is,

$$f(x, y(x)) = c, \forall x \in B_r(x_0).$$

Moreover, also from such a theorem, $y'(x)$ exists in $B_r(x_0)$, so that, from the chain rule, we may obtain

$$\frac{df(x, y(x))}{dx} = \frac{dc}{dx} = 0,$$

that is,

$$f_x(x, y(x)) + f_y(x, y(x))y'(x) = 0,$$

so that

$$y'(x) = -\frac{f_x(x,y(x))}{f_y(x,y(x))}.$$

Similarly, for the case in which $f : D \subset \mathbb{R}^3 \to \mathbb{R}$ is of C^1 class and $(x_0, y_0, z_0) \in D^0$, if $f_z(x_0, y_0, z_0) \neq 0$, from this same theorem, we may obtain $r, r_1 > 0$ and a function $z : B_r(x_0, y_0) \to B_{r_1}(z_0)$ defined implicitly by

$$f(x,y,z) = f(x_0, y_0, z_0) \equiv c,$$

that is,

$$f(x,y,z(x,y)) = c, \ \forall (x,y) \in B_r(x_0, y_0)$$

Moreover, also as a result of the implicit function theorem, z_x and z_y exist in $B_r(x_0, y_0)$, so that, from the chain rule, we have,

$$\frac{df(x,y,z(x,y))}{dx} = \frac{dc}{dx} = 0,$$

that is,

$$f_x(x,y,z(x,y)) + f_z(x,y,z(x,y))z_x(x,y) = 0,$$

so that

$$z_x(x,y) = -\frac{f_x(x,y,z(x,y))}{f_z(x,y,z(x,y))}.$$

Similarly,

$$\frac{df(x,y,z(x,y))}{dy} = \frac{dc}{dy} = 0,$$

so that

$$f_y(x,y,z(x,y)) + f_z(x,y,z(x,y))z_y(x,y) = 0,$$

that is,

$$z_y(x,y) = -\frac{f_y(x,y,z(x,y))}{f_z(x,y,z(x,y))}.$$

So, with such a discussion in mind, we now introduce the implicit function theorem for the scalar case.

1.2 Implicit function theorem, scalar case

Theorem 1.2.1 *Let $D \to \mathbb{R}$ be an open set and let $f : D \to \mathbb{R}$ be a function. Assume f is such that, for $(\mathbf{x}_0, y_0) \in D \subset \mathbb{R}^n$, where $\mathbf{x}_0 \in \mathbb{R}^{n-1}$ and $y_0 \in \mathbb{R}$, we have:*

$$f(\mathbf{x}_0, y_0) = c \in \mathbb{R}$$

and

$$f_y(\mathbf{x}_0, y_0) \neq 0.$$

Suppose also f and its first partial derivatives are continuous on $B_r(\mathbf{x}_0, y_0)$, for some $r > 0$. Under such hypotheses, there exists $\delta > 0$ and $\varepsilon > 0$ so that, denoting $J = (y_0 - \varepsilon, y_0 + \varepsilon)$, for each $\mathbf{x} \in B_\delta(\mathbf{x}_0)$ there exists a unique $y_0 \in J$ such that

$$f(\mathbf{x}, y) = c.$$

Moreover, denoting

$$y = \xi(\mathbf{x})$$

we have that

$$\xi : B_\delta(\mathbf{x}_0) \to J,$$

is a C^1 class function such that, for each $\mathbf{x} \in B_\delta(\mathbf{x}_0)$, we have

$$f(\mathbf{x}, \xi(\mathbf{x})) = c,$$

and

$$\frac{\partial \xi(\mathbf{x})}{\partial x_j} = \frac{-f_{x_j}(\mathbf{x}, \xi(\mathbf{x}))}{f_y(\mathbf{x}, \xi(\mathbf{x}))},$$

$\forall j \in \{1, \cdots, n-1\}$.

The function $y = \xi(\mathbf{x})$ is said to be implicitly defined by the equation $f(\mathbf{x}, y) = c$ in a neighborhood of $(\mathbf{x}_0, y_0)$.

Proof 1.1   With no loss in generality, we may assume $f_y(\mathbf{x}_0, y_0) > 0$.
Since f_y is continuous on D, there exist $\delta_1 > 0$ and $\varepsilon > 0$ such that

$$f_y(\mathbf{x}, y) > 0, \ \forall (\mathbf{x}, y) \in B_{\delta_1}(\mathbf{x}_0)) \times \bar{J} \subset D.$$

Hence, the function $f(\mathbf{x}_0, y)$ is strictly increasing in $\bar{J}$ and therefore:

$$f(\mathbf{x}_0, y_0 - \varepsilon) < c = f(\mathbf{x}_0, y_0) < f(\mathbf{x}_0, y_0 + \varepsilon).$$

From this, since f is continuous, there exists $\delta \in \mathbb{R}$ such that $0 < \delta < \delta_1$ and also such that

$$f(\mathbf{x}, y_0 - \varepsilon) < c < f(\mathbf{x}, y_0 + \varepsilon),$$

$\forall \mathbf{x} \in B_\delta(\mathbf{x}_0)$.

Fix $\mathbf{x} \in B_\delta(\mathbf{x}_0)$.

From the intermediate value theorem and the last chain of inequalities, there exists $y \in (y_0 - \varepsilon, y_0 + \varepsilon) = J$ such that

$$f(\mathbf{x}, y) = c.$$

We claim that such a $y \in J$ is unique.

Suppose, to obtain contradiction, that there exists $y_1 \in \bar{J}$ such that $y_1 \neq y$ and

$$f(\mathbf{x}, y_1) = 0$$

Hence,

$$f(\mathbf{x}, y) = f(\mathbf{x}, y_1) = c.$$

From this and the mean value theorem, there exists y_2 between y and y_1 such that

$$f_y(\mathbf{x}, y_2) = 0.$$

This contradicts $f_y(\mathbf{x}, y) > 0$ in $B_\delta(\mathbf{x}_0) \times \bar{J}$.
Hence, the y in question is unique, and we shall denote it by

$$y = \xi(\mathbf{x}).$$

Thus,

$$f(\mathbf{x}, \xi(\mathbf{x})) = c, \ \forall \mathbf{x} \in B_\delta(\mathbf{x}_0).$$

Now, we will prove that the function $\xi : B_\delta(\mathbf{x}_0) \to J$ is continuous.
Let $\mathbf{x} \in B_\delta(\mathbf{x}_0)$.
Let $\{\mathbf{x}_n\} \subset B_\delta(\mathbf{x}_0)$ be such that

$$\lim_{n \to \infty} \mathbf{x}_n = \mathbf{x}.$$

It suffices to show that

$$\lim_{n \to \infty} \xi(\mathbf{x}_n) = \xi(\mathbf{x}) = y.$$

Suppose, to obtain contradiction, we do not have

$$\lim_{n \to \infty} \xi(\mathbf{x}_n) = y.$$

Thus, there exists $\varepsilon_0 > 0$ such that, for each $k \in \mathbb{N}$, there exists $n_k \in \mathbb{N}$ such that $n_k > k$ and

$$|\xi(\mathbf{x}_{n_k}) - y| \geq \varepsilon_0.$$

Observe that $\{\xi(\mathbf{x}_{n_k})\} \subset J$ and such a set is bounded.
Thus, there exists a subsequence of $\{\xi(\mathbf{x}_{n_k})\}$ which we shall also denote by $\{\xi(x_{n_k})\}$ and $y_1 \in \bar{J}$ such that $y_1 \neq y$ and

$$\lim_{k \to \infty} \xi(\mathbf{x}_{n_k}) = y_1,$$

Hence, since f is continuous, we obtain:

$$f(x, y_1) = \lim_{k \to \infty} f(\mathbf{x}_{n_k}, \xi(\mathbf{x}_{n_k})) = \lim_{k \to \infty} c = c.$$

Therefore,

$$f(\mathbf{x}, y) = f(\mathbf{x}, y_1) = c.$$

From this and the mean value theorem, there exists y_2 between y and y_1 such that

$$f_y(\mathbf{x}, y_2) = 0.$$

This contradicts $f_y(\mathbf{x}, y) > 0$ in $B_\delta(\mathbf{x}_0) \times \bar{J}$.
Therefore

$$\lim_{n \to \infty} \xi(\mathbf{x}_n) = y = \xi(\mathbf{x}).$$

Since $\{\mathbf{x}_n\} \subset B_\delta(\mathbf{x}_0)$ such that

$$\lim_{n\to\infty} \mathbf{x}_n = \mathbf{x},$$

is arbitrary, we may conclude that ξ is continuous at $\mathbf{x}$, $\forall \mathbf{x} \in B_\delta(\mathbf{x}_0)$.

For the final part, choose $j \in \{1,...,n-1\}$.

Let $\mathbf{x} \in B_\delta(\mathbf{x}_0)$ and $h \in \mathbb{R}$ be such that

$$\mathbf{x} + h\mathbf{e}_j \in B_\delta(\mathbf{x}_0).$$

Denote

$$v = \xi(\mathbf{x}+h\mathbf{e}_j) - \xi(\mathbf{x}),$$

that is,

$$\xi(\mathbf{x}+h\mathbf{e}_j) = \xi(\mathbf{x}) + v.$$

From the continuity of ξ we have:

$$v \to 0, \text{ as } h \to 0.$$

Observe that

$$f(\mathbf{x}, \xi(\mathbf{x})) = c,$$

and

$$f(\mathbf{x}+h\mathbf{e}_j, \xi(\mathbf{x}+h\mathbf{e}_j)) = f(\mathbf{x}+h\mathbf{e}_j, \xi(\mathbf{x})+v) = c,$$

so that

$$f(\mathbf{x}+h\mathbf{e}_j, \xi(\mathbf{x})+v) - f(\mathbf{x}, \xi(\mathbf{x})) = c - c = 0,$$

and therefore, from the mean value theorem, there exists $\tilde{t} \in (0,1)$ such that

$$
\begin{aligned}
0 &= f(\mathbf{x}+h\mathbf{e}_j, \xi(\mathbf{x})+v) - f(\mathbf{x}, \xi(\mathbf{x})) \\
&= f_{x_j}(\mathbf{x}+\tilde{t}h\mathbf{e}_j, \xi(\mathbf{x})+\tilde{t}v)h + f_y(\mathbf{x}+\tilde{t}h\mathbf{e}_j, \xi(\mathbf{x})+\tilde{t}v)v, \quad (1.1)
\end{aligned}
$$

so that,

$$
\begin{aligned}
\frac{\xi(\mathbf{x}+h\mathbf{e}_j) - \xi(\mathbf{x})}{h} &= \frac{v}{h} \\
&= \frac{-f_{x_j}(\mathbf{x}+\tilde{t}h\mathbf{e}_j, \xi(\mathbf{x})+\tilde{t}v)}{f_y(\mathbf{x}+\tilde{t}h\mathbf{e}_j, \xi(\mathbf{x})+\tilde{t}v)} \\
&\to \frac{-f_{x_j}(\mathbf{x}, \xi(\mathbf{x}))}{f_y(\mathbf{x}, \xi(\mathbf{x}))}, \text{ as } h \to 0. \quad (1.2)
\end{aligned}
$$

Summarizing the result obtained,

$$
\begin{aligned}
\frac{\partial \xi(\mathbf{x})}{\partial x_j} &= \lim_{h\to 0} \frac{\xi(\mathbf{x}+h\mathbf{e}_j) - \xi(\mathbf{x})}{h} \\
&= \frac{-f_{x_j}(\mathbf{x}, \xi(\mathbf{x}))}{f_y(\mathbf{x}, \xi(\mathbf{x}))}. \quad (1.3)
\end{aligned}
$$

The proof is complete.

Example 1.2.2 *Consider the example in which $F : \mathbb{R}^2 \to \mathbb{R}$ is given by*

$$F(x,y) = x^2 y^2 - x\cos(y) - 1, \ \forall (x,y) \in \mathbb{R}^2$$

Observe that

$$F(1, \pi/2) = \frac{\pi^2}{4} - 1$$

and

$$F_y(x,y) = 2x^2 y + x\sin(y),$$

so that

$$F_y(1, \pi/2) = \pi + 1 > 0.$$

hence, from the implicity function theorem, the equation

$$F(x,y) = F(1, \pi/2) = \frac{\pi^2}{4} - 1,$$

implicitly defines a function $y(x)$ in a neighborhood of $x = 1$.

Moreover, from this same theorem, the derivative of y in such a neighborhood is given by,

$$\frac{dy(x)}{dx} = -\frac{F_x(x, y(x))}{F_y(x, y(x))}.$$

Thus, informally,

$$\frac{dy(x)}{dx} = \frac{-2xy^2 + \cos(y)}{2x^2 y + x\sin(x)},$$

where it the dependence $y(x)$ is understood, so that

$$\frac{dy(1)}{dx} = \frac{-2(1)(\pi/2)^2 + \cos(\pi/2)}{2(1)(\pi/2) + 1\sin(\pi/2)} = \frac{-\pi^2}{2(\pi+1)}.$$

Example 1.2.3 *Let $f, g : \mathbb{R}^n \to \mathbb{R}$ be such that*

$$g(\mathbf{x}) = f(\mathbf{x})[1 + f(\mathbf{x})^6 + 5f(\mathbf{x})^8], \ \forall \mathbf{x} \in \mathbb{R}^n.$$

Assume g is of C^1 class. We are going to show that f is of C^1 class as well. Define $F : \mathbb{R}^n \times \mathbb{R} \to \mathbb{R}$ by

$$F(\mathbf{x}, y) = g(\mathbf{x}) - y(1 + y^6 + 5y^8).$$

Let $\mathbf{x} \in \mathbb{R}^n$. Thus

$$F(\mathbf{x}, f(\mathbf{x})) = 0.$$

Observe that

$$F_y(\mathbf{x}, y) = -1 - 7y^6 - 45y^8 < 0, \ \forall y \in \mathbb{R}.$$

From the implicity function theorem, the equation

$$F(\mathbf{x}, y) = 0,$$

defines a unique function $y(\mathbf{x})$ *in an neighborhood of* $\mathbf{x}$. *Since*

$$F(\mathbf{x}, f(\mathbf{x})) = 0, \ \forall \mathbf{x} \in \mathbb{R}^n,$$

we must have

$$y(\mathbf{x}) = f(\mathbf{x})$$

in the concerning neighborhood of $\mathbf{x}$.

Also, from the implicit function theorem,

$$\frac{\partial y(\mathbf{x})}{\partial x_j} = \frac{\partial f(\mathbf{x})}{\partial x_j} = -\frac{F_{x_j}(\mathbf{x}, f(\mathbf{x}))}{F_y(\mathbf{x}, f(\mathbf{x}))},$$

so that,

$$\frac{\partial f(\mathbf{x})}{\partial x_j} = \frac{-g_{x_j}(\mathbf{x})}{-1 - 7f(\mathbf{x})^6 - 45f(\mathbf{x})^8},$$

Since $\mathbf{x} \in \mathbb{R}^n$ *and* $j \in \{1, \cdots, n\}$ *are arbitrary, we may infer that* f *is of* C^1 *class.*

1.3   Vectorial functions in $\mathbb{R}^n$

In this section, we develop some fundamental results concerning vectorial functions.

Definition 1.3.1 (Vectorial functions) *Let* $D \subset \mathbb{R}^n$ *be a set. A binary relation* $\mathbf{f} : D \to \mathbb{R}^m$ *where* $m \geq 2$ *is said to be a vectorial function, if for each* $\mathbf{x} \in D$ *there exists only one* $\mathbf{z} \in \mathbb{R}^m$ *such that* $(\mathbf{x}, \mathbf{z}) \in \mathbf{f}$.

In such a case we denote,

$$\mathbf{z} = \mathbf{f}(\mathbf{x}).$$

Example 1.3.2 *Let* $\mathbf{f} : \mathbb{R}^2 \to \mathbb{R}^3$ *be such that*

$$\mathbf{f}(x, y) = x^2 \mathbf{e}_1 + (y + x)\mathbf{e}_2 + \sin(x + y)\mathbf{e}_3, \ \forall (x, y) \in \mathbb{R}^2.$$

we may also denote

$$\mathbf{f}(\mathbf{x}) = \begin{pmatrix} f_1(x, y) \\ f_2(x, y) \\ f_3(x, y) \end{pmatrix} \tag{1.4}$$

where

$$f_1(x, y) = x^2, \ f_2(x, y) = x + y, \ f_3(x, y) = \sin(x + y).$$

Definition 1.3.3 (Limits for vectorial functions) *Let $D \subset \mathbb{R}^n$ be a non-empty set. Let $\mathbf{f} : D \to \mathbb{R}^m$ be a vectorial function and let $\mathbf{x}_0 \in D'$.*

We say that $\mathbf{L} \in \mathbb{R}^m$ is the limit of $\mathbf{f}$ as $\mathbf{x}$ approaches $\mathbf{x}_0$, as for each $\varepsilon > 0$, there exists $\delta > 0$ such that if $\mathbf{x} \in D$ and $0 < |\mathbf{x} - \mathbf{x}_0| < \delta$, then

$$|\mathbf{f}(\mathbf{x}) - \mathbf{L}| < \varepsilon.$$

In such a case, we denote,

$$\lim_{\mathbf{x} \to \mathbf{x}_0} \mathbf{f}(\mathbf{x}) = \mathbf{L} \in \mathbb{R}^m.$$

Theorem 1.3.4 *Let $D \subset \mathbb{R}^n$ be a non-empty set. Let $\mathbf{f} : D \to \mathbb{R}^m$ be a vectorial function and let $\mathbf{x}_0 \in D'$. Denoting*

$$\mathbf{f}(\mathbf{x}) = \begin{pmatrix} f_1(\mathbf{x}) \\ f_2(\mathbf{x}) \\ \vdots \\ f_m(\mathbf{x}) \end{pmatrix} \tag{1.5}$$

we have that

$$\lim_{\mathbf{x} \to \mathbf{x}_0} \mathbf{f}(\mathbf{x}) = \mathbf{L} = \sum_{k=1}^{m} l_k \mathbf{e}_k \in \mathbb{R}^m,$$

if and only if,

$$\lim_{\mathbf{x} \to \mathbf{x}_0} f_k(\mathbf{x}) = l_k, \ \forall k \in \{1, \cdots, m\}.$$

Proof 1.2 Suppose that

$$\lim_{\mathbf{x} \to \mathbf{x}_0} \mathbf{f}(\mathbf{x}) = \mathbf{L} = \sum_{k=1}^{m} l_k \mathbf{e}_k \in \mathbb{R}^m.$$

Let $\varepsilon > 0$ be given. Thus, there exists $\delta > 0$ such that, if $\mathbf{x} \in D$ and $0 < |\mathbf{x} - \mathbf{x}_0| < \delta$, then

$$|\mathbf{f}(\mathbf{x}) - \mathbf{L}| < \varepsilon.$$

Select $k \in \{1, \cdots, m\}$.
Therefore,

$$\begin{aligned} |f_k(\mathbf{x}) - l_k| \ &\leq \ \sqrt{\sum_{j=1}^{N} |f_j(\mathbf{x}) - l_j|^2} \\ &= \ |\mathbf{f}(\mathbf{x}) - \mathbf{L}| \\ &< \ \varepsilon, \end{aligned} \tag{1.6}$$

if $\mathbf{x} \in D$ and $0 < |\mathbf{x} - \mathbf{x}_0| < \delta$, so that

$$\lim_{\mathbf{x} \to \mathbf{x}_0} f_k(\mathbf{x}) = l_k, \ \forall k \in \{1, \cdots, m\}.$$

Reciprocally, suppose

$$\lim_{\mathbf{x}\to\mathbf{x}_0} \mathbf{f}_k(\mathbf{x}) = l_k, \ \forall k \in \{1,\cdots,m\}.$$

Thus, for each $k \in \{1,\cdots,m\}$ there exists $\delta_k > 0$ such that, if $\mathbf{x} \in D$ and $0 < |\mathbf{x} - \mathbf{x}_0| < \delta_k$, then

$$|f_k(\mathbf{x}) - l_k| < \frac{\varepsilon}{\sqrt{m}}.$$

Define $\delta = \min\{\delta_k, \ k \in \{1,\cdots,m\}\}$.

Thus, if $\mathbf{x} \in D$ and $0 < |\mathbf{x} - \mathbf{x}_0| < \delta$, then

$$|\mathbf{f}(\mathbf{x}) - \mathbf{L}| = \sqrt{\sum_{k=1}^{m} |f_k(\mathbf{x}) - l_k|^2} < \sqrt{\sum_{k=1}^{m} \frac{\varepsilon^2}{m}} = \varepsilon,$$

so that

$$\lim_{\mathbf{x}\to\mathbf{x}_0} \mathbf{f}(\mathbf{x}) = \mathbf{L}.$$

The proof is complete.

1.3.1   Limit proprieties

Theorem 1.3.5 *Let $D \subset \mathbb{R}^m$ be a non-empty set. Let $\mathbf{f}, \mathbf{g} : D \to \mathbb{R}^m$ be vectorial functions and let $\mathbf{x}_0 \in D'$.*

Suppose that

$$\lim_{\mathbf{x}\to\mathbf{x}_0} \mathbf{f}(\mathbf{x}) = \mathbf{L} \in \mathbb{R}^m$$

and

$$\lim_{\mathbf{x}\to\mathbf{x}_0} \mathbf{g}(\mathbf{x}) = \mathbf{M} \in \mathbb{R}^m.$$

Under such assumptions, we have

1.

$$\lim_{\mathbf{x}\to\mathbf{x}_0} \alpha\mathbf{f}(\mathbf{x}) = \alpha\mathbf{L}, \ \forall \alpha \in \mathbb{R}.$$

2.

$$\lim_{\mathbf{x}\to\mathbf{x}_0} \mathbf{f}(\mathbf{x}) + \mathbf{g}(\mathbf{x}) = \mathbf{L} + \mathbf{M},$$

3.

$$\lim_{\mathbf{x}\to\mathbf{x}_0} \mathbf{f}(\mathbf{x}) \cdot \mathbf{g}(\mathbf{x}) = \mathbf{L} \cdot \mathbf{M}.$$

The proof of such results are similar to those of the case of scalar functions and it is left as an exercise.

Definition 1.3.6 (Continuous vectorial function) *Let $D \subset \mathbb{R}^n$ be a non-empty set and let $\mathbf{f} : D \to \mathbb{R}^m$ be a vectorial function. Let $\mathbf{x}_0 \in D$. we say that $\mathbf{f}$ is continuous at $\mathbf{x}_0$ as for for each $\varepsilon > 0$ there exists $\delta > 0$ such that if $\mathbf{x} \in D$ and $|\mathbf{x} - \mathbf{x}_0| < \delta$, then*

$$|\mathbf{f}(\mathbf{x}) - \mathbf{f}(\mathbf{x}_0)| < \varepsilon.$$

Therefore, if $\mathbf{x}_0$ is an isolated point of D, then $\mathbf{f}$ will be automatically continuous at $\mathbf{x}_0$, considering that in such a case, there exists $r > 0$ such that

$$B_r(\mathbf{x}_0) \cap D = \{\mathbf{x}_0\}.$$

thus, for a given $\varepsilon > 0$, it suffices to take $\delta = \frac{r}{2}$, so that if $\mathbf{x} \in D$ and $|\mathbf{x} - \mathbf{x}_0| < \delta$, then $\mathbf{x} = \mathbf{x}_0$ and hence

$$|\mathbf{f}(\mathbf{x}) - \mathbf{f}(\mathbf{x}_0)| = |\mathbf{f}(\mathbf{x}_0) - \mathbf{f}(\mathbf{x}_0)| = 0 < \varepsilon.$$

On the other hand, if $\mathbf{x}_0 \in D \cap D'$, the definition of continuity coincides with the one of limit, so that, $\mathbf{f}$ will be continuous at $\mathbf{x}_0$, if and only if,

$$\lim_{\mathbf{x} \to \mathbf{x}_0} \mathbf{f}(\mathbf{x}) = \mathbf{f}(\mathbf{x}_0).$$

Finally, if $\mathbf{f}$ is continuous on all its domain D, we simply say that $\mathbf{f}$ is a continuous function.

Theorem 1.3.7 *Let $D \subset \mathbb{R}^m$ be a non-empty set. Let $\mathbf{f}, \mathbf{g} : D \to \mathbb{R}^m$ be vectorial continuous functions at $\mathbf{x}_0 \in D \cap D'$.*
   Upon such assumptions,

1. *$\alpha \mathbf{f}$ is continuous at $\mathbf{x}_0$, $\forall \alpha \in \mathbb{R}$.*

2. *$\mathbf{f} + \mathbf{g}$ is continuous at $\mathbf{x}_0$.*

3. *$\mathbf{f} \cdot \mathbf{g}$ is continuous at $\mathbf{x}_0$.*

The proof results from the limit properties and it is left as an exercise.

Definition 1.3.8 (Directional derivative) *Let $D \subset \mathbb{R}^n$ be a non-empty set and let $\mathbf{f} : D \to \mathbb{R}^m$ be a function. Suppose $\mathbf{x}_0 \in D^\circ$ and let $\mathbf{v} \in \mathbb{R}^n$. We define the derivative of $\mathbf{f}$ relating $\mathbf{v}$ at the point $\mathbf{x}_0$, denoted by*

$$\frac{\partial \mathbf{f}(\mathbf{x}_0)}{\partial \mathbf{v}},$$

by

$$\frac{\partial \mathbf{f}(\mathbf{x}_0)}{\partial \mathbf{v}} = \lim_{h \to 0} \frac{\mathbf{f}(\mathbf{x}_0 + h\mathbf{v}) - \mathbf{f}(\mathbf{x}_0)}{h},$$

so that, denoting

$$\mathbf{f}(\mathbf{x}) = \begin{pmatrix} f_1(\mathbf{x}) \\ f_2(\mathbf{x}) \\ \vdots \\ f_m(\mathbf{x}) \end{pmatrix} \tag{1.7}$$

we have

$$\frac{\partial \mathbf{f}(\mathbf{x})}{\partial \mathbf{v}} = \begin{pmatrix} \frac{\partial f_1(\mathbf{x})}{\partial \mathbf{v}} \\ \frac{\partial f_2(\mathbf{x})}{\partial \mathbf{v}} \\ \vdots \\ \frac{\partial f_m(\mathbf{x})}{\partial \mathbf{v}} \end{pmatrix} \tag{1.8}$$

and, in particular,

$$\frac{\partial \mathbf{f}(\mathbf{x})}{\partial x_j} = \begin{pmatrix} \frac{\partial f_1(\mathbf{x})}{\partial x_j} \\ \frac{\partial f_2(\mathbf{x})}{\partial x_j} \\ \vdots \\ \frac{\partial f_m(\mathbf{x})}{\partial x_j} \end{pmatrix} \tag{1.9}$$

where,

$$\frac{\partial \mathbf{f}(\mathbf{x})}{\partial x_j} = \lim_{h \to 0} \frac{\mathbf{f}(\mathbf{x} + h\mathbf{e}_j) - \mathbf{f}(\mathbf{x})}{h}, \ \forall j \in \{1, \cdots, n\}.$$

if such limits exist.

Definition 1.3.9 (Differentiable vectorial functions) *Let $D \subset \mathbb{R}^n$ be a non-empty set and $\mathbf{f} : D \to \mathbb{R}^m$ a function. Suppose $\mathbf{x}_0 \in D^\circ$ and let $r > 0$ be such that $B_r(\mathbf{x}_0) \subset D$. We say that $\mathbf{f}$ is differentiable at $\mathbf{x}_0$, if there exists a matrix $m \times n \ M(\mathbf{x}_0)$ such that the function $\mathbf{r}(\mathbf{v})$ defined through the relation*

$$\mathbf{f}(\mathbf{x}_0 + \mathbf{v}) = f(\mathbf{x}_0) + M(\mathbf{x}_0) \cdot \mathbf{v} + \mathbf{r}(\mathbf{v}), \ \forall \mathbf{v} \in B_r(\mathbf{0})$$

that is

$$\mathbf{r}(\mathbf{v}) = \mathbf{f}(\mathbf{x}_0 + \mathbf{v}) - f(\mathbf{x}_0) - M(\mathbf{x}_0) \cdot \mathbf{v},$$

is such that

$$\lim_{\mathbf{v} \to \mathbf{0}} \frac{\mathbf{r}(\mathbf{v})}{|\mathbf{v}|} = \mathbf{0}.$$

Remark 1.3.10 *In the context of this last definition, for all $0 < |h| < r$, we have*

$$f(\mathbf{x}_0 + h\mathbf{e}_j) - f(\mathbf{x}_0) = M(\mathbf{x}_0) \cdot \mathbf{e}_j h + r(h\mathbf{e}_j),$$

so that

$$\frac{f(\mathbf{x}_0 + h\mathbf{e}_j) - f(\mathbf{x}_0)}{h} = M(\mathbf{x}_0) \cdot \mathbf{e}_j + \frac{r(h\mathbf{e}_j)}{h} \to M(\mathbf{x}_0) \cdot \mathbf{e}_j, \ as \ h \to 0.$$

Thus,

$$M(\mathbf{x}_0) \cdot \mathbf{e}_j = \frac{\partial \mathbf{f}(\mathbf{x}_0)}{\partial x_j}.$$

Hence, denoting

$$M(\mathbf{x}_0) = \begin{pmatrix} M_{11} & M_{12} & \cdots & M_{1n} \\ M_{21} & M_{22} & \cdots & M_{2n} \\ \vdots & \vdots & \ddots & \vdots \\ M_{m1} & M_{m2} & \cdots & M_{mn} \end{pmatrix} \tag{1.10}$$

we obtain

$$M(\mathbf{x}_0) \cdot \mathbf{e}_j = \begin{pmatrix} M_{1j} \\ M_{2j} \\ \vdots \\ M_{mj} \end{pmatrix} = \begin{pmatrix} \frac{\partial f_1(\mathbf{x})}{\partial x_j} \\ \frac{\partial f_2(\mathbf{x})}{\partial x_j} \\ \vdots \\ \frac{\partial f_m(\mathbf{x})}{\partial x_j} \end{pmatrix} = \frac{\partial \mathbf{f}(\mathbf{x})}{\partial x_j} \tag{1.11}$$

so that

$$M(\mathbf{x}_0) = \{M_{ij}\}_{m \times n} = \left(\frac{\partial \mathbf{f}(\mathbf{x}_0)}{\partial x_j} \right)_{m \times n}.$$

Therefore, we define

$$\mathbf{f}'(\mathbf{x}_0) = \{M_{ij}\}_{m \times n} = \left(\frac{\partial \mathbf{f}(\mathbf{x}_0)}{\partial x_j} \right)_{m \times n},$$

where $\mathbf{f}'(\mathbf{x}_0)$ is to be the derivative matrix (or Jacobian matrix)) of $\mathbf{f}$ at $\mathbf{x}_0$, that is

$$\mathbf{f}'(\mathbf{x}_0) = \begin{pmatrix} M_{11} & M_{12} & \cdots & M_{1n} \\ M_{21} & M_{22} & \cdots & M_{2n} \\ \vdots & \vdots & \ddots & \vdots \\ M_{m1} & M_{m2} & \cdots & M_{mn} \end{pmatrix} = \begin{pmatrix} \frac{\partial f_1(\mathbf{x})}{\partial x_1} & \frac{\partial f_1(\mathbf{x})}{\partial x_2} & \cdots & \frac{\partial f_1(\mathbf{x})}{\partial x_n} \\ \frac{\partial f_2(\mathbf{x})}{\partial x_1} & \frac{\partial f_2(\mathbf{x})}{\partial x_2} & \cdots & \frac{\partial f_2(\mathbf{x})}{\partial x_n} \\ \vdots & \vdots & \ddots & \vdots \\ \frac{\partial f_m(\mathbf{x})}{\partial x_1} & \frac{\partial f_m(\mathbf{x})}{\partial x_2} & \cdots & \frac{\partial f_m(\mathbf{x})}{\partial x_n} \end{pmatrix}. \tag{1.12}$$

Theorem 1.3.11 *Let $D \subset \mathbb{R}^n$ be a non-empty set and $\mathbf{f} : D \to \mathbb{R}^m$ a vectorial function, where we denote*

$$\mathbf{f}(\mathbf{x}) = \begin{pmatrix} f_1(\mathbf{x}) \\ f_2(\mathbf{x}) \\ \vdots \\ f_m(\mathbf{x}) \end{pmatrix} \tag{1.13}$$

for $f_k : D \to \mathbb{R}$, $\forall k \in \{1, \cdots, m\}$.

Suppose $\mathbf{x}_0 \in D^\circ$. Under such hypotheses, $\mathbf{f}$ is differentiable at $\mathbf{x}_0$ if, and only if, f_k is differentiable at $\mathbf{x}_0$, $\forall k \in \{1, \cdots, m\}$.

Proof 1.3   Suppose **f** is differentiable at $\mathbf{x}_0$.

Let $r_0 > 0$ be such that $B_{r_0}(\mathbf{x}_0) \subset D$. Thus, $\mathbf{r} : B_{r_0}(\mathbf{0}) \to \mathbb{R}^m$ defined through the relation

$$\mathbf{f}(\mathbf{x}_0 + \mathbf{v}) - \mathbf{f}(\mathbf{x}_0) = \mathbf{f}'(\mathbf{x}_0)\mathbf{v} + \mathbf{r}(\mathbf{v}), \ \forall \mathbf{v} \in B_{r_0}(\mathbf{0}),$$

that is,

$$\mathbf{r}(\mathbf{v}) = \mathbf{f}(\mathbf{x}_0 + \mathbf{v}) - \mathbf{f}(\mathbf{x}_0) - \mathbf{f}'(\mathbf{x}_0)\mathbf{v},$$

is such that

$$\frac{\mathbf{r}(\mathbf{v})}{|\mathbf{v}|} \to \mathbf{0}, \text{ as } \mathbf{v} \to \mathbf{0}.$$

From this, we obtain

$$\begin{pmatrix} f_1(\mathbf{x}_0 + \mathbf{v}) \\ f_2(\mathbf{x}_0 + \mathbf{v}) \\ \vdots \\ f_m(\mathbf{x}_0 + \mathbf{v}) \end{pmatrix} - \begin{pmatrix} f_1(\mathbf{x}_0) \\ f_2(\mathbf{x}_0) \\ \vdots \\ f_m(\mathbf{x}_0) \end{pmatrix} = \left[\frac{\partial f_i}{\partial x_j} \right] \mathbf{v} + \begin{pmatrix} r_1(\mathbf{v}) \\ r_2(\mathbf{v}) \\ \vdots \\ r_m(\mathbf{v}) \end{pmatrix} \qquad (1.14)$$

so that from the k-th line of this equation, we have

$$f_k(\mathbf{x}_0 + \mathbf{v}) - f_k(\mathbf{x}_0) = \sum_{j=1}^m \frac{\partial f_k(\mathbf{x}_0)}{\partial x_j} v_j + r_k(\mathbf{v}),$$

where

$$\frac{|r_k(\mathbf{v})|}{|\mathbf{v}|} \le \left| \frac{\mathbf{r}(\mathbf{v})}{|\mathbf{v}|} \right| \to 0, \text{ as } \mathbf{v} \to \mathbf{0}.$$

Therefore, f_k is differentiable at $\mathbf{x}_0, \ \forall k \in \{1, \cdots, m\}$.

Reciprocally, suppose f_k is differentiable at $\mathbf{x}_0, \ \forall k \in \{1, \cdots, m\}$.

Choose $k \in \{1, \cdots, m\}$. Thus, $r_k : B_{r_0}(\mathbf{0}) \to \mathbb{R}$ defined through the relation

$$f_k(\mathbf{x}_0 + \mathbf{v}) - f_k(\mathbf{x}_0) = \sum_{j=1}^m \frac{\partial f_k(\mathbf{x}_0)}{\partial x_j} v_j + r_k(\mathbf{v}), \qquad (1.15)$$

is such that,

$$\frac{r_k(\mathbf{v})}{|\mathbf{v}|} \to 0, \text{ as } \mathbf{v} \to \mathbf{0}.$$

Therefore, from (1.15), we obtain

$$\begin{pmatrix} f_1(\mathbf{x}_0 + \mathbf{v}) \\ f_2(\mathbf{x}_0 + \mathbf{v}) \\ \vdots \\ f_m(\mathbf{x}_0 + \mathbf{v}) \end{pmatrix} - \begin{pmatrix} f_1(\mathbf{x}_0) \\ f_2(\mathbf{x}_0) \\ \vdots \\ f_m(\mathbf{x}_0) \end{pmatrix} = \left[\frac{\partial f_i}{\partial x_j} \right] \mathbf{v} + \begin{pmatrix} r_1(\mathbf{v}) \\ r_2(\mathbf{v}) \\ \vdots \\ r_m(\mathbf{v}) \end{pmatrix}, \qquad (1.16)$$

that is,

$$\mathbf{r}(\mathbf{v}) = \mathbf{f}(\mathbf{x}_0 + \mathbf{v}) - \mathbf{f}(\mathbf{x}_0) - \mathbf{f}'(\mathbf{x}_0)\mathbf{v},$$

where

$$\mathbf{r}(\mathbf{v}) = \begin{pmatrix} r_1(\mathbf{v}) \\ r_2(\mathbf{v}) \\ \vdots \\ r_m(\mathbf{v}) \end{pmatrix},$$

is such that

$$\frac{\mathbf{r}(\mathbf{v})}{|\mathbf{v}|} = \begin{pmatrix} r_1(\mathbf{v})/|\mathbf{v}| \\ r_2(\mathbf{v})/|\mathbf{v}| \\ \vdots \\ r_m(\mathbf{v})/|\mathbf{v}| \end{pmatrix} \to \begin{pmatrix} 0 \\ 0 \\ \vdots \\ 0 \end{pmatrix} = \mathbf{0} \text{ as } \mathbf{v} \to \mathbf{0}.$$

Hence, $\mathbf{f}$ is differentiable at $\mathbf{x}_0$.

The proof is complete.

Remark 1.3.12 *Specially for the case $n = 1$, we shall define a new different notation for $\mathbf{f}$.*

We denote,

$$\mathbf{f}(t) = \begin{pmatrix} f_1(t) \\ f_2(t) \\ \vdots \\ f_m(t) \end{pmatrix} = \begin{pmatrix} x_1(t) \\ x_2(t) \\ \vdots \\ x_m(t) \end{pmatrix} \equiv \mathbf{r}(t)^T \qquad (1.17)$$

so that

$$\mathbf{r}(t) = (x_1(t), \dots, x_m(t)), \forall t \in [a,b],$$

for a one variable vectorial function

$$\mathbf{r} : [a,b] \to \mathbb{R}^m.$$

We are going to define the derivative of $\mathbf{r}$, denoted by $\mathbf{r}'$, by

$$\mathbf{r}'(t) = (x_1'(t), \dots, x_m'(t)),$$

if the derivatives in question exist at $t \in [a,b]$.

Theorem 1.3.13 (Mean value inequality for one variable vectorial functions)
Let $\mathbf{r} : [a,b] \to \mathbb{R}^m$ be a continuous one variable vectorial function, such that $\mathbf{r}'$ is continuous on (a,b), where $a < b$. Under such hypotheses, there exists $\tilde{t} \in (a,b)$ such that

$$|\mathbf{r}(b) - \mathbf{r}(a)| \leq (b-a)|\mathbf{r}'(\tilde{t})|.$$

Proof 1.4   Define $\mathbf{z} = \mathbf{r}(b) - \mathbf{r}(a)$, and

$$\phi(t) = \mathbf{z} \cdot \mathbf{r}(t), \ \forall t \in [a,b].$$

Thus, ϕ is continuous on $[a,b]$ and it is differentiable on (a,b). From the mean value theorem, there exists, $\tilde{t} \in (a,b)$ such that

$$\phi(b) - \phi(a) = (b-a)\phi'(\tilde{t}).$$

Thus,

$$
\begin{aligned}
\mathbf{z} \cdot (\mathbf{r}(b) - \mathbf{r}(a)) = |\mathbf{z}|^2 \;\; &= \;\; (b-a)(\mathbf{z} \cdot \mathbf{r}'(\tilde{t})) \\
&= \;\; (b-a)|\mathbf{z} \cdot \mathbf{r}'(\tilde{t})| \\
&\leq \;\; (b-a)|\mathbf{z}||\mathbf{r}'(\tilde{t})|,
\end{aligned}
\tag{1.18}
$$

so that

$$|\mathbf{r}(b) - \mathbf{r}(a)| = |\mathbf{z}| \leq (b-a)|\mathbf{r}'(\tilde{t})|.$$

This completes the proof.

Theorem 1.3.14 (Mean value inequality for vectorial functions) *Let $D \subset \mathbb{R}^n$ be an open set and let $\mathbf{f} : D \to \mathbb{R}^m$ be a differentiable function at each point of the set*

$$A = \{\mathbf{x}_0 + t\mathbf{v} \; : \; t \in (0,1)\},$$

for a given $\mathbf{v} \in \mathbb{R}^n$.

Assume also that $\mathbf{f}$ is continuous on

$$\overline{A} = \{\mathbf{x}_0 + t\mathbf{v} \; : \; t \in [0,1]\},$$

$$|\mathbf{f}'(\mathbf{x})| \leq M, \; \forall \mathbf{x} \in A,$$

for some $M \in \mathbb{R}^+$.

Under such hypotheses

$$|\mathbf{f}(\mathbf{x}_0 + \mathbf{v}) - \mathbf{f}(\mathbf{x}_0)| \leq M|\mathbf{v}|.$$

Proof 1.5 Let $\mathbf{r} : [0,1] \to \mathbb{R}^m$ be defined by

$$\mathbf{r}(t) = \mathbf{f}(\mathbf{x}_0 + t\mathbf{v}), \forall t \in [0,1].$$

From the hypotheses, $\mathbf{r}$ is continuous on $[0,1]$ and differentiable in $(0,1)$.

From the mean value inequality for one variable vectorial functions, there exists $\tilde{t} \in (0,1)$ such that

$$|\mathbf{r}(1) - \mathbf{r}(0)| \leq |\mathbf{r}'(\tilde{t})|(1-0),$$

so that

$$
\begin{aligned}
|\mathbf{f}(\mathbf{x}_0 + \mathbf{v}) - \mathbf{f}(\mathbf{x}_0)| \;\; &\leq \;\; |\mathbf{r}'(\tilde{t})| \\
&= \;\; |\mathbf{f}'(\mathbf{x}_0 + \tilde{t}\mathbf{v}) \cdot \mathbf{v}| \\
&\leq \;\; |\mathbf{f}'(\mathbf{x}_0 + \tilde{t}\mathbf{v})||\mathbf{v}| \\
&\leq \;\; M|\mathbf{v}|.
\end{aligned}
\tag{1.19}
$$

The proof is complete.

1.4 Implicit function theorem for the vectorial case

We start with the following auxiliary result. Indeed, its proof is very similar to that of the one-dimensional case, however we present the proof again for the sake of completeness.

Theorem 1.4.1 *Let $0 \leq \lambda < 1$. Suppose that $\{\mathbf{x}_n\} \subset \mathbb{R}^n$ is such that,*

$$|\mathbf{x}_{n+2} - \mathbf{x}_{n+1}| \leq \lambda |\mathbf{x}_{n+1} - \mathbf{x}_n|, \forall n \in \mathbb{N}.$$

Upon such assumptions, $\{\mathbf{x}_n\}$ is a Cauchy sequence so that it is converging.

Proof 1.6   Observe that

$$
\begin{aligned}
|\mathbf{x}_3 - \mathbf{x}_2| &\leq \lambda |\mathbf{x}_2 - \mathbf{x}_1| \\
|\mathbf{x}_4 - \mathbf{x}_3| &\leq \lambda |\mathbf{x}_3 - \mathbf{x}_2| \leq \lambda^2 |\mathbf{x}_2 - \mathbf{x}_1| \\
\cdots \quad & \cdots \quad \cdots \\
|\mathbf{x}_{n+1} - \mathbf{x}_n| &\leq \lambda^{n-1} |\mathbf{x}_2 - \mathbf{x}_1|.
\end{aligned}
\tag{1.20}
$$

Thus, for $n, p \in \mathbb{N}$ we have

$$
\begin{aligned}
|\mathbf{x}_{n+p} - \mathbf{x}_n| &= |\mathbf{x}_{n+p} - \mathbf{x}_{n+p-1} + \mathbf{x}_{n+p-1} - \mathbf{x}_{n+p-2} + \mathbf{x}_{n+p-1} + \cdots - \mathbf{x}_{n+1} + \mathbf{x}_{n+1} - \mathbf{x}_n| \\
&\leq |\mathbf{x}_{n+p} - \mathbf{x}_{n+p-1}| + |\mathbf{x}_{n+p-1} - \mathbf{x}_{n+p-2}| + \cdots + |\mathbf{x}_{n+1} - \mathbf{x}_n| \\
&\leq (\lambda^{n+p-2} + \lambda^{n+p-3} + \cdots + \lambda^{n-1})|\mathbf{x}_2 - \mathbf{x}_1|.
\end{aligned}
\tag{1.21}
$$

Therefore,

$$
\begin{aligned}
|\mathbf{x}_{n+p} - \mathbf{x}_n| &\leq \lambda^{n-1}(\lambda^{p-1} + \lambda^{p-2} + \cdots + 1)|\mathbf{x}_2 - \mathbf{x}_1| \\
&\leq \frac{\lambda^{n-1}(1 - \lambda^p)}{1 - \lambda}|\mathbf{x}_2 - \mathbf{x}_1| \\
&\leq \frac{\lambda^{n-1}}{1 - \lambda}|\mathbf{x}_2 - \mathbf{x}_1|.
\end{aligned}
\tag{1.22}
$$

Observe that

$$\lim_{n \to \infty} \frac{\lambda^{n-1}}{1 - \lambda}|\mathbf{x}_2 - \mathbf{x}_1| = 0.$$

Let $\varepsilon > 0$. Thus, there exists $n_0 \in \mathbb{N}$ such that, if $n > n_0$, then

$$\frac{\lambda^{n-1}}{1 - \lambda}|\mathbf{x}_2 - \mathbf{x}_1| < \varepsilon.$$

From this and (1.22), we get,

$$|\mathbf{x}_{n+p} - \mathbf{x}_n| < \varepsilon, \text{ if } n > n_0.$$

Hence, for $m = p + n$ we obtain

$$|\mathbf{x}_m - \mathbf{x}_n| < \varepsilon, \text{ if } m > n > n_0.$$

Thus, $\{\mathbf{x}_n\}$ is a Cauchy sequence, and is therefore converging.

At this point, we present in detail the Banach contractor function theorem.

Theorem 1.4.2 *Let $D \subset \mathbb{R}^m$ be a non-empty, closed and convex set. Let $\mathbf{f} : D \to D$ be a continuous vectorial function such that*

$$|\mathbf{f}(\mathbf{x}) - \mathbf{f}(\mathbf{y})| \le \lambda |\mathbf{x} - \mathbf{y}|, \ \forall \mathbf{x}, \ \mathbf{y} \in D,$$

for some

$$0 \le \lambda < 1,$$

that is, $\mathbf{f}$ is a contractor function.

Choose $\mathbf{x}_1 \in D$. With such a choice in mind, the sequence defined by

$$\mathbf{x}_{k+1} = \mathbf{f}(\mathbf{x}_k), \ \forall k \in \mathbb{N}$$

is such that there exists $\mathbf{x}_0 \in D$ such that

$$\lim_{k \to \infty} \mathbf{x}_k = \mathbf{x}_0$$

and

$$\mathbf{f}(\mathbf{x}_0) = \mathbf{x}_0,$$

there is, $\mathbf{x}_0$ is a fixed point for $\mathbf{f}$.

Moreover, such a $\mathbf{x}_0 \in D$ is unique.

Proof 1.7   Observe that

$$|\mathbf{x}_{n+2} - \mathbf{x}_{n+1}| = |\mathbf{f}(\mathbf{x}_{n+1}) - \mathbf{f}(\mathbf{x}_n)| \le \lambda |\mathbf{x}_{n+1} - \mathbf{x}_n|,$$

$\forall n \in \mathbb{N}$.

From the last theorem and since D is closed, $\{\mathbf{x}_n\}$ converges to some $\mathbf{x}_0 \in D$. Observe that

$$\mathbf{x}_{n+1} = \mathbf{f}(\mathbf{x}_n), \ \forall n \in \mathbb{N}.$$

From the continuity of $\mathbf{f}$ we obtain,

$$\mathbf{x}_0 = \lim_{n \to \infty} \mathbf{x}_{n+1} = \lim_{n \to \infty} \mathbf{f}(\mathbf{x}_n) = \mathbf{f}(\mathbf{x}_0),$$

so that

$$\mathbf{x}_0 = \mathbf{f}(\mathbf{x}_0).$$

Now, suppose $\mathbf{y} \in D$ is such that

$$\mathbf{f}(\mathbf{y}) = \mathbf{y}.$$

Thus,

$$|\mathbf{y} - \mathbf{x}_0| = |\mathbf{f}(\mathbf{y}) - \mathbf{f}(\mathbf{x}_0)| \le \lambda |\mathbf{y} - \mathbf{x}_0|,$$

so that,

$$(1 - \lambda)|\mathbf{y} - \mathbf{x}_0| \le 0.$$

Since $1 - \lambda > 0$, we have $|\mathbf{y} - \mathbf{x}_0| = 0$, that is,

$$\mathbf{y} = \mathbf{x}_0.$$

From this, we may infer that $\mathbf{x}_0$ is unique.
The proof is complete.

Remark 1.4.3 *For $D \subset \mathbb{R}^m$ closed and convex, if $\mathbf{f} : D \to D$ is a differentiable vectorial function such that*

$$|\mathbf{f}'(\mathbf{x})| < \lambda, \forall \mathbf{x} \in D,$$

for some $0 \leq \lambda < 1$, then from mean value inequality, given $\mathbf{x}, \mathbf{y} \in D$, there exists $t \in (0,1)$ such that

$$|\mathbf{f}(\mathbf{y}) - \mathbf{f}(\mathbf{x})| \leq |\mathbf{f}'(\mathbf{x} + t(\mathbf{y} - \mathbf{x}))||\mathbf{y} - \mathbf{x}| \leq \lambda |\mathbf{y} - \mathbf{x}|.$$

Therefore, in such a case, $\mathbf{f}$ is a contractor function, so that from the last theorem, there exists one and only one $\mathbf{x}_0 \in D$ such that

$$\mathbf{x}_0 = \mathbf{f}(\mathbf{x}_0).$$

Theorem 1.4.4 (Implicit function theorem, vectorial case) *Denote* $(\mathbf{x}, \mathbf{y}) \in \mathbb{R}^{n+m}$ *where* $\mathbf{x} = (x_1, ..., x_n) \in \mathbb{R}^n$ *and* $\mathbf{y} = (y_1, ..., y_m) \in \mathbb{R}^m$. *Let $D \subset \mathbb{R}^{n+m}$ be an non-empty open set and let $\mathbf{f} : D \to \mathbb{R}^m$ be a function of C^1 class on D, that is, $\mathbf{f}$ and its first order partial derivatives are continuous on D, so that $\mathbf{f}$ is differentiable on D. Denote*

$$\mathbf{f}'(\mathbf{x}, \mathbf{y}) = \left\{ \frac{\partial f_k(\mathbf{x}, \mathbf{y})}{\partial x_j} \Big|_{j=1}^n , \frac{\partial f_k(\mathbf{x}, \mathbf{y})}{\partial y_j} \Big|_{j=1}^m \right\},$$

where $k \in \{1, ..., m\}$.
Thus, we may also denote:

$$\mathbf{f}'(\mathbf{x}, \mathbf{y}) = \left\{ \mathbf{f}_x(\mathbf{x}, \mathbf{y}) , \mathbf{f}_y(\mathbf{x}, \mathbf{y}) \right\},$$

where,

$$\mathbf{f}_x(\mathbf{x}, \mathbf{y}) = \left\{ \frac{\partial f_k(\mathbf{x}, \mathbf{y})}{\partial x_j} \right\}_{j=1}^n ,$$

$$\mathbf{f}_y(\mathbf{x}, \mathbf{y}) = \left\{ \frac{\partial f_k(\mathbf{x}, \mathbf{y})}{\partial y_j} \right\}_{j=1}^m ,$$

where,

$$\mathbf{f}(\mathbf{x}, \mathbf{y}) = (f_k(\mathbf{x}, \mathbf{y})) = \begin{pmatrix} f_1(\mathbf{x}, \mathbf{y}) \\ f_2(\mathbf{x}, \mathbf{y}) \\ \vdots \\ f_m(\mathbf{x}, \mathbf{y}) \end{pmatrix} \qquad (1.23)$$

Assume $\mathbf{f}(\mathbf{x}_0, \mathbf{y}_0) = \mathbf{0}$, *and that*

$$\det(\mathbf{f}_y(\mathbf{x}_0, \mathbf{y}_0)) \neq 0,$$

where

$$(\mathbf{x}_0, \mathbf{y}_0) \in D.$$

Let $\mathbf{v} \in \mathbb{R}^n$. *Under such hypotheses, there exists* $\delta_1 > 0$ *and* $\delta_2 > 0$ *such that, for each* $\mathbf{x} \in B_{\delta_1}(\mathbf{x}_0)$, *there exists a unique* $\mathbf{y} \in \overline{B}_{\delta_2}(\mathbf{y}_0)$ *such that*

$$\mathbf{f}(\mathbf{x}, \mathbf{y}) = \mathbf{0}.$$

Moreover, denoting

$$\mathbf{y} = \xi(\mathbf{x})$$

we have

$$\mathbf{f}(\mathbf{x}, \xi(\mathbf{x})) = \mathbf{0}, \forall \mathbf{x} \in B_\delta(\mathbf{x}_0),$$

and also

$$\frac{\partial \xi(\mathbf{x})}{\partial \mathbf{v}} = -[\mathbf{f}_y(\mathbf{x}, \xi(\mathbf{x}))]^{-1} \cdot [\mathbf{f}_x(\mathbf{x}, \xi(\mathbf{x})) \cdot \mathbf{v}].$$

Proof 1.8   Denote $A = \mathbf{f}_y(\mathbf{x}_0, \mathbf{y}_0)$.
From the hypotheses A^{-1} exists. Define

$$\lambda = \frac{1}{2|A^{-1}|} > 0. \qquad (1.24)$$

Since $\mathbf{f}_y$ is continuous on D, there exist $\tilde{\delta}_1 > 0$ and $\delta_2 > 0$ such that, if $x \in B_{\tilde{\delta}_1}(\mathbf{x}_0)$ and $\mathbf{y} \in \overline{B}_{\delta_2}(\mathbf{y}_0)$, then

$$\det(\mathbf{f}'(\mathbf{x})) \neq 0$$

and

$$|\mathbf{f}_y(\mathbf{x}, \mathbf{y}) - A| = |\mathbf{f}_y(\mathbf{x}, \mathbf{y}) - \mathbf{f}_y(\mathbf{x}_0, \mathbf{y}_0)| < \lambda \qquad (1.25)$$

Define

$$\varepsilon = \frac{\delta_2}{2|A^{-1}|} > 0.$$

From the continuity of $\mathbf{f}$, there exists $\delta_1 > 0$ such that $\delta_1 < \tilde{\delta}_1$ and also such that, if $\mathbf{x} \in B_{\delta_1}(\mathbf{x}_0)$, then

$$|\mathbf{f}(\mathbf{x}, \mathbf{y}_0)| = |\mathbf{f}(\mathbf{x}, \mathbf{y}_0) - \mathbf{f}(\mathbf{x}_0, \mathbf{y}_0)| < \varepsilon.$$

Let $\mathbf{x} \in B_{\delta_1}(\mathbf{x}_0)$.
Define

$$\phi(\mathbf{y}) = \mathbf{y} - A^{-1}(\mathbf{f}(\mathbf{x}, \mathbf{y})).$$

Thus,

$$\begin{aligned}
\phi'(\mathbf{y}) &= I - A^{-1}(\mathbf{f}_y(\mathbf{x}, \mathbf{y})) \\
&= A^{-1}A - A^{-1}(\mathbf{f}_y(\mathbf{x}, \mathbf{y})) \\
&= A^{-1}(A - \mathbf{f}_y(\mathbf{x}, \mathbf{y})),
\end{aligned} \tag{1.26}$$

so that for $\mathbf{y} \in \overline{B}_{\delta_2}(\mathbf{y}_0)$, we obtain

$$\begin{aligned}
|\phi'(\mathbf{y})| &\leq |A^{-1}||A - \mathbf{f}_y(\mathbf{x}, \mathbf{y})| \\
&\leq \frac{1}{2\lambda}\lambda = \frac{1}{2},
\end{aligned} \tag{1.27}$$

so that

$$|\phi'(\mathbf{y})| < \frac{1}{2}, \ \forall \mathbf{y} \in \overline{B}_{\delta_2}(\mathbf{y}_0).$$

From this and the mean value inequality for vectorial functions, we get

$$|\phi(\mathbf{y}_1) - \phi(\mathbf{y}_2)| \leq \frac{1}{2}|\mathbf{y}_1 - \mathbf{y}_2|, \ \forall \mathbf{y}_1, \mathbf{y}_2 \in \overline{B}_{\delta_2}(\mathbf{y}_0). \tag{1.28}$$

We are going to prove that

$$\phi(\overline{B}_{\delta_2}(\mathbf{y}_0)) \subset B_{\delta_2}(\mathbf{y}_0).$$

Let $\mathbf{y} \in B_{\delta_2}(\mathbf{y}_0)$, thus,

$$\begin{aligned}
|\phi(\mathbf{y}) - \mathbf{y}_0| &= |\phi(\mathbf{y}) - \phi(\mathbf{y}_0) + \phi(\mathbf{y}_0) - \mathbf{y}_0| \\
&\leq |\phi(\mathbf{y}) - \phi(\mathbf{y}_0)| + |\phi(\mathbf{y}_0) - \mathbf{y}_0| \\
&\leq \frac{1}{2}|\mathbf{y} - \mathbf{y}_0| + |\phi(\mathbf{y}_0) - \mathbf{y}_0| \\
&\leq \frac{\delta_2}{2} + |\phi(\mathbf{y}_0) - \mathbf{y}_0|.
\end{aligned} \tag{1.29}$$

On the other hand,

$$\phi(\mathbf{y}_0) = \mathbf{y}_0 - A^{-1}(f(\mathbf{x}, \mathbf{y}_0)),$$

that is

$$\begin{aligned}
|\phi(\mathbf{y}_0) - \mathbf{y}_0| &\leq |A^{-1}||\mathbf{f}(\mathbf{x}, \mathbf{y}_0)| \\
&\leq |A^{-1}|\varepsilon \\
&\leq \frac{|A^{-1}|\delta_2}{2|A^{-1}|} \\
&= \frac{\delta_2}{2}.
\end{aligned} \tag{1.30}$$

From this and (1.29), we obtain

$$\begin{aligned}
|\phi(\mathbf{y}) - \mathbf{y}_0| &\leq \frac{\delta_2}{2} + |\phi(\mathbf{y}_0) - \mathbf{y}_0| \\
&< \frac{\delta_2}{2} + \frac{\delta_2}{2} \\
&= \delta_2
\end{aligned} \tag{1.31}$$

Therefore,

$$\phi(\mathbf{y}) \in \overline{B}_{\delta_2}(\mathbf{y}_0), \ \forall \mathbf{y} \in \overline{B}_{\delta_2}(\mathbf{y}_0).$$

From these last results we may infer that ϕ is a contractor mapping on $\overline{B}_{\delta_2}(\mathbf{y}_0)$.
From the contractor mapping theorem, there exists a unique $\mathbf{y} \in \overline{B}_{\delta_2}(\mathbf{y}_0)$ such that

$$\phi(\mathbf{y}) = \mathbf{y},$$

so that

$$\phi(\mathbf{y}) = \mathbf{y} - A^{-1}(\mathbf{f}(\mathbf{x},\mathbf{y})) = \mathbf{y},$$

so that

$$A^{-1}(\mathbf{f}(\mathbf{x},\mathbf{y})) = \mathbf{0}$$

and since $det(A^{-1}) \neq 0$, we obtain

$$\mathbf{f}(\mathbf{x},\mathbf{y}) = \mathbf{0}.$$

Since, for each $\mathbf{x} \in B_{\delta_1}(\mathbf{x}_0)$ such a $\mathbf{y}$ is unique, we denote

$$\mathbf{y} = \xi(\mathbf{x})$$

so that,

$$\mathbf{f}(\mathbf{x}, \xi(\mathbf{x})) = \mathbf{0}, \ \forall \mathbf{x} \in B_{\delta_1}(\mathbf{x}_0).$$

Next, we show that ξ is continuous on $B_{\delta_1}(\mathbf{x}_0)$.
Let $\mathbf{x} \in B_{\delta_1}(\mathbf{x}_0)$.
Let $\{\mathbf{x}_n\} \subset B_{\delta_1}(\mathbf{x}_0)$ be such that

$$\lim_{n \to \infty} \mathbf{x}_n = \mathbf{x}.$$

It suffices to show that

$$\lim_{n \to \infty} \xi(\mathbf{x}_n) = \xi(\mathbf{x}) = \mathbf{y}.$$

Suppose, to obtain contradiction, we do not have

$$\lim_{n \to \infty} \xi(\mathbf{x}_n) = \mathbf{y}.$$

Thus, there exists $\varepsilon_0 > 0$ such that for each $k \in \mathbb{N}$ there exists $n_k \in \mathbb{N}$ such that $n_k > k$ and

$$|\xi(\mathbf{x}_{n_k}) - \mathbf{y}| \geq \varepsilon_0.$$

Observe that $\{\xi(\mathbf{x}_{n_k})\} \subset \overline{B}_{\delta_2}(\mathbf{y}_0)$ and such a set is compact.
Thus, there exists a subsequence of $\{\xi(\mathbf{x}_{n_k})\}$ which we shall also denote by $\{\xi(\mathbf{x}_{n_k})\}$ and $\mathbf{y}_1 \in \overline{B}_{\delta_2}(\mathbf{y}_0)$ such that $\mathbf{y}_1 \neq \mathbf{y}$ and

$$\lim_{k \to \infty} \xi(\mathbf{x}_{n_k}) = \mathbf{y}_1,$$

Hence, since $\mathbf{f}$ is continuous, we obtain:

$$\mathbf{f}(\mathbf{x}, \mathbf{y}_1) = \lim_{k \to \infty} \mathbf{f}(\mathbf{x}_{n_k}, \xi(\mathbf{x}_{n_k})) = \lim_{k \to \infty} \mathbf{0} = \mathbf{0}.$$

Therefore,

$$\mathbf{f}(\mathbf{x}, \mathbf{y}) = \mathbf{f}(\mathbf{x}, \mathbf{y}_1) = \mathbf{0}.$$

so that from $f(\mathbf{x}, \mathbf{y}_1) = \mathbf{0}$ we obtain

$$\phi(\mathbf{y}_1) = \mathbf{y}_1 - A^{-1}(\mathbf{f}(\mathbf{x}, \mathbf{y}_1)) = \mathbf{y}_1,$$

that is

$$\phi(\mathbf{y}_1) = \mathbf{y}_1.$$

Since the solution of this last equation is unique, we would obtain

$$\mathbf{y}_1 = \mathbf{y},$$

which contradicts

$$\mathbf{y}_1 \neq \mathbf{y}.$$

Therefore

$$\lim_{n \to \infty} \xi(\mathbf{x}_n) = \mathbf{y} = \xi(\mathbf{x}).$$

Since $\{\mathbf{x}_n\} \subset B_\delta(\mathbf{x}_0)$ such that

$$\lim_{n \to \infty} \mathbf{x}_n = \mathbf{x},$$

is arbitrary, we may conclude that ξ is continuous at $\mathbf{x}$, $\forall \mathbf{x} \in B_{\delta_1}(\mathbf{x}_0)$.
For the final part, let $\mathbf{x} \in B_{\delta_1}(\mathbf{x}_0)$ and $h \in \mathbb{R}$ be such that

$$\mathbf{x} + h\mathbf{v} \in B_\delta(\mathbf{x}_0).$$

Denote

$$\mathbf{u} = \xi(\mathbf{x} + h\mathbf{v}) - \xi(\mathbf{x}),$$

that is,

$$\xi(\mathbf{x} + h\mathbf{v}) = \xi(\mathbf{x}) + \mathbf{u}.$$

From the continuity of ξ, we have:

$$\mathbf{u} \to \mathbf{0}, \text{ as } h \to 0.$$

Observe that

$$\mathbf{f}(\mathbf{x}, \xi(\mathbf{x})) = \mathbf{0},$$

and

$$\mathbf{f}(\mathbf{x} + h\mathbf{v}, \xi(\mathbf{x} + h\mathbf{v})) = \mathbf{f}(\mathbf{x} + h\mathbf{v}, \xi(\mathbf{x}) + \mathbf{u}) = \mathbf{0},$$

so that

$$\mathbf{f}(\mathbf{x} + h\mathbf{v}, \xi(\mathbf{x}) + \mathbf{u}) - \mathbf{f}(\mathbf{x}, \xi(\mathbf{x})) = \mathbf{0} - \mathbf{0} = \mathbf{0},$$

Therefore

$$
\begin{aligned}
0 &= \mathbf{f}(\mathbf{x}+h\mathbf{v}, \xi(\mathbf{x})+\mathbf{u}) - \mathbf{f}(\mathbf{x}, \xi(\mathbf{x})) \\
&= \mathbf{f}_x(\mathbf{x}, \xi(\mathbf{x})) \cdot h\mathbf{v} + \mathbf{f}_y(\mathbf{x}, \xi(\mathbf{x})) \cdot \mathbf{u} \\
&\quad + \mathbf{w}(h, \mathbf{u})(|h||\mathbf{v}| + |\mathbf{u}|),
\end{aligned}
\tag{1.32}
$$

Observe that

$$\mathbf{u} \to \mathbf{0}, \text{ as } h \to 0,$$

so that, from the differentiability definition,

$$\mathbf{w}(h, \mathbf{u}) \to \mathbf{0}, \text{ as } h \to 0.$$

From (1.32), we have,

$$
-\mathbf{f}_y(\mathbf{x}, \xi(\mathbf{x})) \cdot \frac{\mathbf{u}}{h} = \mathbf{f}_x(\mathbf{x}, \xi(\mathbf{x})) \cdot \mathbf{v} + \mathbf{w}(h, \mathbf{u}) \left(\frac{|h|}{h} |\mathbf{v}| + \frac{|\mathbf{u}|}{h} \right)
\tag{1.33}
$$

so that

$$
\begin{aligned}
\frac{\mathbf{u}}{h} &= -[\mathbf{f}_y(\mathbf{x}, \xi(\mathbf{x}))]^{-1} \cdot [\mathbf{f}_x(\mathbf{x}, \xi(\mathbf{x})) \cdot \mathbf{v}] + \\
&\quad -[\mathbf{f}_y(\mathbf{x}, \xi(\mathbf{x}))]^{-1} \cdot \left[\mathbf{w}(h, \mathbf{u}) \left(\frac{|h|}{h} |\mathbf{v}| + \frac{|\mathbf{u}|}{h} \right) \right]
\end{aligned}
\tag{1.34}
$$

At this point, we denote

$$B = -[\mathbf{f}_y(\mathbf{x}, \xi(\mathbf{x}))]^{-1},$$

and

$$C = -[\mathbf{f}_y(\mathbf{x}, \xi(\mathbf{x}))]^{-1} \cdot [\mathbf{f}_x(\mathbf{x}, \xi(\mathbf{x})) \cdot \mathbf{v}],$$

so that

$$
\left| \frac{\mathbf{u}}{h} \right| \le |C| + |B| \left[|\mathbf{w}(h, \mathbf{u})| \left(\frac{|h|}{|h|} |\mathbf{v}| + \frac{|\mathbf{u}|}{|h|} \right) \right]
\tag{1.35}
$$

and thus, since

$$\mathbf{w}(h, \mathbf{u}) \to \mathbf{0}, \text{ as } h \to 0,$$

we obtain, for all h sufficiently small,

$$
\begin{aligned}
\left| \frac{\mathbf{u}}{h} \right| &\le \frac{|C| + |B| [|\mathbf{w}(h, \mathbf{u})||\mathbf{v}|]}{(1 - |B||\mathbf{w}(h, \mathbf{u})|)} \\
&\to |C|, \text{ as } h \to 0.
\end{aligned}
\tag{1.36}
$$

Thus, there exists $h_0 > 0$ such that, if $0 < |h| < h_0$, then

$$\left| \frac{\mathbf{u}}{h} \right| \le |C| + 1,$$

so that, from this, we obtain

$$|\mathbf{w}(h, \mathbf{u})| \left(\frac{|\mathbf{u}|}{|h|} \right) \to 0, \text{ as } h \to 0.$$

Hence, from this and (1.35), we have

$$
\begin{aligned}
\frac{\mathbf{u}}{h} &= -[\mathbf{f}_y(\mathbf{x},\xi(\mathbf{x}))]^{-1} \cdot [\mathbf{f}_x(\mathbf{x},\xi(\mathbf{x})) \cdot \mathbf{v}] + \\
&\quad -[\mathbf{f}_y(\mathbf{x},\xi(\mathbf{x}))]^{-1} \cdot \left[\mathbf{w}(h,\mathbf{u})\left(\frac{|h|}{h}|\mathbf{v}| + \frac{|\mathbf{u}|}{h}\right)\right] \\
&\to -[\mathbf{f}_y(\mathbf{x},\xi(\mathbf{x}))]^{-1} \cdot [\mathbf{f}_x(\mathbf{x},\xi(\mathbf{x})) \cdot \mathbf{v}]
\end{aligned}
\tag{1.37}
$$

as $h \to 0$.

Therefore,

$$
\begin{aligned}
\frac{\partial \xi(\mathbf{x})}{\partial \mathbf{v}} &= \lim_{h\to 0} \frac{\xi(\mathbf{x}+h\mathbf{v}) - \xi(\mathbf{x})}{h} \\
&= \lim_{h\to 0} \left(\frac{\mathbf{u}}{h}\right) \\
&= -[\mathbf{f}_y(\mathbf{x},\xi(\mathbf{x}))]^{-1} \cdot [\mathbf{f}_x(\mathbf{x},\xi(\mathbf{x})) \cdot \mathbf{v}]
\end{aligned}
\tag{1.38}
$$

The proof is complete.

1.5   Lagrange multipliers

In this section, we develop necessary conditions for extremals with equality restrictions by using the implicit function theorem for vectorial functions.

Let $D \subset \mathbb{R}^4$ be an open set. Let $f, g, h : D \to \mathbb{R}$ be functions of C^1 class. Suppose $(x_0, y_0, u_0, v_0) \in D$ is a point of local minimum of f subject to

$$
\begin{cases}
g(x, y, u, v) = 0 \\
h(x, y, u, v) = 0
\end{cases}
\tag{1.39}
$$

Suppose that

$$
det \begin{bmatrix} g_u(\mathbf{x}_0) & g_v(\mathbf{x}_0) \\ h_u(\mathbf{x}_0) & h_v(\mathbf{x}_0) \end{bmatrix} \neq 0
\tag{1.40}
$$

Under such hypotheses, since

$$
\begin{cases}
g(x_0, y_0, u_0, v_0) = 0 \\
h(x_0, y_0, u_0, v_0) = 0
\end{cases}
\tag{1.41}
$$

from the implicit theorem for vectorial functions, the equations

$$
\begin{cases}
g(x, y, u, v) = 0 \\
h(x, y, u, v) = 0
\end{cases}
\tag{1.42}
$$

implicitly define the functions $u(x,y)$, $v(x,y)$ at $B_\delta(x_0, y_0)$ for some $\delta > 0$, so that

$$
\begin{cases}
g(x, y, u(x,y), v(x,y)) = 0 \\
h(x, y, u(x,y), v(x,y)) = 0, \ \forall(x,y) \in B_\delta(x_0, y_0).
\end{cases}
\tag{1.43}
$$

Therefore, the original problem of restricted optimization will correspond to the local minimization of the function $F : B_\delta(x_0, y_0) \to \mathbb{R}$ given by

$$F(x,y) = f(x,y,u(x,y),v(x,y)).$$

Hence, the necessary conditions for a local minimum will be:

$$F_x(x_0, y_0) = 0 \text{ and } F_y(x_0, y_0) = 0.$$

Observe that, from $F_x(x_0, y_0) = 0$, we obtain:

$$f_x(\mathbf{x}_0) + f_u(\mathbf{x}_0)u_x(x_0, y_0) + f_v(\mathbf{x}_0)v_x(x_0, y_0) = 0, \tag{1.44}$$

and from $F_y(x_0, y_0) = 0$, we obtain:

$$f_y(\mathbf{x}_0) + f_u(\mathbf{x}_0)u_y(x_0, y_0) + f_v(\mathbf{x}_0)v_y(x_0, y_0) = 0. \tag{1.45}$$

On the other hand, from

$$\begin{cases} g(x,y,u(x,y),v(x,y)) = 0 \\ h(x,y,u(x,y),v(x,y)) = 0, \ \forall(x,y) \in B_\delta(x_0, y_0). \end{cases} \tag{1.46}$$

we get

$$\frac{dg(\mathbf{x}_0)}{dx} = g_x(\mathbf{x}_0) + g_u(\mathbf{x}_0)u_x(x_0, y_0) + g_v(\mathbf{x}_0)v_x(x_0, y_0) = 0,$$

and

$$\frac{dh(\mathbf{x}_0)}{dx} = h_x(\mathbf{x}_0) + h_u(\mathbf{x}_0)u_x(x_0, y_0) + h_v(\mathbf{x}_0)v_x(x_0, y_0) = 0,$$

so that

$$\frac{dg(\mathbf{x}_0)}{dy} = g_y(\mathbf{x}_0) + g_u(\mathbf{x}_0)u_y(x_0, y_0) + g_v(\mathbf{x}_0)v_y(x_0, y_0) = 0,$$

and

$$\frac{dh(\mathbf{x}_0)}{dy} = h_y(\mathbf{x}_0) + h_u(\mathbf{x}_0)u_y(x_0, y_0) + ghv(\mathbf{x}_0)v_y(x_0, y_0) = 0,$$

so that

$$\begin{bmatrix} u_x(x_0, y_0) \\ v_x(x_0, y_0) \end{bmatrix} = - \begin{bmatrix} g_u(\mathbf{x}_0) & g_v(\mathbf{x}_0) \\ h_u(\mathbf{x}_0) & h_v(\mathbf{x}_0) \end{bmatrix}^{-1} \begin{bmatrix} g_x(\mathbf{x}_0) \\ h_x(\mathbf{x}_0) \end{bmatrix} \tag{1.47}$$

and

$$\begin{bmatrix} u_y(x_0, y_0) \\ v_y(x_0, y_0) \end{bmatrix} = - \begin{bmatrix} g_u(\mathbf{x}_0) & g_v(\mathbf{x}_0) \\ h_u(\mathbf{x}_0) & h_v(\mathbf{x}_0) \end{bmatrix}^{-1} \begin{bmatrix} g_y(\mathbf{x}_0) \\ h_y(\mathbf{x}_0) \end{bmatrix}. \tag{1.48}$$

Thus,

$$f_x(\mathbf{x}_0) + [f_u \quad f_v] \begin{bmatrix} u_x(x_0, y_0) \\ v_x(x_0, y_0) \end{bmatrix} = 0, \tag{1.49}$$

so that

$$f_x(\mathbf{x}_0) - [f_u \ \ f_v] \begin{bmatrix} g_u(\mathbf{x}_0) & g_v(\mathbf{x}_0) \\ h_u(\mathbf{x}_0) & h_v(\mathbf{x}_0) \end{bmatrix}^{-1} \begin{bmatrix} g_x(\mathbf{x}_0) \\ h_x(\mathbf{x}_0) \end{bmatrix} = 0. \qquad (1.50)$$

and thus

$$f_x(\mathbf{x}_0) + \lambda_1 g_x(\mathbf{x}_0) + \lambda_2 h_x(\mathbf{x}_0) = 0$$

where:

$$[\lambda_1 \ \ \lambda_2] = -[f_u \ \ f_v] \begin{bmatrix} g_u(\mathbf{x}_0) & g_v(\mathbf{x}_0) \\ h_u(\mathbf{x}_0) & h_v(\mathbf{x}_0) \end{bmatrix}^{-1}. \qquad (1.51)$$

Similarly, we may obtain:

$$f_y(\mathbf{x}_0) + \lambda_1 g_y(\mathbf{x}_0) + \lambda_2 h_y(\mathbf{x}_0) = 0.$$

Finally, observe that

$$
\begin{aligned}
& f_u(\mathbf{x}_0) + \lambda_1 g_u(\mathbf{x}_0) + \lambda_2 h_u(\mathbf{x}_0) \\
= \ & f_u + [\lambda_1 \ \ \lambda_2] \begin{bmatrix} g_u(\mathbf{x}_0) \\ h_u(\mathbf{x}_0) \end{bmatrix} \\
= \ & f_u - [f_u \ \ f_v] \begin{bmatrix} g_u(\mathbf{x}_0) & g_v(\mathbf{x}_0) \\ h_u(\mathbf{x}_0) & h_v(\mathbf{x}_0) \end{bmatrix}^{-1} \begin{bmatrix} g_u(\mathbf{x}_0) \\ h_u(\mathbf{x}_0) \end{bmatrix} \\
= \ & f_u - [f_u \ \ f_v] \begin{bmatrix} h_v(\mathbf{x}_0) & -g_v(\mathbf{x}_0) \\ -h_u(\mathbf{x}_0) & g_u(\mathbf{x}_0) \end{bmatrix} \begin{bmatrix} g_u(\mathbf{x}_0) \\ h_u(\mathbf{x}_0) \end{bmatrix} / (g_u h_v - h_u g_v) \\
= \ & f_u - [f_u \ \ f_v] \begin{bmatrix} 1 \\ 0 \end{bmatrix} \\
= \ & f_u - f_u = 0, \qquad (1.52)
\end{aligned}
$$

Similarly, we obtain:

$$f_v + \lambda_1 g_v + \lambda_2 h_v = 0.$$

We may summarize the set of necessary conditions as indicated below:

$$\begin{cases} f_x + \lambda_1 g_x + \lambda_2 h_x = 0, \\ f_y + \lambda_1 g_y + \lambda_2 h_y = 0, \\ f_u + \lambda_1 g_u + \lambda_2 h_u = 0, \\ f_v + \lambda_1 g_v + \lambda_2 h_v = 0, \\ g = 0 \\ h = 0, \end{cases} \qquad (1.53)$$

with all functions in question considered at the point $\mathbf{x}_0 = (x_0, y_0, u_0, v_0)$.

1.6 Lagrange multipliers, the general case

Let $D \subset \mathbb{R}^{n+m}$ be a non-empty open set, where we denote

$$(\mathbf{x}, \mathbf{y}) \in D \subset \mathbf{R}^{n+m}$$

and where

$$\mathbf{x} = (x_1, \cdots, x_n) \in \mathbb{R}^n$$

and

$$\mathbf{y} = (y_1, \cdots, y_m) \in \mathbb{R}^m.$$

Let $f, g_1, \cdots, g_m : D \to \mathbb{R}$ be functions of C^1 class. Suppose that

$$(\mathbf{x}_0, \mathbf{y}_0) \in D$$

is a point of local minimum of f subject to

$$\begin{cases} g_1(\mathbf{x}, \mathbf{y}) = 0, \\ g_2(\mathbf{x}, \mathbf{y}) = 0, \\ \quad \vdots \\ g_m(\mathbf{x}, \mathbf{y}) = 0. \end{cases} \tag{1.54}$$

Thus, there exists $\delta > 0$ such that

$$f(\mathbf{x}, \mathbf{y}) \geq f(\mathbf{x}_0, \mathbf{y}_0),$$

$\forall (\mathbf{x}, \mathbf{y}) \in B_\delta(\mathbf{x}_0, \mathbf{y}_0)$ such that ,

$$\begin{cases} g_1(\mathbf{x}, \mathbf{y}) = 0, \\ g_2(\mathbf{x}, \mathbf{y}) = 0, \\ \quad \vdots \\ g_m(\mathbf{x}, \mathbf{y}) = 0. \end{cases} \tag{1.55}$$

We shall define, $G : D \to \mathbb{R}^m$ by

$$G(\mathbf{x}, \mathbf{y}) = \begin{bmatrix} g_1(\mathbf{x}, \mathbf{y}) = 0 \\ g_2(\mathbf{x}, \mathbf{y}) = 0 \\ \vdots \\ g_m(\mathbf{x}, \mathbf{y}) = 0 \end{bmatrix} \tag{1.56}$$

Therefore, the restriction in question is equivalent to

$$G(\mathbf{x}, \mathbf{y}) = \mathbf{0} \in \mathbb{R}^m.$$

Denoting

$$G_y(\mathbf{x}_0, \mathbf{y}_0) = \left\{ \frac{\partial g_i(\mathbf{x}_0, \mathbf{y}_0)}{\partial y_j} \right\},$$

where $i \in \{1, \cdots, m\}$ and $j \in \{1, \cdots, m\}$, assume

$$\det(G_y(\mathbf{x}_0, \mathbf{y}_0)) \neq 0.$$

From the implicit theorem, vectorial case, there exist $\delta_1 > 0$ and $\delta_2 > 0$ such that, for each $\mathbf{x} \in B_{\delta_1}(\mathbf{x}_0)$, there exists a unique $\mathbf{y} \in \overline{B}_{\delta_2}(\mathbf{y}_0)$ such that

$$G(\mathbf{x}, \mathbf{y}) = \mathbf{0}.$$

Denoting such a $\mathbf{y}$ by $\mathbf{y} = \mathbf{y}(\mathbf{x})$, we obtain

$$G(\mathbf{x}, \mathbf{y}(\mathbf{x})) = \mathbf{0}, \ \forall \mathbf{x} \in B_{\delta_1}(\mathbf{x}_0).$$

Therefore, the original constrained optimization problem may be seen as the non-constrained local minimization of $F : B_{\delta_1}(\mathbf{x}_0)$, where

$$F(\mathbf{x}) = f(\mathbf{x}, \mathbf{y}(\mathbf{x})).$$

Observe that $\mathbf{x}_0$ is a point of local minimum for F so that the first order optimality conditions stand for,

$$F_{x_j}(\mathbf{x}_0) = 0, \forall j \in \{1, \cdots, n\}.$$

Thus,

$$\frac{df(\mathbf{x}_0, \mathbf{y}(\mathbf{x}_0))}{dx_j} = 0,$$

that is,

$$\frac{\partial f(\mathbf{x}_0, \mathbf{y}_0)}{\partial x_j} + \sum_{k=1}^{m} \frac{\partial f(\mathbf{x}_0, \mathbf{y}_0)}{\partial y_k} \frac{\partial y_k(\mathbf{x}_0)}{\partial x_j} = 0, \tag{1.57}$$

$\forall j \in \{1, \cdots, n\}$.

On the other hand, from implicit function theorem, vectorial case, we have

$$\left\{ \frac{\partial y_k(\mathbf{x}_0)}{\partial x_j} \right\} = -[G_y(\mathbf{x}_0, \mathbf{y}_0)]^{-1} [G_{x_j}(\mathbf{x}_0, \mathbf{y}_0)]$$

$$= -\left\{ \frac{\partial g_i(\mathbf{x}_0, \mathbf{y}_0)}{\partial y_p} \right\}_{m \times m}^{-1} \begin{bmatrix} (g_1)_{x_j}(\mathbf{x}_0, \mathbf{y}_0) \\ (g_2)_{x_j}(\mathbf{x}_0, \mathbf{y}_0) \\ \vdots \\ (g_m)_{x_j}(\mathbf{x}_0, \mathbf{y}_0) \end{bmatrix}_{m \times 1}, \tag{1.58}$$

where $i \in \{1, \cdots, m\}$, $p \in \{1, \cdots, m\}$, $\forall i \in \{1, \cdots, n\}$. From this and (1.57), we obtain

$$
\frac{\partial f(\mathbf{x}_0, \mathbf{y}_0)}{\partial x_j} - \begin{bmatrix} \frac{\partial f(\mathbf{x}_0, \mathbf{y}_0)}{\partial y_1} & \cdots & \frac{\partial f(\mathbf{x}_0, \mathbf{y}_0)}{\partial y_m} \end{bmatrix}_{1 \times m} \begin{bmatrix} (y_1)_{x_j}(\mathbf{x}_0) \\ (y_2)_{x_j}(\mathbf{x}_0) \\ \vdots \\ (y_m)_{x_j}(\mathbf{x}_0) \end{bmatrix}_{m \times 1}
$$

$$
= \frac{\partial f(\mathbf{x}_0, \mathbf{y}_0)}{\partial x_j}
$$

$$
- \begin{bmatrix} \frac{\partial f(\mathbf{x}_0, \mathbf{y}_0)}{\partial y_1} & \cdots & \frac{\partial f(\mathbf{x}_0, \mathbf{y}_0)}{\partial y_m} \end{bmatrix}_{1 \times m} \begin{bmatrix} \frac{\partial g_i(\mathbf{x}_0, \mathbf{y}_0)}{\partial y_p} \end{bmatrix}^{-1}_{m \times m} \begin{bmatrix} (g_1)_{x_j}(\mathbf{x}_0, \mathbf{y}_0) \\ (g_2)_{x_j}(\mathbf{x}_0, \mathbf{y}_0) \\ \vdots \\ (g_m)_{x_j}(\mathbf{x}_0, \mathbf{y}_0) \end{bmatrix}_{m \times 1}
$$

$$
= \frac{\partial f(\mathbf{x}_0, \mathbf{y}_0)}{\partial x_j} +
$$

$$
\begin{bmatrix} \lambda_1 & \cdots & \lambda_m \end{bmatrix}_{1 \times m} \begin{bmatrix} (g_1)_{x_j}(\mathbf{x}_0, \mathbf{y}_0) \\ (g_2)_{x_j}(\mathbf{x}_0, \mathbf{y}_0) \\ \vdots \\ (g_m)_{x_j}(\mathbf{x}_0, \mathbf{y}_0) \end{bmatrix}_{m \times 1}
$$

$$
= 0, \tag{1.59}
$$

Indeed, this last equation stands for,

$$
\frac{\partial f(\mathbf{x}_0, \mathbf{y}_0)}{\partial x_j} + \sum_{k=1}^{m} \lambda_k \frac{\partial g_k(\mathbf{x}_0, \mathbf{y}_0)}{\partial x_j} = 0, \tag{1.60}
$$

$\forall j \in \{1, \cdots, n\}$, where,

$$
[\lambda_1, \cdots, \lambda_m] = - \begin{bmatrix} \frac{\partial f(\mathbf{x}_0, \mathbf{y}_0)}{\partial y_1} & \cdots & \frac{\partial f(\mathbf{x}_0, \mathbf{y}_0)}{\partial y_m} \end{bmatrix}_{1 \times m} \begin{bmatrix} \frac{\partial g_i(\mathbf{x}_0, \mathbf{y}_0)}{\partial y_p} \end{bmatrix}^{-1}_{m \times m}.
$$

On the other hand,

$$\frac{\partial f(\mathbf{x}_0, \mathbf{y}_0)}{\partial y_j} + \sum_{k=1}^{m} \lambda_k \frac{\partial g_k(\mathbf{x}_0, \mathbf{y}_0)}{\partial y_j}$$

$$= \frac{\partial f(\mathbf{x}_0, \mathbf{y}_0)}{\partial y_j}$$

$$+ \begin{bmatrix} \lambda_1 & \cdots & \lambda_m \end{bmatrix}_{1 \times m} \begin{bmatrix} (g_1)_{y_j}(\mathbf{x}_0, \mathbf{y}_0) \\ (g_2)_{y_j}(\mathbf{x}_0, \mathbf{y}_0) \\ \vdots \\ (g_m)_{y_j}(\mathbf{x}_0, \mathbf{y}_0) \end{bmatrix}_{m \times 1}$$

$$= \frac{\partial f(\mathbf{x}_0, \mathbf{y}_0)}{\partial x_j}$$

$$- \begin{bmatrix} \frac{\partial f(\mathbf{x}_0, \mathbf{y}_0)}{\partial y_1} & \cdots & \frac{\partial f(\mathbf{x}_0, \mathbf{y}_0)}{\partial y_m} \end{bmatrix}_{1 \times m} \begin{bmatrix} \frac{\partial g_i(\mathbf{x}_0, \mathbf{y}_0)}{\partial y_p} \end{bmatrix}_{m \times m}^{-1} \begin{bmatrix} (g_1)_{y_j}(\mathbf{x}_0, \mathbf{y}_0) \\ (g_2)_{y_j}(\mathbf{x}_0, \mathbf{y}_0) \\ \vdots \\ (g_m)_{y_j}(\mathbf{x}_0, \mathbf{y}_0) \end{bmatrix}_{m \times 1}.$$

$$(1.61)$$

At this point, define

$$\mathbf{z} = \begin{bmatrix} \frac{\partial g_i(\mathbf{x}_0, \mathbf{y}_0)}{\partial y_p} \end{bmatrix}_{m \times m}^{-1} \begin{bmatrix} (g_1)_{y_j}(\mathbf{x}_0, \mathbf{y}_0) \\ (g_2)_{y_j}(\mathbf{x}_0, \mathbf{y}_0) \\ \vdots \\ (g_m)_{y_j}(\mathbf{x}_0, \mathbf{y}_0) \end{bmatrix}_{m \times 1},$$

so that

$$\begin{bmatrix} (g_1)_{y_1}(\mathbf{x}_0, \mathbf{y}_0) & (g_1)_{y_2}(\mathbf{x}_0, \mathbf{y}_0) & \cdots & (g_1)_{y_m}(\mathbf{x}_0, \mathbf{y}_0) \\ (g_2)_{y_1}(\mathbf{x}_0, \mathbf{y}_0) & (g_2)_{y_2}(\mathbf{x}_0, \mathbf{y}_0) & \cdots & (g_2)_{y_m}(\mathbf{x}_0, \mathbf{y}_0) \\ \vdots & \vdots & \ddots & \vdots \\ (g_m)_{y_1}(\mathbf{x}_0, \mathbf{y}_0) & (g_m)_{y_2}(\mathbf{x}_0, \mathbf{y}_0) & \cdots & (g_m)_{y_m}(\mathbf{x}_0, \mathbf{y}_0) \end{bmatrix}_{m \times m} \begin{bmatrix} z_1 \\ z_2 \\ \vdots \\ z_m \end{bmatrix}_{m \times 1}$$

$$= \begin{bmatrix} (g_1)_{y_j}(\mathbf{x}_0, \mathbf{y}_0) \\ (g_2)_{y_j}(\mathbf{x}_0, \mathbf{y}_0) \\ \vdots \\ (g_m)_{y_j}(\mathbf{x}_0, \mathbf{y}_0) \end{bmatrix}_{m \times 1}.$$

$$(1.62)$$

From this last equation, it is clear that $z_j = 1$ and $z_k = 0$, if $k \neq j$. Therefore, we have obtained,

$$
\begin{aligned}
&\frac{\partial f(\mathbf{x}_0, \mathbf{y}_0)}{\partial y_j} + \sum_{k=1}^{m} \lambda_k \frac{\partial g_k(\mathbf{x}_0, \mathbf{y}_0)}{\partial y_j} \\
&= \frac{\partial f(\mathbf{x}_0, \mathbf{y}_0)}{\partial y_j} - \left[\frac{\partial f(\mathbf{x}_0, \mathbf{y}_0)}{\partial y_1} \cdots \frac{\partial f(\mathbf{x}_0, \mathbf{y}_0)}{\partial y_m} \right]_{1 \times m} \mathbf{z} \\
&= \frac{\partial f(\mathbf{x}_0, \mathbf{y}_0)}{\partial y_j} - \frac{\partial f(\mathbf{x}_0, \mathbf{y}_0)}{\partial y_j} \\
&= 0, \forall j \in \{1, \cdots, m\}.
\end{aligned} \tag{1.63}
$$

In summary, the first order optimality conditions for the local minimization of f subject to $G(\mathbf{x}, \mathbf{y}) = \mathbf{0}$ will be

$$
\frac{\partial f(\mathbf{x}_0, \mathbf{y}_0)}{\partial x_i} + \sum_{k=1}^{m} \lambda_k \frac{\partial g_k(\mathbf{x}_0, \mathbf{y}_0)}{\partial x_i} = 0, \ \forall i \in \{1, \cdots, n\},
$$

$$
\frac{\partial f(\mathbf{x}_0, \mathbf{y}_0)}{\partial y_j} + \sum_{k=1}^{m} \lambda_k \frac{\partial g_k(\mathbf{x}_0, \mathbf{y}_0)}{\partial y_j} = 0, \ \forall j \in \{1, \cdots, m\},
$$

$$
g_k(\mathbf{x}_0, \mathbf{y}_0) = 0, \ \forall k \in \{1, \cdots, m\}.
$$

We have $n + 2m$ variables and $n + 2m$ equations.

Observe that such a system corresponds to a critical point of the Lagrangian L, where

$$
L(\mathbf{x}, \mathbf{y}, \lambda_1, \cdots, \lambda_m) = f(\mathbf{x}, \mathbf{y}) + \sum_{k=1}^{m} \lambda_k g_k(\mathbf{x}, \mathbf{y}).
$$

Finally, $\lambda_1, \cdots, \lambda_m$ are said to be the Lagrange multipliers relating the corresponding constraints of the original problem.

1.7 Inverse function theorem

Theorem 1.7.1 *Let $D \subset \mathbb{R}^n$ be an open set and let $\mathbf{f} : D \to \mathbb{R}^n$ be a C^1 class function on D.*

Let $\mathbf{x}_0 \in D$ be such that $\det(\mathbf{f}'(\mathbf{x}_0)) \neq 0$.

Under such hypotheses, denoting $\mathbf{y}_0 = \mathbf{f}(\mathbf{x}_0)$, we have

1. *There exist open sets $U, V \subset \mathbb{R}^n$ such that $\mathbf{x}_0 \in U$ and $\mathbf{y}_0 \in V$ and such that $\mathbf{f}(U) = V$. Moreover, $\mathbf{f}$ is injective on U.*

2. *Defining the local inverse of $\mathbf{f}$, denoted by*

$$
\mathbf{f}^{-1} = \mathbf{g} : V \to U,
$$

by

$$\mathbf{g}(\mathbf{y}) = \mathbf{x} \Leftrightarrow \mathbf{y} = \mathbf{f}(\mathbf{x}), \, \forall \mathbf{x} \in U, \, \mathbf{y} \in V$$

we have that

$$\mathbf{g}'(\mathbf{y}) = [\mathbf{f}'(\mathbf{g}(\mathbf{y}))]^{-1}.$$

Moreover, $\mathbf{g}$ *is also of* C^1 *class.*

Proof 1.9 Denote $\mathbf{f}'(\mathbf{x}_0) = A$. From the hypotheses, $det(A) \neq 0$.
Define $\lambda = \frac{1}{2|A^{-1}|}$.

Since $\mathbf{f}'$ is continuous on D, there exists $\delta_1 > 0$ such that if $\mathbf{x} \in \overline{B}_{\delta_1}(\mathbf{x}_0)$ then

$$det(\mathbf{f}'(\mathbf{x})) \neq 0,$$

and

$$|\mathbf{f}'(\mathbf{x}) - A| = |\mathbf{f}'(\mathbf{x}) - \mathbf{f}'(\mathbf{x}_0)| < \lambda.$$

Fix $\mathbf{y} \in \mathbb{R}^n$ and define

$$\phi(\mathbf{x}) = \mathbf{x} + A^{-1}(\mathbf{y} - \mathbf{f}(\mathbf{x})).$$

Hence,

$$
\begin{aligned}
\phi'(\mathbf{x}) &= I + A^{-1}(-\mathbf{f}'(\mathbf{x})) \\
&= A^{-1}A - A^{-1}(\mathbf{f}'(\mathbf{x})) \\
&= A^{-1}(A - \mathbf{f}'(\mathbf{x})). \quad\quad (1.64)
\end{aligned}
$$

Therefore,

$$|\phi'(\mathbf{x})| \leq |A^{-1}||A - \mathbf{f}'(\mathbf{x})| \leq \frac{1}{2\lambda}\lambda = \frac{1}{2}.$$

Thus,

$$|\phi'(\mathbf{x})| \leq \frac{1}{2}, \, \forall \mathbf{x} \in \overline{B}_{\delta_1}(\mathbf{x}), \, \mathbf{y} \in \mathbb{R}^n.$$

From the mean value inequality, we obtain

$$|\phi(\mathbf{x}_2) - \phi(\mathbf{x}_1)| \leq \frac{1}{2}|\mathbf{x}_2 - \mathbf{x}_1|,$$

$\forall \mathbf{x}_1, \mathbf{x}_2 \in \overline{B}_{\delta_1}(\mathbf{x}_0), \, \mathbf{y} \in \mathbb{R}^n$.

Define $\delta_2 = \frac{\delta_1}{2|A^{-1}|}$.

Let $\mathbf{y} \in B_{\delta_2}(\mathbf{y}_0)$.

Observe that, for such a specific $\mathbf{y}$, we have

$$
\begin{aligned}
|\phi(\mathbf{x}_0) - \mathbf{x}_0| &= |A^{-1}(\mathbf{y} - \mathbf{f}(\mathbf{x}_0))| \\
&\leq |A^{-1}||\mathbf{y} - \mathbf{y}_0| \\
&\leq \frac{|A^{-1}|\delta_1}{2|A^{-1}|} \\
&= \frac{\delta_1}{2}, \quad\quad (1.65)
\end{aligned}
$$

so that

$$|\phi(\mathbf{x}_0) - \mathbf{x}_0| < \frac{\delta_1}{2}.$$

Thus, for $\mathbf{x} \in B_{\delta_1}(\mathbf{x}_0)$, we have that

$$
\begin{aligned}
|\phi(\mathbf{x}) - \mathbf{x}_0| &= |\phi(\mathbf{x}) - \phi(\mathbf{x}_0) + \phi(\mathbf{x}_0) - \mathbf{x}_0| \\
&\leq |\phi(\mathbf{x}) - \phi(\mathbf{x}_0)| + \frac{\delta_1}{2} \\
&\leq \frac{1}{2}|\mathbf{x} - \mathbf{x}_0| + \frac{\delta_1}{2} \\
&< \frac{\delta_1}{2} + \frac{\delta_1}{2} = \delta_1.
\end{aligned}
\tag{1.66}
$$

Therefore, we may infer that

$$|\phi(\mathbf{x}) - \mathbf{x}_0| < \delta_1, \ \forall \mathbf{x} \in \overline{B}_{\delta_1}(\mathbf{x}_0),$$

so that,

$$\phi(\overline{B}_{\delta_1}(\mathbf{x}_0)) \subset \overline{B}_{\delta_1}(\mathbf{x}_0),$$

in fact, from the exposed above,

$$\phi(\overline{B}_{\delta_1}(\mathbf{x}_0)) \subset B_{\delta_1}(\mathbf{x}_0).$$

We may conclude that ϕ is a contractor mapping on $\overline{B}_{\delta_1}(\mathbf{x}_0)$.

Thus, from the contractor mapping theorem, there exists a unique $\mathbf{x} \in B_\delta(\mathbf{x}_0)$ such that $\phi(\mathbf{x}) = \mathbf{x}$.

From

$$\phi(\mathbf{x}) = \mathbf{x} + A^{-1}(\mathbf{y} - \mathbf{f}(\mathbf{x})),$$

we obtain

$$A^{-1}(\mathbf{y} - \mathbf{f}(\mathbf{x})) = \mathbf{0},$$

and since $det(A^{-1}) \neq 0$, we get

$$\mathbf{y} - \mathbf{f}(\mathbf{x}) = \mathbf{0},$$

that is,

$$\mathbf{y} = \mathbf{f}(\mathbf{x}),$$

where such a $\mathbf{x} \in B_{\delta_1}(\mathbf{x}_0)$ is unique, $\forall \mathbf{y} \in B_{\delta_2}(\mathbf{y}_0)$.

Hence, $\mathbf{g}(\mathbf{y}) = \mathbf{x}$.

Define

$$V = B_{\delta_2}(\mathbf{y}_0).$$

We claim that $\mathbf{g}$ is injective on V. Suppose $\mathbf{y}_1, \mathbf{y}_2 \in V$ are such that

$$\mathbf{g}(\mathbf{y}_1) = \mathbf{g}(\mathbf{y}_2) = \mathbf{x}.$$

From above
$$\mathbf{y}_1 = \mathbf{y}_2 = \mathbf{f}(\mathbf{x}),$$
so we may conclude that $\mathbf{g}$ is injective on V.

Define $U = f^{-1}(V) \cap B_{\delta_1}(x_0)$. From the injectivity of $\mathbf{g}$ on V, we infer that $\mathbf{f}$ is injective on U.

Observe that U and V are open sets and
$$\mathbf{f}(U) = V.$$

Finally, $\mathbf{g} : V \to U$ is such that
$$\mathbf{g}(\mathbf{y}) = \mathbf{x} \Leftrightarrow \mathbf{y} = \mathbf{f}(\mathbf{x}).$$

For the next item, let $\mathbf{y}, \mathbf{v} \in \mathbb{R}^n$ be such that $\mathbf{y}$ and $\mathbf{y} + \mathbf{v} \in V$. Thus, there exists $\mathbf{x}, \mathbf{x}_1 \in U$ such that
$$\mathbf{y} = \mathbf{f}(\mathbf{x}) \text{ and } \mathbf{y} + \mathbf{v} = \mathbf{f}(\mathbf{x}_1).$$

Here, we denote
$$\mathbf{h} = \mathbf{x}_1 - \mathbf{x},$$
so that $\mathbf{x}_1 = \mathbf{x} + \mathbf{h}$, and therefore
$$\mathbf{y} + \mathbf{v} = \mathbf{f}(\mathbf{x} + \mathbf{h}),$$
moreover,
$$\mathbf{g}(\mathbf{y}) = \mathbf{x},$$
and
$$\mathbf{g}(\mathbf{y} + \mathbf{v}) = \mathbf{x} + \mathbf{h}.$$

Observe that, from the exposed above,
$$|\phi(\mathbf{x}+\mathbf{h}) - \phi(\mathbf{x})| \leq \frac{1}{2}|\mathbf{h}|, \tag{1.67}$$
where
$$\phi(\mathbf{x}+\mathbf{h}) = \mathbf{x} + \mathbf{h} + A^{-1}(\mathbf{y} - \mathbf{f}(\mathbf{x}+\mathbf{h})),$$
$$\phi(\mathbf{x}) = \mathbf{x} + A^{-1}(\mathbf{y} - \mathbf{f}(\mathbf{x})),$$
therefore, from this and (1.67), we obtain,
$$\begin{aligned}
\frac{1}{2}|\mathbf{h}| &\geq |\phi(\mathbf{x}+\mathbf{h}) - \phi(\mathbf{x})| \\
&= |\mathbf{h} + A^{-1}(\mathbf{f}(\mathbf{x}) - \mathbf{f}(\mathbf{x}+\mathbf{h}))| \\
&= |\mathbf{h} + A^{-1}(\mathbf{y} - (\mathbf{y}+\mathbf{v}))| \\
&= |\mathbf{h} - A^{-1}\mathbf{v}| \\
&\geq |\mathbf{h}| - |A^{-1}\mathbf{v}|, \tag{1.68}
\end{aligned}$$

so that

$$\frac{1}{2}|\mathbf{h}| \geq |\mathbf{h}| - |A^{-1}\mathbf{v}|,$$

and thus,

$$|A^{-1}\mathbf{v}| \geq \frac{1}{2}|\mathbf{h}|.$$

From this, we may infer that

$$|\mathbf{h}| \leq 2|A^{-1}||\mathbf{v}| \leq \lambda^{-1}|\mathbf{v}|, \tag{1.69}$$

so that

$$\frac{1}{|\mathbf{v}|} \leq \frac{1}{\lambda|\mathbf{h}|}.$$

Observe that, from (1.69), we have, $\mathbf{h} \to \mathbf{0}$ as $\mathbf{v} \to \mathbf{0}$, so that $\mathbf{g}$ is continuous on V.

Observe also that $det(\mathbf{f}'(\mathbf{x})) \neq 0$ and define

$$T = [\mathbf{f}'(\mathbf{x})]^{-1}.$$

Therefore,

$$\begin{aligned}
\mathbf{g}(\mathbf{y}+\mathbf{v}) - \mathbf{g}(\mathbf{y}) - T\mathbf{v} &= \mathbf{h} - T\mathbf{v} \\
&= TT^{-1}\mathbf{h} - T\mathbf{v} \\
&= -T(\mathbf{v} - T^{-1}\mathbf{h}) \\
&= -T(\mathbf{f}(\mathbf{x}+\mathbf{h}) - \mathbf{f}(\mathbf{x}) - \mathbf{f}'(\mathbf{x})\mathbf{h}), \tag{1.70}
\end{aligned}$$

so that

$$\begin{aligned}
\frac{|\mathbf{g}(\mathbf{y}+\mathbf{v}) - \mathbf{g}(\mathbf{y}) - T\mathbf{v}|}{|\mathbf{v}|} &\leq \frac{|-T||\mathbf{f}(\mathbf{x}+\mathbf{h}) - \mathbf{f}(\mathbf{x}) - \mathbf{f}'(\mathbf{x})\mathbf{h}|}{|\mathbf{v}|} \\
&\leq \frac{|T|}{\lambda} \frac{|\mathbf{f}(\mathbf{x}+\mathbf{h}) - \mathbf{f}(\mathbf{x}) - \mathbf{f}'(\mathbf{x})\mathbf{h}|}{|\mathbf{h}|} \\
&\to 0 \quad \text{as } \mathbf{v} \to \mathbf{0}. \tag{1.71}
\end{aligned}$$

From this, we may conclude that g is differentiable at $\mathbf{y}$ and also

$$\mathbf{g}'(\mathbf{y}) = T = [\mathbf{f}'(\mathbf{x})]^{-1} = [\mathbf{f}'(\mathbf{g}(\mathbf{y}))]^{-1}.$$

Moreover, from $det[\mathbf{f}'(\mathbf{x})]^{-1} \neq 0$ on $B_{\delta_1}(\mathbf{x}_0)$, and with $\mathbf{g}$ continuous on V and $\mathbf{f}'$ continuous on $B_{\delta_1}(\mathbf{x}_0)$, we may conclude that $\mathbf{g}$ is of C^1 class on V.

The proof is complete.

Example 1.7.2 *In this example, we define* $\mathbf{f} : \mathbb{R}^2 \to \mathbb{R}^2$ *by*

$$\mathbf{f}(x,y) = \begin{pmatrix} f_1(\mathbf{x}) \\ f_2(\mathbf{x}) \end{pmatrix} \tag{1.72}$$

where

$$f_1(x,y) = \sin^2[\pi(x+y)] + y,$$

and

$$f_2(x,y) = \ln(x^2 + y^2 + 1).$$

Let $\mathbf{x}_0 = (1,0) \in \mathbb{R}^2$.
Observe that

$$\mathbf{f}'(x,y) = \begin{pmatrix} (f_1)_x(x,y) & (f_1)_y(x,y) \\ (f_2)_x(x,y) & (f_2)_y(x,y) \end{pmatrix}$$

$$= \begin{bmatrix} 2\sin[\pi(x+y)]\cos[\pi(x+y)]\pi & 2\sin[\pi(x+y)]\cos[\pi(x+y)]\pi + 1 \\ \frac{2x}{x^2+y^2+1} & \frac{2y}{x^2+y^2+1} \end{bmatrix} \quad (1.73)$$

Hence,

$$\mathbf{f}'(1,0) = \begin{bmatrix} 0 & 1 \\ 1 & 0 \end{bmatrix},$$

so that

$$\det(\mathbf{f}'(1,0)) = -1 \neq 0.$$

Thus, from the inverse function theorem, the inverse $\mathbf{g} = \mathbf{f}^{-1}$ *of f is well defined in a neighborhood of*

$$\mathbf{f}(\mathbf{x}_0) = \mathbf{f}(1,0) = (0, \ln 2)^T.$$

Also, from such a theorem,

$$\mathbf{g}'(\mathbf{f}(\mathbf{x}_0)) = \mathbf{g}'(0, \ln 2) = [\mathbf{f}'(\mathbf{x}_0)]^{-1} = [\mathbf{f}'(1,0)]^{-1},$$

that is,

$$\mathbf{g}'(0, \ln 2) = \begin{bmatrix} 0 & 1 \\ 1 & 0 \end{bmatrix}^{-1} = \begin{bmatrix} 0 & -1 \\ -1 & 0 \end{bmatrix} / (-1) = \begin{bmatrix} 0 & 1 \\ 1 & 0 \end{bmatrix}.$$

Exercises 1.7.3

1. *Let* $A, B \subset \mathbb{R}^n$ *be open sets.*
   Prove that $A \cup B$ *and* $A \cap B$ *are open.*

2. *Let* $\{A_\alpha, \ \alpha \in L\} \subset \mathbb{R}^n$ *be a collection of sets such that* A_α *is open* $\forall \alpha \in L$.
   Show that $\cup_{\alpha \in L} A_\alpha$ *is open.*

3. *Let* $\{A_\alpha, \ \alpha \in L\} \subset \mathbb{R}^n$ *be a collection of sets such that* A_α *is closed* $\forall \alpha \in L$.
   Show that $\cap_{\alpha \in L} A_\alpha$ *is closed.*

4. *Let* $A, B \subset \mathbb{R}^n$ *be closed sets.*
   Prove that $A \cup B$ *is closed.*

5. *Through the proof of the Heine-Borel for $\mathbb{R}$, prove a version of such a theorem for $\mathbb{R}^n$.*

   That is, prove that for $E \subset \mathbb{R}^n$, the following three properties are equivalent.

   (a) *E is compact.*

   (b) *E is closed and bounded.*

   (c) *Every infinite set contained in E has a limit point in E.*

6. *Given $A, B \subset \mathbb{R}^n$, we define the distance between A and B, denoted by $d(A,B)$, as*

   $$d(A,B) = \inf\{|u - v| \, : \, u \in A \text{ and } v \in B\}.$$

   Let $K, F \subset \mathbb{R}^n$ be sets such that K is compact, F is closed and

   $$K \cap F = \emptyset.$$

   Prove that $d(K,F) > 0$ and there exist $u_0 \in K$ and $v_0 \in F$ such that

   $$d(K,F) = |u_0 - v_0|.$$

7. *Let $K, V \subset \mathbb{R}^n$ be sets such that K is compact, V is open and*

   $$K \subset V.$$

   Prove that there exists an open set $W \subset \mathbb{R}^n$ such that

   $$K \subset W \subset \overline{W} \subset V.$$

8. *Calculate the limits:*

   (a)

   $$\lim_{(x,y) \to (0,0)} \frac{x - y}{\sqrt{x} - \sqrt{y}},$$

   (b)

   $$\lim_{(x,y) \to (2,2)} \frac{x + y - 4}{\sqrt{x + y} - 2},$$

9. *Prove formally that:*

   (a) $\lim_{(x,y) \to (1,3)} 3x - 5y + 7 = -5,$

   (b) $\lim_{(x,y) \to (-2,1)} -x + 4y + 4 = 10,$

   (c) $\lim_{(x,y) \to (1,3)} x^2 + y^2 - 2x + 1 = 9,$

   (d) $\lim_{(x,y) \to (-1,2)} 3x^2 - 2y^2 - 2x + 3y + 5 = 8.$

10. *For the functions given below, show that* $\lim_{(x,y)\to(0,0)} f(x,y)$ *does not exist, where*

(a)
$$f(x,y) = \frac{x}{\sqrt{x^2+y^2}},$$

(b)
$$f(x,y) = \frac{xy}{|xy|},$$

(c)
$$f(x,y) = \frac{x^2-y^2}{x^2+y^2},$$

(d)
$$f(x,y) = \frac{x^4+3x^2y^2+2xy^3}{(x^2+y^2)^2},$$

(e)
$$f(x,y) = \frac{x^9y}{(x^6+y^2)^2}.$$

11. *For the functions* $f : \mathbb{R}^2 \setminus \{(0,0)\} \to \mathbb{R}$ *below, show that* $\lim_{(x,y)\to(0,0)} f(x,y)$ *exists and calculate its value, where,*

(a)
$$f(x,y) = x\cos\left(\frac{1}{x^2+y^2}\right),$$

(b)
$$f(x,y) = \frac{x^2+3xy}{\sqrt{x^2+y^2}},$$

(c)
$$f(x,y) = \frac{x^2y+xy^2}{x^2+y^2},$$

(d)
$$f(x,y) = \cos\left(\frac{x^3-y^3}{x^2+y^2}\right).$$

12. *For the functions* $f : \mathbb{R}^2 \setminus \{(0,0)\} \to \mathbb{R}$ *below, calculate, if they exist, the limits* $\lim_{(x,y)\to(0,0)} f(x,y)$ *and discuss the possibility or not of such functions to be continuously extended to* $(0,0)$, *by appropriately defining* $f(0,0)$, *where*

(a)

$$f(x,y) = \ln\left(\frac{3x^4 - x^2y^2 + 3y^4}{x^2 + y^2} + 2\right),$$

(b)

$$f(x,y) = \ln\left(x^2\cos^2\left(\frac{1}{x^2+y^2}\right) + 3\right).$$

13. *Let $a, b, c \in \mathbb{R}$, where $a \neq 0$ or $b \neq 0$.*

   Prove formally that,

   $$\lim_{(x,y)\to(x_0,y_0)} ax + by + c = ax_0 + by_0 + c.$$

14. *Let $a, b, c, d, e, f \in \mathbb{R}$ where $a \neq 0$, $b \neq 0$ or $c \neq 0$.*

   Prove formally that,

   $$\lim_{(x,y)\to(x_0,y_0)} ax^2 + by^2 + cxy + dx + ey + f = ax_0^2 + by_0^2 + cx_0y_0 + dx_0 + ey_0 + f.$$

15. *Let $D \subset \mathbb{R}^n$ be an open set, let $f, g : D \to \mathbb{R}$ be real functions and let $\mathbf{x_0} \in D$.*

   Suppose there exists $K > 0$ and $\delta > 0$ such that $|g(\mathbf{x})| < K$, if $0 < |\mathbf{x} - \mathbf{x_0}| < \delta$.

   Assume

   $$\lim_{\mathbf{x}\to\mathbf{x_0}} f(\mathbf{x}) = 0.$$

   Under such hypotheses, prove that

   $$\lim_{\mathbf{x}\to\mathbf{x_0}} f(\mathbf{x})g(\mathbf{x}) = 0.$$

16. *Use the item 15 to prove that*

   $$\lim_{(x,y)\to(0,0)} f(x,y) = 0,$$

   where $f : \mathbb{R}^2 \to \mathbb{R}$ is defined by

   $$f(x,y) = \begin{cases} (x^2 + y^2 + x - y)\sin\left(\frac{1}{x^2+y^2}\right), & \text{if } (x,y) \neq (0,0) \\ 5, & \text{if } (x,y) = (0,0). \end{cases}$$

17. *Let $D \subset \mathbb{R}^n$ be an open set and let $\mathbf{x_0} \in D$.*

   Assume $f, g : D \to \mathbb{R}$ are such that

   $$\lim_{\mathbf{x}\to\mathbf{x_0}} f(\mathbf{x}) = L \in \mathbb{R}$$

and

$$\lim_{\mathbf{x}\to\mathbf{x_0}} g(\mathbf{x}) = M \in \mathbb{R},$$

where $L < M$.

Prove that there exists delta > 0 such that, if $0 < |\mathbf{x} - \mathbf{x_0}| < \delta$, then

$$f(\mathbf{x}) < \frac{L+M}{2} < g(\mathbf{x}).$$

Hint: Define $\varepsilon = \frac{M-L}{2}$.

18. *Let $D \subset \mathbb{R}^n$ be an open set and let $f : D \to \mathbb{R}$ be a continuous function. Assume $A \subset \mathbb{R}$ is open. Prove that $f^{-1}(A)$ is open, where,*

$$f^{-1}(A) = \{\mathbf{x} \in D \ : \ f(\mathbf{x}) \in A\}.$$

19. *Let $f : \mathbb{R}^n \to \mathbb{R}$ be a continuous function and let $c \in \mathbb{R}$.*

    Prove that the sets B and C are closed, where

    (a) *$B = \{\mathbf{x} \in \mathbb{R}^n \ : \ f(\mathbf{x}) \le c\}$.*
    (b) *$C = \{\mathbf{x} \in \mathbb{R}^n \ : \ f(\mathbf{x}) = c\}$.*

20. *Let $A, F \subset \mathbb{R}^n$ be such that A is open and F is closed. Prove that $A \setminus F$ is open and $F \setminus A$ is closed.*

21. *Let $D \subset \mathbb{R}^n$ be an open set and let $f : D \to \mathbb{R}$ be a continuous function.*

    Assume $F \subset \mathbb{R}$ is closed. Prove that there exists a closed set $F_1 \subset \mathbb{R}^n$ such that $f^{-1}(F) = D \cap F_1$, where

    $$f^{-1}(F) = \{\mathbf{x} \in D \ : \ f(\mathbf{x}) \in F\}.$$

22. *Let $f : \mathbb{R}^2 \to \mathbb{R}$ be such that*

    $$f(x,y) = \begin{cases} \frac{\sin(x+y)}{x+y}, & \text{if } x+y \ne 0 \\ 1, & \text{if } x+y = 0. \end{cases}$$

    Prove that f is continuous on $\mathbb{R}^2$.

23. *Let $f : \mathbb{R}^2 \to \mathbb{R}$ be such that*

    $$f(x,y) = \begin{cases} \frac{xy}{|x|+|y|}, & \text{if } (x,y) \ne (0,0) \\ 0, & \text{if } (x,y) = (0,0). \end{cases}$$

    Prove that f is continuous on $\mathbb{R}^2$.

24. *Through the definition of partial derivative, calculate*

$$\frac{\partial f(x,y)}{\partial x} \text{ and } \frac{\partial f(x,y)}{\partial y},$$

where

$$f(x,y) = x^2 - 3y^2 + x.$$

25. *Through the definition of partial derivative, for* $(x,y) \in \mathbb{R}^2$ *such that* $x + 2y > 0$, *calculate*

$$\frac{\partial f(x,y)}{\partial x} \text{ and } \frac{\partial f(x,y)}{\partial y},$$

where

$$f(x,y) = \frac{1}{\sqrt{x+2y}}.$$

26. *Through the definition of partial derivative, for* $(x,y) \in \mathbb{R}^2$ *such that* $x^2 - y \neq 0$, *calculate*

$$\frac{\partial f(x,y)}{\partial y},$$

where

$$f(x,y) = \frac{x+2y}{x^2 - y}.$$

27. *Through the definition of differentiability, prove that the functions below are differentiable on the respective domains,*

(a) $f(x,y) = 3x^2 - 2xy + 5y^2$,

(b) $f(x,y) = 2xy^2 - 3xy$,

(c) $f(x,y) = \frac{x^2}{y}$.

28. *Let* $f : \mathbb{R}^2 \to \mathbb{R}$ *be defined by*

$$f(x,y) = \begin{cases} \frac{(x^3 + y^3)}{x^2 + y^2}, & \text{if } (x,y) \neq (0,0) \\ 0, & \text{if } (x,y) = (0,0). \end{cases}$$

Calculate $f_x(0,0)$ *e* $f_y(0,0)$.

29. *Let* $f : \mathbb{R}^2 \to \mathbb{R}$ *defined by*

$$f(x,y) = \begin{cases} \frac{3x^2 y^2}{x^4 + y^4}, & \text{if } (x,y) \neq (0,0) \\ 0, & \text{if } (x,y) = (0,0). \end{cases}$$

Prove that $f_x(0,0)$ *and* $f_y(0,0)$ *exist, however* f *is not differentiable at* $(0,0)$.

30. *Let $f : \mathbb{R}^2 \to \mathbb{R}$ be defined by*

$$f(x,y) = \begin{cases} \frac{xy(x^2-y^2)}{x^2+y^2}, & \text{if } (x,y) \neq (0,0) \\ 0, & \text{if } (x,y) = (0,0). \end{cases}$$

Prove that $f_x(0,0)$ and $f_y(0,0)$ exist and f is differentiable at $(0,0)$.

31. *Let $f : \mathbb{R}^3 \to \mathbb{R}$ be defined by*

$$f(x,y,z) = \begin{cases} \frac{xyz^2}{x^2+y^2+z^2}, & \text{if } (x,y,z) \neq (0,0,0) \\ 0, & \text{if } (x,y,z) = (0,0,0). \end{cases}$$

Prove that f is differentiable at $(0,0,0)$.

32. *Let $f : \mathbb{R}^2 \to \mathbb{R}$ be defined by*

$$f(x,y) = \begin{cases} (x^2+y^2)\sin\left(\frac{1}{\sqrt{x^2+y^2}}\right), & \text{if } (x,y) \neq (0,0) \\ 0, & \text{if } (x,y) = (0,0). \end{cases}$$

   (a) *Obtain $\Delta f(0,0,\Delta x, \Delta y)$.*
   (b) *Calculate $f_x(0,0)$ e $f_y(0,0)$.*
   (c) *Through the definition of differentiability, show that f is differentiable at $(0,0)$.*

33. *For the functions below, obtain the respective domains and prove that they are differentiable (on the domains in question):*

   (a) $f(x,y) = \frac{x+y}{x^2+5y}$
   (b) $f(x,y) = y\ln x - x/y,$
   (c) $f(x,y) = \arctan(x^2 - y) + \frac{1}{\sqrt{x^2-y}},$

34. *Let $D \subset \mathbb{R}^n$ be an open connected set. Suppose all partial derivatives of f are zero on D.*

   Prove that f is constant on D.

35. *Let $D \subset \mathbb{R}^2$ be an open rectangle and let $f : D \to \mathbb{R}$ be a function. Assume f has partial derivatives well defined on D. Let (x,y) and $(x+u, y+v) \in D$.*

   Prove that there exists $\lambda \in (0,1)$ such that

$$f(x+u, y+v) - f(x,y) = f_x(x+\lambda u, y+v)u + f_y(x, y+\lambda v)v.$$

36. *Let $D \subset \mathbb{R}^n$ be an open convex set and let $f : D \to \mathbb{R}$ be a function. Suppose there exists $K > 0$ such that*

$$\left| \frac{\partial f(\mathbf{x})}{\partial x_j} \right| \leq K, \ \forall \mathbf{x} \in D, \ j \in \{1, .., n\}.$$

Prove that

$$|f(\mathbf{x}) - f(\mathbf{y})| \leq Kn|\mathbf{x} - \mathbf{y}|, \ \forall \mathbf{x}, \mathbf{y} \in D.$$

37. *Let $D \subset \mathbb{R}^n$ be an open set and let $f : D \to \mathbb{R}$ be a differentiable function at $\mathbf{x_0} \in D$. Prove that there exist $\delta > 0$ and $K > 0$ such that, if $|\mathbf{h}| < \delta$, then $\mathbf{x_0} + \mathbf{h} \in D$ and*

$$|f(\mathbf{x_0} + \mathbf{h}) - f(\mathbf{x_0})| < K|\mathbf{h}|.$$

38. *Let $f : \mathbb{R}^n \setminus \{\mathbf{0}\} \to \mathbb{R}$ be defined by $f(\mathbf{x}) = |\mathbf{x}|^c$, where $c \in \mathbb{R}$. Let $\mathbf{x} = (x_1, ..., x_n)$ and $\mathbf{v} = (v_1, ..., v_n) \in \mathbb{R}^n$.*

Calculate

$$\nabla f(\mathbf{x}) \cdot \mathbf{v}.$$

39. *let $f : \mathbb{R}^3 \to \mathbb{R}$ be defined by*

$$f(x, y, t) = \frac{t^2 + y}{e^t + x^2 + t^2}.$$

Suppose the functions $x : \mathbb{R} \to \mathbb{R}$ and $y : \mathbb{R} \to \mathbb{R}$ be defined by

$$x(t) = \cos^2(t^3),$$

and

$$y(t) = e^{t^2}.$$

Through the chain rule, calculate $g'(t)$, where $g(t) = f(x(t), y(t), t), \ \forall t \in \mathbb{R}$.

Finally, obtain the equation of the tangent line to the graph of g at the points $t = 0$ and $t = \pi$.

40. *Let $g : \mathbb{R}^3 \to \mathbb{R}$ be defined by*

$$g(x, y, z) = \frac{x^2 + y^2 + xy}{z^2 + e^x + \cos^2(y)}.$$

Let $z(x, y) = \cos^2(x^2 + y^2)$ and define $h : \mathbb{R}^2 \to \mathbb{R}$ by

$$h(x, y) = g(x, y, z(x, y)).$$

Through the chain rule, calculate $h_x(x, y)$ and $h_y(x, y)$.

Find the equation of the normal line and the equation of the tangent plane, to the graph of h at the point $(1, 0)$.

41. Let $f : \mathbb{R} \to \mathbb{R}$ be a differentiable function. Let $u(x,y) = bx - ay$. Show that $z(x,y) = f(u(x,y))$ satisfies the equation,

$$a\frac{\partial z}{\partial x} + b\frac{\partial z}{\partial y} = 0.$$

42. Let $f : \mathbb{R}^2 \to \mathbb{R}$ be a differentiable function.

    Denoting $u(r,\theta) = f(x,y)$, where $x = r\cos\theta$ and $y = r\sin\theta$, show that

$$\frac{\partial u}{\partial x} = \frac{\partial u}{\partial r}\cos\theta - \frac{\partial u}{\partial \theta}\frac{\sin\theta}{r},$$

and

$$\frac{\partial u}{\partial y} = \frac{\partial u}{\partial r}\sin\theta + \frac{\partial u}{\partial \theta}\frac{\cos\theta}{r}.$$

43. Consider the ellipsoid of equation

$$\frac{x^2}{a^2} + \frac{y^2}{b^2} + \frac{z^2}{c^2} = 1,$$

    where $a,b,c > 0$.

    Find the closest points on such surface to the origin $(0,0,0)$.

44. Let A be a matrix $m \times n$. Let $\mathbf{y}_0 \in \mathbb{R}^m$ and let $f : \mathbb{R}^n \to \mathbb{R}$ be defined by

$$f(\mathbf{x}) = \langle (A\mathbf{x}), \mathbf{y}_0 \rangle,$$

    where $\langle \cdot, \cdot \rangle : \mathbb{R}^m \times \mathbb{R}^m \to \mathbb{R}$ denotes the usual inner product in $\mathbb{R}^m$. Through the method of Lagrange multipliers, find the points of minimum and maximum of $f(\mathbf{x})$ subject to $|\mathbf{x}| = 1$.

45. A function $f : \mathbb{R}^n \to \mathbb{R}$ is said to be convex if

$$f(\lambda\mathbf{x} + (1-\lambda)\mathbf{y}) \leq \lambda f(\mathbf{x}) + (1-\lambda)f(\mathbf{y}), \ \forall \mathbf{x}, \mathbf{y} \in \mathbb{R}^n, \ \lambda \in [0,1].$$

    (a) Suppose that $f : \mathbb{R}^n \to \mathbb{R}$ is differentiable. Show that f is convex if, and only if,

$$f(\mathbf{y}) - f(\mathbf{x}) \geq \nabla f(\mathbf{x}) \cdot (\mathbf{y} - \mathbf{x}), \ \forall \mathbf{x}, \mathbf{y} \in \mathbb{R}^n.$$

    (b) Prove that if f is convex, differentiable and $\nabla f(\mathbf{x}) = \mathbf{0}$ then $\mathbf{x} \in \mathbb{R}^n$ is a point of global minimum for f.

46. Let $f : \mathbb{R}^n \to \mathbb{R}$ be a twice differentiable function such that

$$H(\mathbf{x}) = \left\{ \frac{\partial^2 f(\mathbf{x})}{\partial x_i \partial x_j} \right\}$$

    is a positive definite matrix. $\forall \mathbf{x} \in \mathbb{R}^n$.

    Show that f is convex on $\mathbb{R}^n$.

47. Let $F,G : \mathbb{R}^4 \to \mathbb{R}$ be defined by $F(x,y,u,v) = x^2 + y^3 - u + v^2$ and $G(x,y,u,v) = e^{2x} + e^{3y} + 2uv + 3v^2$. Assuming the hypotheses of the vectorial case of implicit function theorem, consider the functions $u(x,y)$ and $v(x,y)$ implicitly defined on a neighborhood of a point $(x,y,u,v) \in \mathbb{R}^4$ such that

$$F(x,y,u,v) = 0 \quad and \quad G(x,y,u,v) = 0.$$

Find u_x, u_y, v_x and v_y on such neighborhood.

1.8   Some topics on differential geometry

Definition 1.8.1 (Limit) *Let $C \subset \mathbb{R}^3$ be a curve defined by a one variable vectorial function $\mathbf{r} : [a,b] \to \mathbb{R}^3$. Let $t_0 \in (a,b)$. We say that $\mathbf{A}$ is the limit of $\mathbf{r}$ as t approaches t_0, as, for each $\varepsilon > 0$, there exists $\delta > 0$ such that, if $t \in [a,b]$ and $0 < |t - t_0| < \delta$, then $|\mathbf{r}(t) - \mathbf{A}| < \varepsilon$.*

In such a case, we denote

$$\lim_{t \to t_0} \mathbf{r}(t) = \mathbf{A}.$$

Theorem 1.8.2 *Let $\mathbf{r} : [a,b] \to \mathbb{R}^3$ be point-wisely expressed by*

$$\mathbf{r}(t) = x_1(t)\mathbf{e}_1 + x_2(t)\mathbf{e}_2 + x_3(t)\mathbf{e}_3,$$

where $\{\mathbf{e}_1, \mathbf{e}_2, \mathbf{e}_3\}$ is the canonical basis for $\mathbb{R}^3$.
Let $t_0 \in [a,b]$.
Under such hypotheses,

$$\lim_{t \to t_0} \mathbf{r}(t) = A_1\mathbf{e}_1 + A_2\mathbf{e}_2 + A_3\mathbf{e}_3 \equiv \mathbf{A},$$

if, and only if,

$$\lim_{t \to t_0} x_j(t) = A_j, \ \forall j \in \{1,2,3\}.$$

Proof 1.10   Assume first

$$\lim_{t \to t_0} \mathbf{r}(t) = A_1\mathbf{e}_1 + A_2\mathbf{e}_2 + A_3\mathbf{e}_3.$$

Let $\varepsilon > 0$. Thus, there exists $\delta > 0$ such that if $t \in [a,b]$ and $0 < |t - t_0| < \delta$, then

$$|\mathbf{r}(t) - \mathbf{A}| < \varepsilon.$$

In particular

$$
\begin{aligned}
|x_k(t) - A_k| \ &\leq \ \sqrt{\sum_{k=1}^{3} |x_k(t) - A_k|^2} \\
&= \ |\mathbf{r}(t) - \mathbf{A}| < \varepsilon,
\end{aligned}
\tag{1.74}
$$

$\forall t \in [a,b]$, such that $0 < |t - t_0| < \delta$.

Thus,

$$\lim_{t \to t_0} x_k(t) = A_k, \ \forall k \in \{1,2,3\}.$$

Reciprocally, suppose that

$$\lim_{t \to t_0} x_k(t) = A_k, \ \forall k \in \{1,2,3\}.$$

Let $\varepsilon > 0$. Let $k \in \{1,2,3\}$.

Thus, there exists $\delta_k > 0$ such that, if $0 < |t - t_0| < \delta_k$, then

$$|x_k(t) - A_k| < \varepsilon/\sqrt{3}.$$

Therefore, denoting $\delta = \min\{\delta_1, \delta_2, \delta_3\}$, if $t \in [a,b]$ and $0 < |t - t_0| < \delta$, we have,

$$
\begin{aligned}
|\mathbf{r}(t) - \mathbf{A}| &= \sqrt{\sum_{k=1}^{3} |x_k(t) - A_k|^2} \\
&< \sqrt{\sum_{k=1}^{3} \varepsilon^2/3} \\
&= \sqrt{\varepsilon^2} \\
&= \varepsilon,
\end{aligned}
\tag{1.75}
$$

Therefore,

$$\lim_{t \to t_0} \mathbf{r}(t) = \mathbf{A}.$$

The proof is complete.

In the next lines, we present the definition of derivative for a one variable vectorial function.

Definition 1.8.3 (Derivative) *Let* $\mathbf{r} : [a,b] \to \mathbb{R}^3$ *be a one variable vectorial function.*

Let $t_0 \in (a,b)$. *We define the derivative of* $\mathbf{r}$ *relating* t *at* t_0, *denoted by* $\mathbf{r}'(t_0)$, *by*

$$\mathbf{r}'(t_0) = \lim_{h \to 0} \frac{\mathbf{r}(t_0 + h) - \mathbf{r}(t_0)}{h},$$

if such a limit exists.

Similarly, if $t_0 = a$ *or* $t_0 = b$ *we define the derivative of* $\mathbf{r}$ *at these values for* t_0, *through one-sided limits.*

Remark 1.8.4 *Considering this last definition and the last theorem, if*

$$\mathbf{r}(t) = \sum_{k=1}^{3} x_k(t) \mathbf{e}_k,$$

on $[a,b]$, for some $t \in [a,b]$, we may infer that $\mathbf{r}'(t)$ exists, if and only if, $x'_k(t)$ exists, $\forall k \in \{1,2,3\}$, and in such a case

$$\mathbf{r}'(t) = \sum_{k=1}^{3} x'_k(t)\mathbf{e}_k.$$

1.8.1 Arc length

Consider a curve $C \subset \mathbb{R}^3$ defined by the one variable vectorial differentiable function $\mathbf{r} : [a,b] \to \mathbb{R}^3$ point-wisely represented by

$$\mathbf{r}(t) = \sum_{k=1}^{3} x_k(t)\mathbf{e}_k.$$

Consider a partition P of $[a,b]$, given by

$$P = \{t_0 = a, t_1, \cdots, t_n = b\}.$$

Thus, a first approximation for the length of C, denoted by L, is given by

$$L \approx \sum_{j=1}^{n} |\Delta \mathbf{r}_j|,$$

where,

$$\begin{aligned} \Delta \mathbf{r}_j &= \mathbf{r}(t_j) - \mathbf{r}(t_{j-1}) \\ &= \frac{d\mathbf{r}(t_j)}{dt}\Delta t_j + \mathscr{O}(\Delta t_j^2), \end{aligned} \tag{1.76}$$

where $\Delta t_j = t_j - t_{j-1}$, $\forall j \in \{1, \cdots, n-1\}$.
If $|P| = \max\{\Delta t_j, \; j \in \{1, \cdots, n-1\}\}$ is small enough, we have

$$\begin{aligned} L &\approx \sum_{j=1}^{n} |\Delta \mathbf{r}_j| \\ &\approx \sum_{j=1}^{m} \left|\frac{d\mathbf{r}(t_j)}{dt}\right|\Delta t_j \\ &\equiv S_{\mathbf{r}}^P. \end{aligned} \tag{1.77}$$

With such an approximation in mind, assuming $\mathbf{r}$ is of C^1 class or at least piecewise continuous, we define the length of C, by

$$L = \lim_{|P| \to 0} S_{\mathbf{r}}^P = \int_a^b \left|\frac{d\mathbf{r}(t)}{dt}\right| dt,$$

so that

$$L = \int_a^b \sqrt{\sum_{j=1}^{3} (x'_j(t))^2} \, dt,$$

1.8.2 The arc length function

Definition 1.8.5 *consider a C^1 class curve defined by the function $\mathbf{r} : [a,b] \to \mathbb{R}^3$.*
We define the relating arc length function to C, denoted by $s : [a,b] \to \mathbb{R}^+$,
point-wisely by

$$s(t) = \int_a^t \sqrt{\mathbf{r}'(u) \cdot \mathbf{r}'(u)} \, du, \ \forall t \in [a,b].$$

Hence, $s(t)$ provides a measure of the arc length of C between a and t.

Observe that

$$\frac{ds(t)}{dt} = \sqrt{\mathbf{r}'(t) \cdot \mathbf{r}'(t)},$$

so that

$$
\begin{aligned}
ds(t) &= \sqrt{\mathbf{r}'(t) \cdot \mathbf{r}'(t)} dt \\
&= \sqrt{d\mathbf{r}(t) \cdot d\mathbf{r}(t)},
\end{aligned}
\tag{1.78}
$$

and thus, in a differential form, we may denote,

$$ds(t)^2 = d\mathbf{r}(t) \cdot d\mathbf{r}(t).$$

Moreover, we denote,

$$
\begin{aligned}
s(t) &= \int_a^t \sqrt{\overline{\mathbf{r}'(u) \cdot \mathbf{r}'(u)}} \, du \\
&= f(t),
\end{aligned}
\tag{1.79}
$$

so that

$$t = f^{-1}(s) \equiv t(s).$$

With such results in mind, we define the parametrization of C by its arc length, denoted point-wisely by $\hat{\mathbf{r}}(s)$, by

$$\hat{\mathbf{r}}(s) = \mathbf{r}(t(s)).$$

When the meaning is clear, we denote

$$\dot{\mathbf{r}} = \frac{d\mathbf{r}(t(s))}{ds},$$

$$\ddot{\mathbf{r}} = \frac{d^2\mathbf{r}(t(s))}{ds^2},$$

$$\mathbf{r}' = \frac{d\mathbf{r}(t)}{dt},$$

$$\mathbf{r}'' = \frac{d^2\mathbf{r}(t)}{dt^2}.$$

Definition 1.8.6 (Unit tangent vector) *Let C be a C^1 class curve defined by* $\mathbf{r}:$ $[a,b] \to \mathbb{R}^3$.

Let $t \in [a,b]$. We define the unit tangent vector to C at t, denoted by $\mathbf{T}(t)$, *by*

$$\mathbf{T}(t) = \frac{\mathbf{r}'(t)}{|\mathbf{r}'(t)|},$$

where we recall to have assumed through the C^1 class definition, $\mathbf{r}'(t) \neq \mathbf{0}$, for all $t \in [a,b]$.

Observe that

$$
\begin{aligned}
\mathbf{T}(t) &= \frac{\mathbf{r}'(t)}{|\mathbf{r}'(t)|} \\
&= \frac{\mathbf{r}'(t)}{s'(t)} \\
&= \frac{\frac{d\mathbf{r}(t)}{dt}}{\frac{ds(t)}{dt}} \\
&= \frac{d\mathbf{r}(t(s))}{ds} \\
&= \hat{\mathbf{T}}(s).
\end{aligned}
\tag{1.80}
$$

Therefore

$$\hat{\mathbf{T}} = \dot{\mathbf{r}},$$

and

$$\dot{\hat{\mathbf{T}}} = \ddot{\mathbf{r}}.$$

Definition 1.8.7 *Let C be a curve defined by a C^1 class function $\mathbf{r}:[a,b] \to \mathbb{R}^3$. Let $t \in [a,b]$.*

We define the curvature vector of C at t, denoted by $\mathbf{K}(t)$, *by*

$$\mathbf{K}(t) = \frac{d\mathbf{T}(t)}{ds}.$$

Observe that

$$
\begin{aligned}
\mathbf{K}(t) &= \frac{d\mathbf{T}(t)}{ds} \\
&= \frac{d\mathbf{T}(t)}{dt}\frac{dt}{ds} \\
&= \frac{\mathbf{T}'(t)}{\frac{ds(t)}{dt}} \\
&= \frac{\mathbf{T}'(t)}{|\mathbf{r}'(t)|}.
\end{aligned}
\tag{1.81}
$$

We may also define

$$
\begin{aligned}
\hat{\mathbf{K}}(s) &= \frac{d\hat{\mathbf{T}}(s)}{ds} \\
&= \dot{\hat{\mathbf{T}}}(s) \\
&= \frac{d^2\mathbf{r}(t(s))}{ds^2}.
\end{aligned}
\tag{1.82}
$$

Observe that $|\mathbf{T}(t)| = 1$, $\forall t \in [a,b]$, that is,

$$
\mathbf{T}(t) \cdot \mathbf{T}(t) = 1,
$$

so that

$$
\mathbf{T}'(t) \cdot \mathbf{T}(t) = 0.
$$

Hence $\mathbf{T}'(t)$ has a direction orthogonal to the one of $\mathbf{T}(t)$.
So, at this point we introduce the next definition.

Definition 1.8.8 *Let C be a curve defined by a C^1 class function $\mathbf{r} : [a,b] \to \mathbb{R}^3$. Let $t \in [a,b]$. We define the normal vector to C at t, denoted by $\mathbf{N}(t)$, by*

$$
\mathbf{N}(t) = \frac{\mathbf{T}'(t)}{|\mathbf{T}'(t)|}.
$$

Also, we define the bi-normal vector to C, denoted by $\mathbf{B}(t)$, by

$$
\mathbf{B}(t) = \mathbf{T}(t) \times \mathbf{N}(t).
$$

In the context of the last definition, from

$$
\mathbf{B} = \mathbf{T} \times \mathbf{N},
$$

we obtain

$$
\begin{aligned}
\frac{d\mathbf{B}}{ds} &= \frac{d\mathbf{T}}{ds} \times \mathbf{N} + \mathbf{T} \times \frac{d\mathbf{N}}{ds} \\
&= \frac{\mathbf{T}'(t)}{|\mathbf{r}'(t)|} \times \mathbf{N} + \mathbf{T} \times \frac{d\mathbf{N}}{ds} \\
&= \mathbf{T} \times \frac{d\mathbf{N}}{ds}.
\end{aligned}
\tag{1.83}
$$

Therefore, $\frac{d\mathbf{B}}{ds}$ is orthogonal to $\mathbf{T}$.
Also, from $\mathbf{B} \cdot \mathbf{B} = 1$, we obtain,

$$
\frac{d\mathbf{B}}{ds} \cdot \mathbf{B} = 0,
$$

and hence,

$$
\frac{d\mathbf{B}}{ds}
$$

is orthogonal to $\mathbf{T}$ and $\mathbf{B}$, so that it has the direction of $\mathbf{N}$. So, we may denote

$$\frac{d\mathbf{B}}{ds} = \tau \mathbf{N},$$

where the scalar τ is said to be the torsion of the curve C at the point corresponding to $t \in [a,b]$.

Moreover, we define the osculating plane to C at t, as the one which has as normal directional the vector $\mathbf{B}(t)$ and contains the point $\mathbf{r}(t)$. The normal plane to C at t, is the one which has as normal direction the vector $\mathbf{T}(t)$ and contains the point $\mathbf{r}(t)$. Finally, the rectifying plane to C at t is the one which has $\mathbf{N}(t)$ as the normal direction and it contains $\mathbf{r}(t)$.

Finally, observe also that, from

$$\mathbf{N} = \mathbf{B} \times \mathbf{T},$$

we obtain

$$\begin{aligned} \frac{d\mathbf{N}}{ds} &= \frac{d\mathbf{B}}{ds} \times \mathbf{T} + \mathbf{B} \times \frac{d\mathbf{T}}{ds} \\ &= \tau \mathbf{N} \times \mathbf{T} + \mathbf{B} \times \kappa \mathbf{N} \\ &= -\kappa \mathbf{T} - \tau \mathbf{B}. \end{aligned} \tag{1.84}$$

Here, we have denoted

$$\kappa = |\mathbf{K}(t)| = \left| \frac{d\mathbf{T}}{ds} \right| = \left| \frac{\mathbf{T}'(t)}{|r'(t)|} \right|.$$

Since $\mathbf{T}'$ and $\mathbf{N}$ has the same direction, we may write,

$$\mathbf{K}(t) = \frac{d\mathbf{T}}{ds} = \kappa(t)\mathbf{N}(t).$$

$\kappa(t)$ is said to be the scalar curvature of C at the point corresponding to t.

1.9 Some notes about the scalar curvature in a surface in $\mathbb{R}^3$

Consider a surface $S \subset \mathbb{R}^3$, where

$$S = \{\mathbf{r}(\mathbf{u}) \ : \ \mathbf{u} \in D\},$$

where $\mathbf{r} : D \to \mathbb{R}^3$ is a C^1 class function and $D \subset \mathbb{R}^2$ is an open, bounded, simply connect set with a C^1 class boundary ∂D. Here $\mathbf{u} = (u_1, u_2) \in \mathbb{R}^2$, and

$$\mathbf{r}(u_1, u_2) = X_1(u_1, u_2)\mathbf{e}_1 + X_2(u_1, u_2)\mathbf{e}_2 + X_3(u_1, u_2)\mathbf{e}_3,$$

and $\{\mathbf{e}_1, \mathbf{e}_2, \mathbf{e}_3\}$ is the canonical basis of $\mathbb{R}^3$.

Generically, we shall denote

$$\mathbf{r}_\alpha = \frac{\partial \mathbf{r}}{\partial u_\alpha},$$

$$\mathbf{r}_{\alpha\beta} = \frac{\partial^2 \mathbf{r}}{\partial u_\alpha \partial u_\beta}.$$

Let $(\mathbf{r} \circ \mathbf{u}) : [a,b] \to S$, be a curve on S, that is,

$$\mathbf{r}(\mathbf{u}(t)) = \mathbf{r}(u_1(t), u_2(t)), \ \forall t \in [a,b].$$

For

$$s(t) = \int_a^t |(\mathbf{r} \circ \mathbf{u})'(v)| \, dv,$$

we shall consider this same curve parameterized by its arc length, where, as the meaning is clear, we denote simply

$$\hat{\mathbf{r}}(s) = (\mathbf{r} \circ \mathbf{u})(t(s)) \equiv \mathbf{r},$$

so that

$$\dot{\mathbf{r}} = \frac{\partial \mathbf{r}}{\partial u_1} \frac{du_1}{ds} + \frac{\partial \mathbf{r}}{\partial u_2} \frac{du_2}{ds} = \mathbf{r}_\alpha \dot{u}_\alpha.$$

Thus,

$$\ddot{\mathbf{r}} = \mathbf{r}_{\alpha\beta} \dot{u}_\alpha \dot{u}_\beta + \mathbf{r}_\alpha \ddot{u}_\alpha.$$

Hence, for $\mathbf{n} = \frac{\mathbf{r}_\alpha \times \mathbf{r}_\beta}{|\mathbf{r}_\alpha \times \mathbf{r}_\beta|}$, we get

$$\begin{aligned} \ddot{\mathbf{r}} \cdot \mathbf{n} &= (\mathbf{r}_{\alpha\beta} \cdot \mathbf{n}) \dot{u}_\alpha \dot{u}_\beta \\ &= b_{\alpha\beta} \dot{u}_\alpha \dot{u}_\beta, \end{aligned} \tag{1.85}$$

where

$$\{b_{\alpha\beta}\} = \{\mathbf{r}_{\alpha\beta} \cdot \mathbf{n}\}$$

is the curvature tensor of S at the point corresponding to s, relating C.

Observe that $\mathbf{r}_\alpha \cdot \mathbf{n} = 0$, so that

$$\mathbf{r}_{\alpha\beta} \cdot \mathbf{n} + \mathbf{r}_\alpha \cdot \mathbf{n}_\beta = 0,$$

and thus,

$$\begin{aligned} b_{\alpha\beta} &= \mathbf{r}_{\alpha\beta} \cdot \mathbf{n} \\ &= -\mathbf{r}_\alpha \cdot \mathbf{n}_\beta, \end{aligned} \tag{1.86}$$

where

$$\mathbf{n}_\beta = \frac{\partial \mathbf{n}}{\partial u_\beta}.$$

At this point we recall that:

$$
\begin{aligned}
\ddot{\mathbf{r}} &= \hat{\dot{\mathbf{T}}}(s) \\
&= \hat{\mathbf{K}}(s) \\
&= \kappa(s)\mathbf{N}(t(s)),
\end{aligned}
\tag{1.87}
$$

Let γ be the angle between

$$
\mathbf{n} = \frac{\mathbf{r}_\alpha \times \mathbf{r}_\beta}{|\mathbf{r}_\alpha \times \mathbf{r}_\beta|},
$$

and

$$
\mathbf{N}(t(s)) \equiv \mathbf{N}.
$$

so that

$$
\begin{aligned}
\cos(\gamma) &= \mathbf{N} \cdot \mathbf{n} \\
&= \frac{\ddot{\mathbf{r}}}{\kappa} \cdot \mathbf{n},
\end{aligned}
\tag{1.88}
$$

and thus

$$
\ddot{\mathbf{r}} \cdot \mathbf{n} = \kappa\cos(\gamma),
$$

and hence

$$
\begin{aligned}
\ddot{\mathbf{r}} \cdot \mathbf{n} &= (\mathbf{r}_{\alpha\beta} \cdot \mathbf{n})\dot{u}_\alpha \dot{u}_\beta \\
&= b_{\alpha\beta}\dot{u}_\alpha \dot{u}_\beta \\
&= \kappa\cos(\gamma).
\end{aligned}
\tag{1.89}
$$

On the other hand,

$$
\begin{aligned}
ds^2 &= d\mathbf{r} \cdot d\mathbf{r} \\
&= \left(\frac{\partial \mathbf{r}}{\partial u_\alpha}\,du_\alpha\right) \cdot \left(\frac{\partial \mathbf{r}}{\partial u_\beta}\,du_\beta\right) \\
&= \mathbf{r}_\alpha \cdot \mathbf{r}_\beta\,du_\alpha du_\beta \\
&= g_{\alpha\beta}\,du_\alpha du_\beta,
\end{aligned}
\tag{1.90}
$$

where,

$$
\mathbf{g}_\alpha = \mathbf{r}_\alpha,
$$

$$
g_{\alpha\beta} = \mathbf{g}_\alpha \cdot \mathbf{g}_\beta,
$$

so that, from this and

$$
b_{\alpha\beta} = -\mathbf{r}_\alpha \cdot \mathbf{n}_\beta,
$$

we obtain

$$
\begin{aligned}
b_{\alpha\beta}du_{\alpha}du_{\beta} &= -\mathbf{r}_{\alpha} \cdot \mathbf{n}_{\beta} du_{\alpha}du_{\beta} \\
&= -\left(\frac{\partial \mathbf{r}}{\partial u_{\alpha}}du_{\alpha}\right) \cdot \left(\frac{\partial \mathbf{n}}{\partial u_{\beta}}du_{\beta}\right) \\
&= -d\mathbf{r} \cdot \mathbf{n}.
\end{aligned} \tag{1.91}
$$

Observe also that

$$
\begin{aligned}
\kappa\cos(\gamma) &= b_{\alpha\beta}\dot{u}_{\alpha}\dot{u}_{\beta} \\
&= b_{\alpha\beta}\frac{\partial u_{\alpha}}{\partial s}\frac{\partial u_{\beta}}{\partial s} \\
&= b_{\alpha\beta}\frac{u'_{\alpha}u'_{\beta}}{s's'} \\
&= b_{\alpha\beta}\frac{u'_{\alpha}u'_{\beta}}{g_{\alpha\beta}u'_{\alpha}u'_{\beta}},
\end{aligned} \tag{1.92}
$$

so that denoting

$$
\kappa_n = \kappa\cos(\gamma),
$$

we obtain

$$
\kappa_n = \frac{b_{\alpha\beta}u'_{\alpha}u'_{\beta}}{g_{\alpha\beta}u'_{\alpha}u'_{\beta}},
$$

so that,

$$
(b_{\alpha\beta} - \kappa_n g_{\alpha\beta})c_{\alpha}c_{\beta} = 0, \tag{1.93}
$$

where we have denoted $c_{\alpha} = u'_{\alpha}$, for $\alpha \in \{1,2\}$.

Observe that c_{α}, κ_n depend on the curve in question, but $\{b_{\alpha\beta}\}$ and $\{g_{\alpha\beta}\}$ do not, so we shall differentiate (1.93) looking for the conditions for which κ_n has extremal values, that is

$$
\frac{\partial \kappa_n}{\partial c_{\alpha}} = 0.
$$

So, denoting

$$
a_{\alpha\beta} = b_{\alpha\beta} - \kappa_n g_{\alpha\beta},
$$

we obtain

$$
\frac{\partial}{\partial c_{\gamma}}[a_{\alpha\beta}c_{\alpha}c_{\beta}] = 0,
$$

so that

$$
\begin{aligned}
0 &= a_{\alpha\beta}\left(\frac{\partial c_{\alpha}}{\partial c_{\gamma}}c_{\beta} + c_{\alpha}\frac{\partial c_{\beta}}{\partial c_{\gamma}}\right) \\
&= a_{\alpha\beta}(\delta_{\gamma}^{\alpha}c_{\beta} + c_{\alpha}\delta_{\gamma}^{\beta}) \\
&= a_{\gamma\beta}c_{\beta} + a_{\alpha\gamma}c_{\alpha} \\
&= (a_{\gamma\alpha} + a_{\alpha\gamma})c_{\alpha}
\end{aligned} \tag{1.94}
$$

Since $a_{\alpha\gamma} = a_{\gamma\alpha}$, from this and the equation above, we get,

$$a_{\alpha\gamma}c_\alpha = 0,$$

so that in particular

$$(b_{\alpha\gamma} - k_n g_{\alpha\gamma})c_\gamma = 0,$$

so that

$$b_\alpha^\beta c_\alpha - k_n c_\beta = 0,$$

that is,

$$\left[\begin{array}{cc} b_1^1 - \kappa_n & b_2^1 \\ b_1^2 & b_2^2 - \kappa_n \end{array} \right] \left[\begin{array}{c} c_1 \\ c_2 \end{array} \right] = \left[\begin{array}{c} 0 \\ 0 \end{array} \right],$$

so that, for having non zero solutions, the determinant concerning this last system must be zero, and hence

$$\kappa_n^2 - \kappa_n b_\alpha^\alpha + det(\{b_\alpha^\beta\}) = 0.$$

Since

$$det\{b_\alpha^\beta\} = \frac{b}{g},$$

we obtain,

$$\kappa_n^2 - b_{\alpha\beta}g^{\alpha\beta}\kappa_n + b/g = 0,$$

so that denoting by κ_1 and κ_2 the solutions of this last equation, we have,

$$\kappa_1\kappa_2 = b/g.$$

Finally,

$$K = \kappa_1\kappa_2 = b/g$$

is said to be the Gaussian curvature of S at the point in question.

Moreover,

$$H = \frac{1}{2}(\kappa_1 + \kappa_2) = \frac{1}{2}b_{\alpha\beta}g^{\alpha\beta} = \frac{1}{2}b_\alpha^\alpha,$$

is said to be the mean curvature of S at the same point.

About the notation, we recall that

$$\mathbf{r}_{\alpha\beta} = \frac{\partial^2\mathbf{r}}{\partial u_\alpha u_\beta},$$

and we shall denote

$$\mathbf{r}_{\alpha\beta} = \Gamma_{\alpha\beta}^\gamma \mathbf{r}_\gamma + a_{\alpha\beta}\mathbf{n}.$$

Thus,

$$b_{\alpha\beta} = \mathbf{r}_{\alpha\beta} \cdot \mathbf{n} = a_{\alpha\beta}\mathbf{n} \cdot \mathbf{n} = a_{\alpha\beta}.$$

At this point, we start to specify the coefficients $\Gamma_{\alpha\beta}^\gamma$.

Observe that,

$$\begin{aligned}
\mathbf{r}_{\alpha\beta} \cdot \mathbf{r}_\lambda &= \Gamma^\gamma_{\alpha\beta}\mathbf{r}_\gamma \cdot \mathbf{r}_\lambda + b_{\alpha\beta}\mathbf{n} \cdot \mathbf{r}_\lambda \\
&= \Gamma^\gamma_{\alpha\beta}\mathbf{r}_\gamma \cdot \mathbf{r}_\lambda \\
&= \Gamma^\gamma_{\alpha\beta}g_{\gamma\lambda}.
\end{aligned} \tag{1.95}$$

Hence,

$$\Gamma^\rho_{\alpha\beta} = \mathbf{r}_{\alpha\beta} \cdot \mathbf{r}_\lambda g^{\lambda\rho}.$$

Let us denote,

$$\Gamma_{\alpha\beta\gamma} = \mathbf{r}_{\alpha\beta} \cdot \mathbf{r}_\lambda,$$

so that

$$\Gamma_{\alpha\beta\lambda} = g_{\gamma\lambda}\Gamma^\lambda_{\alpha\beta},$$

and also,

$$\Gamma^\rho_{\alpha\beta} = g^{\lambda\rho}\Gamma_{\alpha\beta\lambda}.$$

Observe that from

$$\mathbf{r}_{\alpha\beta} = \mathbf{r}_{\beta\alpha},$$

we get

$$\Gamma_{\alpha\beta\lambda} = \Gamma_{\beta\alpha\lambda},$$

and

$$\Gamma^\rho_{\alpha\beta} = \Gamma^\rho_{\beta\alpha}.$$

Observe also that,

$$g_{\alpha\lambda} = \mathbf{r}_\alpha \cdot \mathbf{r}_\lambda,$$

so that

$$\frac{\partial g_{\alpha\lambda}}{\partial u_\beta} = \mathbf{r}_{\alpha\beta} \cdot \mathbf{r}_\lambda + \mathbf{r}_\alpha \cdot \mathbf{r}_{\lambda\beta},$$

so that

$$\frac{\partial g_{\alpha\lambda}}{\partial u_\beta} = \Gamma_{\alpha\beta\lambda} + \Gamma_{\lambda\beta\alpha}, \tag{1.96}$$

$$\frac{\partial g_{\lambda\beta}}{\partial u_\alpha} = \Gamma_{\lambda\alpha\beta} + \Gamma_{\beta\alpha\lambda}, \tag{1.97}$$

and,

$$\frac{\partial g_{\beta\alpha}}{\partial u_\lambda} = \Gamma_{\beta\lambda\alpha} + \Gamma_{\alpha\lambda\beta}, \tag{1.98}$$

so that, from (1.96), (1.97) and (1.98), we obtain

$$\Gamma_{\alpha\beta\lambda} = \frac{1}{2}\left[\frac{\partial g_{\beta\lambda}}{\partial u_\alpha} + \frac{\partial g_{\lambda\alpha}}{\partial u_\beta} - \frac{\partial g_{\alpha\beta}}{\partial u_\lambda}\right].$$

We recall also that,

$$\mathbf{r}_{\alpha\beta} = \Gamma^{\gamma}_{\alpha\beta}\mathbf{r}_{\gamma} + b_{\alpha\beta}\mathbf{n}.$$

Observe that for a C^3 class surface, we have

$$\mathbf{r}_{\alpha\beta\lambda} = \mathbf{r}_{\alpha\lambda\beta},$$

Hence,

$$\mathbf{r}_{\alpha\beta\lambda} = \frac{\partial \Gamma^{\gamma}_{\alpha\beta}}{\partial u_{\lambda}}\mathbf{r}_{\gamma} + \Gamma^{\gamma}_{\alpha\beta}\mathbf{r}_{\gamma\lambda} + \frac{\partial b_{\alpha\beta}}{\partial u_{\lambda}}\mathbf{n} + b_{\alpha\beta}\mathbf{n}_{\lambda}.$$

Observe that

$$\mathbf{n}\cdot\mathbf{n} = 1,$$

and hence

$$\mathbf{n}_{\alpha}\cdot\mathbf{n} = 0,$$

so that we denote,

$$\mathbf{n}_{\alpha} = c^{\gamma}_{\alpha}\mathbf{r}_{\gamma}.$$

Observe that,

$$\mathbf{n}_{\alpha}\cdot\mathbf{r}_{\rho} = c^{\gamma}_{\alpha}\mathbf{r}_{\gamma}\cdot\mathbf{r}_{\rho} = c^{\gamma}_{\alpha}g_{\gamma\rho},$$

and

$$\mathbf{n}_{\alpha}\cdot\mathbf{r}_{\rho} = -b_{\alpha\rho},$$

and from this and

$$g_{\gamma\rho}g^{\rho\tau} = \delta^{\tau}_{\gamma},$$

we get

$$\begin{aligned}
\mathbf{n}_{\alpha}\cdot\mathbf{r}_{\rho}g^{\rho\tau} &= -b_{\alpha\rho}g^{\rho\tau} \\
&= -b^{\tau}_{\alpha} \\
&= c^{\lambda}_{\alpha}g_{\lambda\rho}g^{\rho\tau} \\
&= c^{\tau}_{\alpha},
\end{aligned} \tag{1.99}$$

so that

$$c^{\tau}_{\alpha} = -b^{\tau}_{\alpha}.$$

Thus, we obtain the formula of Weingarten:

$$\mathbf{n}_{\alpha} = -b^{\beta}_{\alpha}\mathbf{r}_{\beta},$$

where

$$b^{\beta}_{\alpha} = g^{\rho\beta}b_{\alpha\rho}.$$

Retaking the earlier expression we had obtained,

$$\mathbf{r}_{\alpha\beta\lambda} = \frac{\partial \Gamma^{\gamma}_{\alpha\beta}}{\partial u_{\lambda}}\mathbf{r}_{\gamma} + \Gamma^{\gamma}_{\alpha\beta}\mathbf{r}_{\gamma\lambda} + \frac{\partial b_{\alpha\beta}}{\partial u_{\lambda}}\mathbf{n} + b_{\alpha\beta}\mathbf{n}_{\lambda}.$$

so that,

$$\mathbf{r}_{\alpha\beta\lambda} = \frac{\partial \Gamma^{\gamma}_{\alpha\beta}}{\partial u_{\lambda}} \mathbf{r}_{\gamma}$$
$$+ \Gamma^{\gamma}_{\alpha\beta} \left[\Gamma^{\rho}_{\gamma\lambda} \mathbf{r}_{\rho} + b_{\gamma\lambda} \mathbf{n} \right]$$
$$+ \frac{\partial b_{\alpha\beta}}{\partial u_{\lambda}} \mathbf{n}$$
$$- b_{\alpha\beta} b^{\tau}_{\lambda} \mathbf{r}_{\tau}, \tag{1.100}$$

that is,

$$\mathbf{r}_{\alpha\beta\lambda} = \left[\frac{\partial \Gamma^{\rho}_{\alpha\beta}}{\partial u_{\lambda}} + \Gamma^{\gamma}_{\alpha\beta}\Gamma^{\rho}_{\gamma\lambda} - b_{\alpha\beta}b^{\rho}_{\lambda} \right] \mathbf{r}_{\rho}$$
$$+ \left[\Gamma^{\rho}_{\alpha\beta} b_{\rho\lambda} + \frac{\partial b_{\alpha\beta}}{\partial u_{\lambda}} \right] \mathbf{n}. \tag{1.101}$$

Interchanging β and λ, we obtain

$$\mathbf{r}_{\alpha\lambda\beta} = \left[\frac{\partial \Gamma^{\rho}_{\alpha\lambda}}{\partial u_{\rho}} + \Gamma^{\gamma}_{\alpha\lambda}\Gamma^{\rho}_{\gamma\beta} - b_{\alpha\lambda}b\beta^{\rho} \right] \mathbf{r}_{\rho}$$
$$+ \left[\Gamma^{\rho}_{\alpha\lambda} b_{\rho\beta} + \frac{\partial b_{\alpha\lambda}}{\partial u_{\beta}} \right] \mathbf{n}. \tag{1.102}$$

The vectors $\mathbf{r}_{\alpha}, \mathbf{r}_{\beta}$, and $\mathbf{n}$ are linearly independent, so that since

$$\mathbf{r}_{\alpha\beta\lambda} = \mathbf{r}_{\alpha\lambda\beta},$$

we must have

$$\Gamma^{\rho}_{\alpha\beta} b_{\rho\lambda} - \Gamma^{\rho}_{\alpha\lambda} b_{\rho\beta} + \frac{\partial b_{\alpha\beta}}{\partial u_{\lambda}} - \frac{\partial b_{\alpha\lambda}}{\partial u_{\beta}} = 0.$$

These are the Mainard-Codazzi formulas.
From

$$\mathbf{r}_{\alpha\beta\lambda} - \mathbf{r}_{\alpha\lambda\beta} = 0,$$

equating to zero the equations in $\mathbf{r}_{\alpha}$ and $\mathbf{r}_{\beta}$, we obtain,

$$R^{\rho}_{\alpha\lambda\beta} = b_{\alpha\beta}b^{\rho}_{\lambda} + b_{\alpha\lambda}b^{\rho}_{\beta},$$

where

$$R^{\rho}_{\alpha\lambda\beta} = \frac{\partial \Gamma^{\rho}_{\alpha\beta}}{\partial u_{\lambda}} - \frac{\partial \Gamma^{\rho}_{\alpha\lambda}}{\partial u_{\beta}}$$
$$+ \Gamma^{\gamma}_{\alpha\beta}\Gamma^{\rho}_{\gamma\lambda} - \Gamma^{\gamma}_{\alpha\lambda}\Gamma^{\rho}_{\gamma\beta}. \tag{1.103}$$

This tensor is called the mixed Riemann curvature tensor.

The tensor

$$R_{\tau\alpha\lambda\beta} = g_{\rho\tau}R^{\rho}_{\alpha\lambda\beta},$$

is called the Riemann covariant curvature tensor.

Observe that

$$R^{\rho}_{\alpha\lambda\beta} = b_{\alpha\beta}b^{\rho}_{\lambda} - b_{\alpha\lambda}b^{\rho}_{\beta},$$

so that

$$
\begin{aligned}
R_{\tau\alpha\lambda\beta} &= g_{\rho\tau}R^{\rho}_{\alpha\lambda\beta} \\
&= g_{\rho\tau}(b_{\alpha\beta}b^{\rho}_{\lambda} - b_{\alpha\lambda}b^{\rho}_{\beta}) \\
&= b_{\alpha\beta}b_{\lambda\tau} - b_{\alpha\lambda}b_{\beta\tau}.
\end{aligned}
\tag{1.104}
$$

In particular,

$$
\begin{aligned}
R_{1212} &= b_{22}b_{11} - b_{21}b_{21} \\
&= b_{11}b_{22} - (b_{12})^2 \\
&= b.
\end{aligned}
\tag{1.105}
$$

Under the context of this last discussion, we have,

Theorem 1.9.1 (Gauss, Egregium Theorem) *Considering the discussion above, the Gaussian curvature of the surface S in question is given by,*

$$K = \frac{R_{1212}}{g}.$$

Proof 1.11   Recall that earlier we had obtained,

$$K = \frac{b}{g},$$

so that from this and

$$R_{1212} = b,$$

we have,

$$K = \frac{R_{1212}}{g}.$$

This completes the proof.

Chapter 2

Manifolds in $\mathbb{R}^n$

2.1 Introduction

This chapter presents basic results on surfaces in $\mathbb{R}^n$. Topics addressed include differential forms and an abstract version of the Stokes theorem in $\mathbb{R}^n$. The main references for this chapter are [23] and [39]. In fact, a substantial part of this chapter is based on [39] (from the first section up to the section on surfaces with boundary), where more details may be found.

We start with some preliminary definitions.

Definition 2.1.1 *Let $U \subset \mathbb{R}^m$ and $V \subset \mathbb{R}^n$. A homeomorphism from U to V is a continuous bijection $\mathbf{f} : U \to V$ for which the respective inverse is also continuous. A diffeomorphism $\mathbf{h} : U \to V$ is a differentiable bijection which the inverse $\mathbf{h}^{-1} : V \to U$ is also differentiable.*

Definition 2.1.2 (Immersion) *Let $U \subset \mathbb{R}^m$ be an open set. We say that a differentiable function $\mathbf{f} : U \to \mathbb{R}^n$ is an immersion, if for each $\mathbf{u} \in U$, $\mathbf{f}'(\mathbf{u}) : \mathbb{R}^m \to \mathbb{R}^n$ is an injective linear transformation. (observe that in such a case $m \leq n$).*

Definition 2.1.3 (Sub-immersion) *Let $U \subset \mathbb{R}^m$ be an open set. We say that a differentiable function $\mathbf{f} : U \to \mathbb{R}^n$ is a sub-immersion, if for each $\mathbf{u} \in U$, $\mathbf{f}'(\mathbf{u}) : \mathbb{R}^m \to \mathbb{R}^n$ is an onto linear transformation. (observe that in such a case $m \geq n$).*

2.2 The local form of sub-immersions

In this section we state and prove the Local Form of Sub-Immersions Theorem.

Theorem 2.2.1 *Let $D \subset \mathbb{R}^{n+m}$ be a non-empty open set and let $\mathbf{f} : D \to \mathbb{R}^m$ be a vectorial function of C^1 class.*

We denote $(\mathbf{x}, \mathbf{y}) \in \mathbb{R}^{n+m}$ *where* $\mathbf{x} = (x_1, \ldots, x_n) \in \mathbb{R}^n$ *and* $\mathbf{y} = (y_1, \ldots, y_m)$ $\in \mathbb{R}^m$.

Suppose $(\mathbf{x}_0, \mathbf{y}_0) \in D$ *is such that*

$$\det[\mathbf{f}_y(\mathbf{x}_0, \mathbf{y}_0)] \neq 0.$$

Under such hypotheses, there exist open sets V, Z, W *such that*

$$(\mathbf{x}_0, \mathbf{y}_0) \in Z \subset \mathbb{R}^{n+m}, \ \mathbf{x}_0 \in V \subset \mathbb{R}^n \ and \ \mathbf{f}(\mathbf{x}_0, \mathbf{y}_0) \in W \subset \mathbb{R}^m,$$

and there exists also a function $h : V \times W \to Z$ *of* C^1 *class, such that*

$$\mathbf{f}(h(\mathbf{x}, \mathbf{w})) = \mathbf{w}, \ \forall (\mathbf{x}, \mathbf{w}) \in V \times W.$$

Proof 2.1   Define $\mathbf{c} = \mathbf{f}(\mathbf{x}_0, \mathbf{y}_0)$. With no loss in generality, assume $\mathbf{c} = \mathbf{0}$. Define $\varphi : D \to \mathbb{R}^n \times \mathbb{R}^m$ by

$$\varphi(\mathbf{x}, \mathbf{y}) = (\mathbf{x}, \mathbf{f}(\mathbf{x}, \mathbf{y})).$$

Observe that

$$\varphi'(\mathbf{x}_0, \mathbf{y}_0) = \begin{bmatrix} I_{n \times n} & 0_{n \times m} \\ \mathbf{f}_x(\mathbf{x}_0, \mathbf{y}_0) & \mathbf{f}_y(\mathbf{x}_0, \mathbf{y}_0) \end{bmatrix} \tag{2.1}$$

where $I_{n \times n}$ denotes the identity matrix $n \times n$ and $0_{n \times m}$ denotes a matrix $n \times m$ with all entries zero.

Observe also that

$$\det(\varphi'(\mathbf{x}_0, \mathbf{y}_0)) = \det[\mathbf{f}_y(\mathbf{x}_0, \mathbf{y}_0)] \neq 0.$$

From the inverse function theorem, we have that $h = \varphi^{-1}$ exists and it is of C^1 class on an open set $V_1 \times W$ so that

$$\varphi(\mathbf{x}_0, \mathbf{y}_0) = (\mathbf{x}_0, \mathbf{f}(\mathbf{x}_0, \mathbf{y}_0)) \in V_1 \times W.$$

Also, from this same theorem, there exists an open set $U_1 \times U_2$ such that $\varphi(U_1 \times U_2) = V_1 \times W$ where $\mathbf{x}_0 \in U_1$ and $\mathbf{y}_0 \in U_2$.

Define $V = U_1 \cap V_1$ and $Z = \varphi^{-1}(V \times W)$.

Observe that φ^{-1} is of C^1 class (also from the inverse function theorem) and moreover, since φ keeps the fist variable fixed, φ^{-1} also keep it fixed, so that there exists a function h_2 such that

$$h(\mathbf{x}, \mathbf{y}) = \varphi^{-1}(\mathbf{x}, \mathbf{y}) = (\mathbf{x}, h_2(\mathbf{x}, \mathbf{y})),$$

Thus, for $(\mathbf{x}, \mathbf{w}) \in V \times W$ we have that

$$\begin{aligned} (\mathbf{x}, \mathbf{w}) &= \varphi[\varphi^{-1}(\mathbf{x}, \mathbf{w})] \\ &= \varphi[h(\mathbf{x}, \mathbf{w})] \\ &= \varphi(\mathbf{x}, h_2(\mathbf{x}, \mathbf{w})) \\ &= (\mathbf{x}, \mathbf{f}(\mathbf{x}, h_2(\mathbf{x}, \mathbf{w}))) \\ &= (\mathbf{x}, \mathbf{f}(h(\mathbf{x}, \mathbf{w}))), \end{aligned} \tag{2.2}$$

that is,

$$(\mathbf{x}, \mathbf{w}) = (\mathbf{x}, \mathbf{f}(h(\mathbf{x}, \mathbf{w}))),$$

so that

$$\mathbf{w} = \mathbf{f}(h(\mathbf{x}, \mathbf{w})), \ \forall (\mathbf{x}, \mathbf{w}) \in V \times W.$$

The proof is complete.

2.3 The local form of immersions

In this section we present and prove the Local Form of Immersions Theorem.

Theorem 2.3.1 *Let $D \subset \mathbb{R}^n$ be a non-empty open set. Let $\mathbf{f} : D \to \mathbb{R}^{n+m}$ be a function of C^1 class such that*

$$\mathbf{f}(\mathbf{x}) = \begin{bmatrix} f_1(\mathbf{x}) \\ f_2(\mathbf{x}) \\ \vdots \\ f_{n+m}(\mathbf{x}) \end{bmatrix}. \tag{2.3}$$

Let $\mathbf{x}_0 \in D$. Define the set A by

$$A = \left\{ \sum_{k=1}^{n} \langle \mathbf{f}'(\mathbf{x}_0)\mathbf{v}, e_k \rangle e_k \ : \ \mathbf{v} \in \mathbb{R}^n \right\},$$

where $\{e_1, e_2, \ldots, e_{n+m}\}$ is the canonical basis for $\mathbb{R}^{n+m}$.
 Assume that there exist linearly independent vectors $\mathbf{v}_1, \mathbf{v}_2, \ldots, \mathbf{v}_m \in \mathbb{R}^{n+m}$ such that

$$\mathbb{R}^{n+m} = A \oplus F,$$

where F is the sub-space spanned by $\{\mathbf{v}_1, \ldots, \mathbf{v}_m\}$.
 Under such hypotheses, there exist open sets V, Z, W such that

$$(\mathbf{x}_0, \mathbf{0}) \in V \times W \subset \mathbb{R}^n \times \mathbb{R}^m,$$

$\mathbf{f}(\mathbf{x}_0) \in Z \subset \mathbb{R}^{n+m}$ *and there exists a function $h : Z \to V \times W$ of C^1 class such that*

$$h(\mathbf{f}(\mathbf{x})) = (\mathbf{x}, \mathbf{0}), \ \forall \mathbf{x} \in V.$$

Proof 2.2 Define $\varphi : D \times \mathbb{R}^m \to \mathbb{R}^{n+m}$, for $\mathbf{y} = (y_1, \ldots, y_m) \in \mathbb{R}^m$ by

$$\varphi(\mathbf{x}, \mathbf{y}) = \begin{bmatrix} f_1(\mathbf{x}) + \sum_{i=1}^{m} y_i \langle \mathbf{v}_i, e_1 \rangle \\ \vdots \\ f_{n+m}(\mathbf{x}) + \sum_{i=1}^{m} y_i \langle \mathbf{v}_i, e_{n+m} \rangle \end{bmatrix} \tag{2.4}$$

Thus,

$\varphi'(\mathbf{x}_0, \mathbf{0})$

$$
= \begin{bmatrix}
(f_1)_{x_1}(\mathbf{x}_0) & (f_1)_{x_2}(\mathbf{x}_0) & \cdots & (f_1)_{x_n}(\mathbf{x}_0) & \langle \mathbf{v}_1, e_1 \rangle & \cdots & \langle \mathbf{v}_m, e_1 \rangle \\
(f_2)_{x_1}(\mathbf{x}_0) & (f_2)_{x_2}(\mathbf{x}_0) & \cdots & (f_2)_{x_n}(\mathbf{x}_0) & \langle \mathbf{v}_1, e_2 \rangle & \cdots & \langle \mathbf{v}_m, e_2 \rangle \\
\vdots & \vdots & \ddots & \vdots & \vdots & \ddots & \vdots \\
(f_{n+m})_{x_1}(\mathbf{x}_0) & (f_{n+m})_{x_2}(\mathbf{x}_0) & \cdots & (f_{n+m})_{x_n}(\mathbf{x}_0) & \langle \mathbf{v}_1, e_{n+m} \rangle & \cdots & \langle \mathbf{v}_m, e_{n+m} \rangle
\end{bmatrix}
$$

From the hypotheses $A \oplus F = \mathbb{R}^{n+m}$ and thus

$$\varphi'(\mathbf{x}_0, \mathbf{0})[\mathbb{R}^{n+m}] = \mathbb{R}^{n+m}.$$

Therefore, $\det(\varphi'(\mathbf{x}_0, \mathbf{0})) \neq 0$ so that from the inverse function theorem there exist open sets V, W, Z such that $V \subset \mathbb{R}^n$, $W \subset \mathbb{R}^m$ and $Z \subset \mathbb{R}^{n+m}$ and such that

$$h \equiv \varphi^{-1} : Z \to V \times W$$

is of C^1 class in Z, $\mathbf{f}(\mathbf{x}_0) \in Z$, $\mathbf{x}_0 \in V$ and $\mathbf{0} \in W$.

Moreover,

$$\varphi(V \times W) = Z.$$

Observe that $\varphi(\mathbf{x}, \mathbf{0}) = \mathbf{f}(\mathbf{x})$.

Thus,

$$h(\mathbf{f}(\mathbf{x})) = \varphi^{-1}(\mathbf{f}(\mathbf{x})) = (\mathbf{x}, \mathbf{0}),$$

that is,

$$h(\mathbf{f}(\mathbf{x})) = (\mathbf{x}, \mathbf{0}), \ \forall \mathbf{x} \in V.$$

The proof is complete.

Remark 2.3.2 *Let $\pi : V \times W \to V$ where*

$$\pi(\mathbf{x}, \mathbf{w}) = \mathbf{x}.$$

Considering the context of the last theorem, let

$$\xi = \pi \circ h : Z \to V.$$

Thus,

$$(\xi \circ f)(\mathbf{x}) = \pi \circ h \circ f(\mathbf{x}) = \pi(\mathbf{x}, \mathbf{0}) = \mathbf{x}.$$

Hence, $\xi|_{f(V)} = (f|_V)^{-1}$. Summarizing, $\xi : Z \to V$ is of C^k class and as it is restricted to $\mathbf{f}(V)$, corresponds to the inverse of

$$\mathbf{f} : V \to \mathbf{f}(V).$$

2.4 Parameterizations and surfaces in $\mathbb{R}^n$

Definition 2.4.1 *Let $U_0 \subset \mathbb{R}^m$ be an open set. An immersion of C^k class $\mathbf{r} : U_0 \to \mathbb{R}^n$ (which is also an homeomorphism on $\mathbf{r}(U_0)$) is said to be a parametrization of C^k class on*

$$U \equiv \mathbf{r}(U_0).$$

Remark 2.4.2 *Considering the injectivity of $\mathbf{r}'(\mathbf{u}) : \mathbb{R}^m \to \mathbb{R}^n$, we have that the following conditions are equivalent.*

1. *$\mathbf{r}'(\mathbf{u}) : \mathbb{R}^m \to \mathbb{R}^n$ is injective.*

2.

$$\frac{\partial \mathbf{r}(\mathbf{u})}{\partial u_1}, \dots, \frac{\partial \mathbf{r}(\mathbf{u})}{\partial u_m}$$

   are linearly independent vectors.

3. *The Jacobian matrix*

$$\left\{ \frac{\partial r_i(\mathbf{u})}{\partial u_j} \right\}$$

   has rank m, that is, the determinant of some of its minors $m \times m$ is different from zero.

Definition 2.4.3 *Let $1 \leq m \leq n$. We say that a non-empty set $M \subset \mathbb{R}^n$ is an m-dimensional surface of C^k class if for each $p \in M$ there exists an open set U in M such that $p \in U$ and there exists a parametrization $\mathbf{r} : U_0 \to U = \mathbf{r}(U_0)$ for some open set $U_0 \subset \mathbb{R}^m$.*

The number $n - m$ is called the co-dimension of M.

2.4.1 Change of coordinates

Let $M \subset \mathbb{R}^n$ be a m-dimensional surface of C^k class, where $(1 \leq m \leq n)$. Let $\mathbf{r} : U_0 \to U$ be a parametrization of the open set in M, $U \subset M$. Let $V_0 \subset \mathbb{R}^m$ be an open set and let $\xi : V_0 \to U_0$ be a diffeomorphism of C^k class.

Thus, $\mathbf{r} \circ \xi : V_0 \to U$ is still a parametrization of U.

Observe that ξ represents a change of coordinates.

Remark 2.4.4 *If $\mathbf{r} : U_0 \to U$ and $\mathbf{s} : V_0 \to V$ are parameterizations of M such that $U \cap V \neq \emptyset$, then*

$$\xi = \mathbf{s}^{-1} \circ \mathbf{r} : \mathbf{r}^{-1}(U \cap V) \to \mathbf{s}^{-1}(U \cap V)$$

is a homeomorphism between open sets of $\mathbb{R}^m$.

Indeed, let us see the next theorem.

Theorem 2.4.5 *Let V_0 be an open subset of $\mathbb{R}^m$ and let $\mathbf{s} : V_0 \to \mathbb{R}^n$ be a parametrization of C^k class of a set $V \subset \mathbb{R}^n$. Given an open set $U_0 \subset \mathbb{R}^r$ and $\mathbf{f} : U_0 \to V \subset \mathbb{R}^n$ of C^k class, then*

1. *The composite function $\mathbf{s}^{-1} \circ \mathbf{f} : U_0 \to V_0 \subset \mathbb{R}^m$ is of C^k class.*

2. *For $\mathbf{x} \in U_0$ and*
$$\mathbf{z} = (\mathbf{s}^{-1} \circ \mathbf{f})(\mathbf{x})$$
   we have that
$$(\mathbf{s}^{-1} \circ \mathbf{f})'(\mathbf{x}) = [\mathbf{s}'(\mathbf{z})]^{-1} \circ \mathbf{f}'(\mathbf{x}).$$

Proof 2.3

1. Since $\mathbf{s} : V_0 \to V$ is an immersion of C^k class, for each point $p \in V$, there exists an open set $Z \subset \mathbb{R}^n$ which contains such a point and a function of C^k class $\mathbf{g} : Z \to \mathbb{R}^m$ (this results from the remark 2.3.2) such that
$$\mathbf{g}|(V \cap Z) = \mathbf{s}^{-1}.$$
   Let $p \in \mathbf{f}(U_0) \subset V$. Thus
$$\mathbf{s}^{-1} \circ \mathbf{f} = \mathbf{g} \circ \mathbf{f} : \mathbf{f}^{-1}(V \cap Z) \subset \mathbb{R}^r \to \mathbb{R}^m$$
   so that $\mathbf{s}^{-1} \circ \mathbf{f}$ is of C^k class, since so are $\mathbf{f}$ and $\mathbf{g}$.

2. Write $h = \mathbf{s}^{-1} \circ \mathbf{f}$ and apply the chain rule to the equality
$$\mathbf{s} \circ h = \mathbf{f}.$$

Corollary 2.4.6 *Let U_0 and $V_0 \subset \mathbb{R}^m$ and let $\mathbf{r} : U_0 \to V$ and $\mathbf{s} : V_0 \to V$ be parameterizations of C^k class from the same set $V \subset \mathbb{R}^n$. Under such hypotheses, the change of coordinates*
$$\xi = \mathbf{s}^{-1} \circ \mathbf{r}$$
is a diffeomorphism of C^k class.

2.4.2 Differentiable functions defined on surfaces

We start this sub-section with some important definitions to be used in some subsequent text parts.

Definition 2.4.7 *Let $M \subset \mathbb{R}^n$ be a surface of C^k class. We say that a function $\mathbf{f} : M \to \mathbb{R}^s$ is differentiable at a point $p \in M$ if there exists a parametrization $\mathbf{r} : U_0 \to U$ of C^k class with $p \in U$, such that*
$$\mathbf{f} \circ \mathbf{r} : U_0 \to \mathbb{R}^s$$
is differentiable at $\mathbf{u}_0 \in U_0$, where $\mathbf{r}(\mathbf{u}_0) = p$.

From the last proposition and corollary, we have that

$$\mathbf{f} \circ \mathbf{s} = (\mathbf{f} \circ \mathbf{r}) \circ (\mathbf{r}^{-1} \circ \mathbf{s})$$

is also differentiable at $\mathbf{u}_0 = \mathbf{r}^{-1}(p)$, whichever is the parametrization $\mathbf{s}$ of C^k in a neighborhood of p.

So, we may infer that the definition in question does not depend on the parametrization chosen.

If we have surfaces $M \subset \mathbb{R}^r$ and $N \subset \mathbb{R}^s$ of dimensions m_1 and m_2 respectively, we say that $\mathbf{f} : M \to N$ is differentiable at the point $p \in M$ when considered as a function on M in $\mathbb{R}^s$, if $\mathbf{f}$ is differentiable at such a point.

Similarly, we say that $\mathbf{f} : M \to N$ is of C^k class if for each $p \in M$ there exists a parametrization $\mathbf{r} : U_0 \to U \subset M$ of C^k class, with $p \in U$ such that

$$\mathbf{f} \circ \mathbf{r} : U_0 \to N \subset \mathbb{R}^s$$

is of C^k class.

Observe that, in such a case, from the last proposition and its corollary, $\mathbf{f} \circ \mathbf{r}$ is of C^k class whichever is the parametrization of C^k class $\mathbf{r} : U_0 \to U$, such that $p \in U$.

Let us then see the next theorem.

Theorem 2.4.8 *Considering the context of the last remarks above, in order that* $\mathbf{f} : M \to N$ *be of C^k class it is necessary and sufficient that, for each $p \in M$, there exist parameterizations of C^k class*

$$\mathbf{s} : V_0 \to V \subset N$$

and

$$\mathbf{r} : U_0 \to U \subset M$$

with $p \in U$, $\mathbf{f}(U) \subset V$ such that

$$\mathbf{s}^{-1} \circ \mathbf{f} \circ \mathbf{r} : U_0 \to V_0 \subset \mathbb{R}^{m_2}$$

is of C^k class.

Proof 2.4 Let $\mathbf{f} : M \to N$ be of C^k class. Let $p \in M$. Thus, there exists a parametrization $s : V_0 \to V \subset N$ of C^k class with $\mathbf{f}(p) \in V$ and $V_0 \subset \mathbb{R}^{m_2}$.

Since $\mathbf{f}$ is continuous, there exists an open set U_1 in M, such that $p \in U_1$ and $\mathbf{f}(U_1) \subset V$. We may obtain a parametrization $\mathbf{r} : \hat{U}_0 \to \mathbf{r}(\hat{U}_0) \subset M$ with $p \in \mathbf{r}(\hat{U}_0)$, where $\hat{U}_0 \subset \mathbb{R}^{m_1}$ is open. Define

$$U_0 = \mathbf{r}^{-1}(U_1 \cap \mathbf{r}(\hat{U}_0))$$

and $U = U_1 \cap \mathbf{r}(\hat{U}_0)$.

Thus

$$\mathbf{f}(U) \subset \mathbf{f}(U_1) \subset V.$$

From the concerning definition, from $\mathbf{f}$ to be of C^k class, we have that

$$\mathbf{f} \circ \mathbf{r} : U_0 \to V \subset \mathbb{R}^s$$

is of C^k class, so that from the last theorem and its corollary

$$\mathbf{s}^{-1} \circ \mathbf{f} \circ \mathbf{r} : U_0 \to V_0$$

is of C^k class as well.

The proof of the reciprocal is left as an exercise.

Corollary 2.4.9 *Let* $M, N, P \subset \mathbb{R}^n$ *be surfaces of* C^k *class of dimensions* m_1, m_2 *and* m_3 *respectively. Let* $\mathbf{f} : M \to N$ *and* $\mathbf{g} : N \to P$ *be functions of* C^k *class.*
Under such hypotheses $\mathbf{g} \circ \mathbf{f} : M \to P$ *is also of* C^k *class.*

The proof of such a corollary is left as an exercise.

2.5 Oriented surfaces

Definition 2.5.1 (Atlas) *An atlas of* C^k *class of a m-dimensional surface* $M \subset \mathbb{R}^n$ *is a collection* $\mathscr{P}$ *of parameterizations* $\mathbf{r} : U_0 \to U \subset M$ *of* C^k *class such that the union of sets* U *covers* M.

Two parameterizations of C^k *class* $\mathbf{r} : U_0 \to U$ *and* $\mathbf{s} : V_0 \to V$ *are said to be coherent if* $U \cap V = \emptyset$ *or, if* $U \cap V \neq \emptyset$, *then* $\xi = \mathbf{r}^{-1} \circ \mathbf{s}$ *has positive Jacobian determinant at each point of* $\mathbf{s}^{-1}(U \cap V)$.

An atlas is said to be coherent if each pair of parameterizations $\mathbf{r}, \mathbf{s}$ *are coherent.*

If M *admits an atlas* $\mathscr{P}$ *coherent, is said to be (positively) oriented by* $\mathscr{P}$.

Theorem 2.5.2 *Let* $M \subset \mathbb{R}^n$ *be a surface of* C^k *class.*
If there exist $n - m$ *continuous vector fields* $\mathbf{n}_1, \ldots, \mathbf{n}_{n-m} : M \to \mathbb{R}^n$ *such that* $\mathbf{n}_1(p), \ldots, \mathbf{n}_{n-m}(p) \in (T_p M)^\perp$ *are linearly independents,* $\forall p \in M$, *then* M *is oriented.*

Proof 2.5 Let $\mathscr{P}$ be a set of parameterizations of C^k class $\mathbf{r} : U_0 \to U \subset M$ such that

1. U_0 is connected.

2. For each $\mathbf{u} \in U_0$, $A(\mathbf{r}(\mathbf{u}))$ is a matrix $n \times n$ in which the columns are

$$\mathbf{r}'(\mathbf{u})e_1, \ldots, \mathbf{r}'(\mathbf{u})e_m, \mathbf{n}_1(\mathbf{r}(\mathbf{u})), \ldots, \mathbf{n}_{n-m}(\mathbf{r}(\mathbf{u})),$$

has positive determinant. Here, $\{e_1, \ldots, e_m\}$ is the canonical basis of $\mathbb{R}^m$.

Let us show that $\mathscr{P}$ is a coherent atlas in M.

Let $p \in M$. Consider a parametrization of C^k class $\mathbf{r} : U_0 \to U \subset M$ with $p \in U$ and U_0 connected.

Thus, by continuity, either $\det A(\mathbf{r}(\mathbf{u})) > 0$, $\forall \mathbf{u} \in U_0$ and in such a case $\mathbf{r} \in \mathscr{P}$, or $\det A(\mathbf{r}(\mathbf{u})) < 0$, $\forall \mathbf{u} \in U_0$ and in such a case, it suffices to replace $\mathbf{r}$ by $\mathbf{r}_1$, where $\mathbf{r}_1(u_1, u_2, \ldots, u_m) = \mathbf{r}(-u_1, u_2, \ldots u_m)$ in order to obtain $\mathbf{r}_1 \in \mathscr{P}$.

Since $p \in M$ is arbitrary, we have shown that the ranges of the parameterizations $\mathscr{P}$ cover M.

Let $\mathbf{r} : U_0 \to U$ and $\mathbf{s} : V_0 \to V$ be elements of $\mathscr{P}$ such that $U \cap V \neq \emptyset$.
We must show that

$$\mathbf{r}^{-1} \circ \mathbf{s} : \mathbf{s}^{-1}(U \cap V) \to \mathbf{r}^{-1}(U \cap V)$$

have positive Jacobian determinants at each point of

$$\mathbf{u} \in \mathbf{s}^{-1}(U \cap V).$$

Let

$$p = \mathbf{r}(\mathbf{u}_1) = \mathbf{s}(\mathbf{u}_2) \in U \cap V \subset M.$$

Observe that

$$\mathbf{s}'(\mathbf{u}_2) e_j = \sum_{i=1}^{m} \alpha_j^i \mathbf{r}'(\mathbf{u}_1) e_i, \ \forall j \in \{1, \ldots, m\}.$$

From this, we have

$$\det A(\mathbf{s}(\mathbf{u}_2)) = \det\{\alpha_j^i\} \det A(\mathbf{r}(\mathbf{u}_1)),$$

so that

$$\det\{\alpha_j^i\} > 0.$$

Observe that the Jacobian matrix of $\mathbf{r}^{-1} \circ \mathbf{s}$ at $\mathbf{u}_2$ is exactly $\{\alpha_j^i\}$.
The proof is complete.

Theorem 2.5.3 *Let* $M \subset \mathbb{R}^n$ *be an m-dimensional of* C^k *class surface, where* $1 \leq m \leq n$. *Let* $p \in M$.

Under such hypotheses, the elements of T_pM *are the velocity vectors at p of the differentiable curves contained in M which pass through p.*

More precisely,

$$T_pM = \{\mathbf{v} = \lambda'(0) : \lambda : (-\varepsilon, \varepsilon) \to M \subset \mathbb{R}^n$$
$$\text{is differentiable and } \lambda(0) = p\}. \tag{2.5}$$

Proof 2.6 Let $\mathbf{v} \in T_pM$. Thus, there exists $\mathbf{w} \in \mathbb{R}^m$ such that

$$\mathbf{v} = \mathbf{r}'(\mathbf{u})\mathbf{w},$$

where $p = \mathbf{r}(\mathbf{u})$, for an appropriate parametrization

$$\mathbf{r} : U_0 \to U.$$

Therefore

$$\mathbf{v} = \lim_{t \to 0} \frac{\mathbf{r}(\mathbf{u} + t\mathbf{w}) - \mathbf{r}(\mathbf{u})}{t}.$$

For $\varepsilon > 0$ sufficiently small, define $\lambda : (-\varepsilon, \varepsilon) \to M$ by

$$\lambda(t) = \mathbf{r}(\mathbf{u} + t\mathbf{w}).$$

Thus,

$$\mathbf{v} = \lambda'(0)$$

and

$$\lambda(0) = \mathbf{r}(\mathbf{u}) = p.$$

Reciprocally, suppose that $\lambda : (-\varepsilon, \varepsilon) \to M$ is a differentiable curve with $\lambda(0) = p$. Let

$$\mathbf{v} = \lambda'(0).$$

Observe that there exists a parametrization $\mathbf{r} : U_0 \to U$, where $p \in \mathbf{r}(\mathbf{u}) \in U$, for $\mathbf{u} \in U_0$, so that for $0 < \varepsilon_1 < \varepsilon$ sufficiently small,

$$\lambda(-\varepsilon_1, \varepsilon_1) \subset U.$$

Observe that

$$\mathbf{r}^{-1} \circ \lambda : (-\varepsilon_1, \varepsilon_1) \to U_0$$

is differentiable.

Define

$$\mathbf{w} = (\mathbf{r}^{-1} \circ \lambda)'(0),$$

thus,

$$\mathbf{w} = [\mathbf{r}'(\mathbf{u})]^{-1} \lambda'(0),$$

so that

$$\mathbf{v} = \lambda'(0) = \mathbf{r}'(\mathbf{u})\mathbf{w} \in T_p M.$$

The proof is complete.

Proposition 2.5.4 *Let $U \subset \mathbb{R}^{m+n}$ be an open set and let $\mathbf{f} : U \to \mathbb{R}^n$ be a function of C^k class.*

Let $\mathbf{c} \in \mathbb{R}^n$.

Define

$$M = \{p \in U : \mathbf{f}(p) = \mathbf{c} \text{ and } \mathbf{f}'(p) : \mathbb{R}^{m+n} \to \mathbb{R}^n \text{ is onto }\}.$$

Under such hypotheses, if $M \neq \emptyset$, then M is a surface of C^k class. Moreover,

$$T_p(M) = Ker\mathbf{f}'(p), \forall p \in M.$$

Proof 2.7 Denote $(\mathbf{x}, \mathbf{y}) \in \mathbb{R}^{m+n}$ and let

$$p = (\mathbf{x}_0, \mathbf{y}_0) \in M.$$

Thus

$$\mathbf{f}(\mathbf{x}_0, \mathbf{y}_0) = \mathbf{c}.$$

Renaming the variables if necessary, we obtain $\det[\mathbf{f}_y(\mathbf{x}_0, \mathbf{y}_0)] \neq 0$ and thus, from the implicit function theorem, there exist $\delta_1 > 0$ and $\delta_2 > 0$ such that for each $\mathbf{x} \in B_{\delta_1}(\mathbf{x}_0)$ there exists a unique $\mathbf{y} \in B_{\delta_2}(\mathbf{y}_0)$ such that

$$\mathbf{f}(\mathbf{x}, \mathbf{y}) = \mathbf{c},$$

where we denote

$$\mathbf{y} = \xi(\mathbf{x}),$$

where such a function is of C^k class, so that

$$\mathbf{f}(\mathbf{x}, \xi(\mathbf{x})) = \mathbf{c}, \ \forall \mathbf{x} \in B_{\delta_1}(\mathbf{x}_0).$$

Thus, the neighborhood $Z = B_{\delta_1}(\mathbf{x}_0) \times B_{\delta_2}(\mathbf{y}_0)$ of p is such that the parametrization $\mathbf{s} : V = B_{\delta_1}(\mathbf{x}_0) \to Z \cap f^{-1}(c)$, where $\mathbf{s}(\mathbf{x}) = (\mathbf{x}, \xi(\mathbf{x}))$ is a bijection.

Summarizing $p \in Z \cap f^{-1}(c) \subset M$.

Since such a p is arbitrary, we have that M is a surface.

Now, let $\mathbf{v} \in T_p M$. Let $\lambda : (-\varepsilon, \varepsilon) \to M$ be such that $\lambda(0) = p$ and $\lambda'(0) = \mathbf{v}$. Thus,

$$
\begin{aligned}
\mathbf{0} &= (\mathbf{f} \circ \lambda)'(0) \\
&= \mathbf{f}'(\lambda(0))\lambda'(0) \\
&= \mathbf{f}'(p)\mathbf{v}
\end{aligned}
\tag{2.6}
$$

Hence $\mathbf{v} \in Ker\mathbf{f}'(p)$.

Since $T_p M$ and $Ker\mathbf{f}'(p)$ are two sub-spaces of dimension m de $\mathbb{R}^{m+n}$ and from above $T_p M \subset Ker\mathbf{f}'(p)$, we get

$$T_p M = Ker\mathbf{f}'(p).$$

The proof is complete.

2.6 Surfaces in $\mathbb{R}^n$ with boundary

Definition 2.6.1 (Semi-space) *Consider a linear function* $\alpha : \mathbb{R}^m \to \mathbb{R}$, *where*

$$\alpha(\mathbf{u}) = \sum_{k=1}^{m} a_k u_k, \ \forall \mathbf{u} \in \mathbb{R}^m,$$

and where $a_k \in \mathbb{R}, \forall k \in \{1, \ldots, m\}$.

Define the semi-space $H \subset \mathbb{R}^m$ by

$$H = \{\mathbf{u} \in \mathbb{R}^m \; : \; \alpha(\mathbf{u}) \leq 0\}.$$

Observe that in such a case, the boundary of H, denoted by ∂H, will be

$$\partial H = \{\mathbf{u} \in \mathbb{R}^m \; : \; \alpha(\mathbf{u}) = 0\}.$$

Remark 2.6.2 *Let $A \subset H \subset \mathbb{R}^m$ be an open set in H. Let $\mathbf{f} : A \to \mathbb{R}^n$ be a differentiable function.*
We shall show that $f'(\mathbf{u}) : \mathbb{R}^m \to \mathbb{R}^n$ is well defined, $\forall \mathbf{u} \in A$.
Let $F : U \to \mathbb{R}^n$, be a differentiable extension of F where $U \supset A$ is open.
Let $\mathbf{u} \in A$. If $\mathbf{u} \in H°$, obviously

$$F'(\mathbf{u}) = \mathbf{f}'(\mathbf{u}).$$

Assume that $\mathbf{u} \in A \cap \partial H$. Let $\{\mathbf{v}_1, \ldots, \mathbf{v}_m\} \subset H$ be a basis of $\mathbb{R}^m$.
Indeed, such a basis exists, since, given any basis of $\mathbb{R}^m$, changing the sign of each element if necessary, we may obtain that each element is in H. (even when changing some signs we still have a basis)
Let $t \geq 0$. Thus

$$\mathbf{u} + t\mathbf{v}_k \in H, \forall k \in \{1, \ldots, m\},$$

since

$$\alpha(\mathbf{u} + t\mathbf{v}_k) = \alpha(\mathbf{u}) + t\alpha(\mathbf{v}_k) \leq 0, \; \forall k \in \{1, \ldots, m\}.$$

Since A is open in H and $\mathbf{u} \in A \cap \partial H$, for all $t \geq 0$ sufficiently small, we have that

$$\mathbf{u} + t\mathbf{v}_k \in A.$$

Let, $k \in \{1, \ldots, m\}$.
Thus,

$$
\begin{aligned}
F'(\mathbf{u})\mathbf{v}_k &= \lim_{t \to 0} \frac{F(\mathbf{u} + t\mathbf{v}_k) - F(\mathbf{u})}{t} \\
&= \lim_{t \to 0^+} \frac{F(\mathbf{u} + t\mathbf{v}_k) - F(\mathbf{u})}{t} \\
&= \lim_{t \to 0^+} \frac{\mathbf{f}(\mathbf{u} + t\mathbf{v}_k) - \mathbf{f}(\mathbf{u})}{t} \\
&= \mathbf{f}'(\mathbf{u})\mathbf{v}_k.
\end{aligned}
\tag{2.7}
$$

Hence,

$$F'(\mathbf{u})\mathbf{v}_k = \mathbf{f}'(\mathbf{u})\mathbf{v}_k, \; \forall k \in \{1, \ldots, m\}.$$

From this, we may infer that $F'(\mathbf{u})$ does not depend on the extension chosen.
The chain rule is also valid, that is, if $\mathbf{f} : A \to \mathbb{R}^n$ and $\mathbf{g} : B \to \mathbb{R}^p$ are differentiable, where $A \subset H$ is open, H and $\mathbf{f}(A) \subset B \subset H_1$, where H is a semi-space of $\mathbb{R}^m$ and H_1 is a semi-space of $\mathbb{R}^n$, then $(\mathbf{g} \circ \mathbf{f}) : A \to \mathbb{R}^p$ is differentiable and

$$(\mathbf{g} \circ \mathbf{f})'(\mathbf{u}) = \mathbf{g}'(\mathbf{f}(\mathbf{u}))\mathbf{f}'(\mathbf{u}), \; \forall \mathbf{u} \in A.$$

Remark 2.6.3 *Consider again a semi-space $H \subset \mathbb{R}^m$ and let $A \subset H$ be an open set in H. We shall define the boundary of A in H, denoted by ∂A, as*

$$\partial A = A \cap \partial H.$$

Observe that the boundary ∂A is an hyper-surface in $\mathbb{R}^m$.
Indeed, since A is open in H, we have that $A = U \cap H$, for some open $U \subset \mathbb{R}^m$.
Therefore,

$$
\begin{aligned}
U \cap \partial H &= U \cap (H \cap \partial H) \\
&= (U \cap H) \cap \partial H \\
&= A \cap \partial H \\
&= \partial A, \qquad\qquad (2.8)
\end{aligned}
$$

so that ∂A is an open subset of the hyper-surface

$$\partial H = \alpha^{-1}(0).$$

Theorem 2.6.4 *Let $A \subset H$ be an open set in H and let $B \subset H_1$ be an open set in H_1, where H, H_1 are semi-spaces in $\mathbb{R}^m$.*
Let $\mathbf{f} : A \to B$ be a diffeomorphism of C^1 class.
Under such hypotheses

$$\mathbf{f}(\partial A) = \partial B.$$

In particular, the restriction $\mathbf{f}|_{\partial A}$ is a diffeomorphism between ∂A and ∂B.

Proof 2.8 *Let $\mathbf{u} \in A \cap H^0 \equiv U$. Thus, $U \subset \mathbb{R}^m$ is an open set such that $\mathbf{u} \in U \subset A \subset H$.*
Restricted to U, $\mathbf{f}$ is a diffeomorphism of C^1 class on $\mathbf{f}(U)$. From the inverse function theorem, $\mathbf{f}(U)$ is open in $\mathbb{R}^m$. Since $\mathbf{f}(U) \subset B \subset H_1$, we may infer that $\mathbf{f}(U) \subset B \cap H_1^\circ$. Thus,

$$\mathbf{f}(A \cap H^\circ) \subset B \cap H_1^\circ.$$

Therefore,

$$\mathbf{f}^{-1}(\partial B) \subset \partial A.$$

Similarly, inter-changing the roles of $\mathbf{f}$, $\mathbf{f}^{-1}$ and, A and B, we may obtain

$$\mathbf{f}(\partial A) \subset \partial B,$$

so that

$$\mathbf{f}(\partial A) = \partial B.$$

2.6.1 Parameterizations for surfaces in $\mathbb{R}^n$ with boundary

Definition 2.6.5 *A parametrization (of C^k class and dimension m) of a set $U \subset \mathbb{R}^n$ is an homeomorphism $\mathbf{r} : U_0 \to U$ of C^k class defined on the open set $U_0 \subset H$ in H, where H is a semi-space, such that $\mathbf{r}'(\mathbf{u}) : \mathbb{R}^m \to \mathbb{R}^n$ is a linear injective transformation for each $\mathbf{u} \in U_0$.*

Definition 2.6.6 *In the context of the last definition, a set $M \subset \mathbb{R}^n$ is said to be a surface of dimension m and class C^k, with a boundary, as for each $p \in M$, there exists a parametrization $\mathbf{r} : U_0 \to U \subset M$, with $p = \mathbf{r}(\mathbf{u})$, for some $\mathbf{u} \in U_0 \subset H$, where U_0 is open in H and $H \subset \mathbb{R}^m$ is a semi-space.*

Theorem 2.6.7 *Let $M \subset \mathbb{R}^n$ be a m-dimensional surface of C^k class with boundary. Let $\mathbf{r} : U_0 \to U$ and $\mathbf{s} : V_0 \to V$ parameterizations of C^k class of open sets (in M) $U, V \subset M$, with $U \cap V \neq \emptyset$. Under such hypotheses,*

$$\mathbf{s}^{-1} \circ \mathbf{r} : \mathbf{r}^{-1}(U \cap V) \to \mathbf{s}^{-1}(U \cap V)$$

is a diffeomorphism of C^k class.

Proof 2.9 Let $\mathbf{u} \in \mathbf{r}^{-1}(U \cap V)$.

Let $p = \mathbf{r}(\mathbf{u})$ and let $\mathbf{v} = \mathbf{s}^{-1}(\mathbf{r}(\mathbf{u}))$.

Observe that, from above, $\mathbf{s}$ may be extended to a function $\mathbf{s}_1 : W \to \mathbb{R}^n$ of C^k class in an open set $W \subset \mathbb{R}^m$ such that $\mathbf{v} \in W$. Since $\mathbf{s}'_1(\mathbf{v}) : \mathbb{R}^m \to \mathbb{R}^n$ is injective, we have that, from the local form of immersions, restricting W if necessary, $\mathbf{s}_1$ is a homeomorphism of W onto its range $\mathbf{s}_1(W)$, so that the inverse homeomorphism $\mathbf{s}_1^{-1} : \mathbf{s}_1(W) \to W$ is the restriction of a function $\mathbf{g}$ of C^k class defined in an open set of $\mathbb{R}^n$.

Defining $A = \mathbf{r}^{-1}(\mathbf{s}_1(W))$, we have that $\mathbf{u} \in A \cap H$.

Observe that

$$(\mathbf{s}^{-1} \circ \mathbf{r})|_{A \cap H} = (\mathbf{g} \circ \mathbf{r})|_{A \cap H}$$

where $\mathbf{g} \circ \mathbf{r}$ is of C^k class.

In particular, $(\mathbf{s}^{-1} \circ \mathbf{r})$ is of C^k class at $\mathbf{u}$, $\forall \mathbf{u} \in \mathbf{r}^{-1}(U \cap V)$.

The proof is complete.

Remark 2.6.8 *For a surface $M \subset \mathbb{R}^n$ with boundary, the boundary of M is the set of points $p \in M$ such that the parametrization $\mathbf{r} : U_0 \to U$ of C^1 class of an open set in M, $U \subset M$ with $p = \mathbf{r}(\mathbf{u})$, necessarily we have $\mathbf{u} \in \partial U_0$.*

From the theorem 2.6.4 and the fact that each change of variables for a parametrization is a diffeomorphism, for $p \in M$, it suffices that there exists a parametrization $\mathbf{r} : U_0 \to U$ of C^1 class on an open set U em M, with $p = \mathbf{r}(\mathbf{u})$ and $\mathbf{u} \in \partial U_0$, in order for us to have that $p \in \partial M$.

Remark 2.6.9 *If $M \subset \mathbb{R}^n$ is a surface with boundary of C^k class and dimension $m+1$, its boundary ∂M is a surface (without boundary) of C^k class and dimension m (since $\partial H \subset \mathbb{R}^{m+1}$ is a sub-space of dimension m).*

The parameterizations which define ∂M as a surface are restrictions to the boundary $\partial U_0 = U_0 \cap \partial H$, of parameterizations $\mathbf{r} : U_0 \to U$ of C^k class, which has as range, an open set U in M, such that $U \cap \partial M \neq \emptyset$. Observe that the restriction $\mathbf{r}|_{\partial U_0} : \partial U_0 \to \partial U$ has $\partial U = U \cap \partial M$ as range. We may parameterize ∂U in the following fashion.

Write $\mathbf{u} = (u_0, u_1, \ldots, u_m) \in H$, and define

$$H_0 = \{\mathbf{u} = (u_0, u_1, \ldots, u_m) \in \mathbb{R}^{m+1} : u_0 \leq 0\}.$$

Thus, we identify ∂H_0 com $\mathbb{R}^m$, where

$$\partial H_0 = \{(0, u_1, \ldots, u_m) : (u_1, \ldots, u_m) \in \mathbb{R}^m\}.$$

Observe that, for each semi-space $H \subset \mathbb{R}^{m+1}$, there exists a linear isomorphism $T : \mathbb{R}^{m+1} \to \mathbb{R}^{m+1}$ such that $T(H_0) = H$.

Hence, for each parametrization $\mathbf{s} : V_0 \to U$ of C^k class in a set V_0 open in H, we define, $U_0 = T^{-1}(V_0)$ and obtain $\mathbf{r} = \mathbf{s} \circ T : U_0 \to U$, which is a standard parametrization defined in an open set H_0, of C^k class and with the same range as that of $\mathbf{s}$.

If $\mathbf{r} : U_0 \to U$ is a standard parametrization and $U \cap \partial M \neq \emptyset$, the restriction $\mathbf{r}|_{\partial U_0} : \partial U_0 \to \partial U$ in the surface ∂M is defined in an open subset $\partial U_0 \subset \mathbb{R}^m$, considering that we have identified ∂H_0 with the set $\mathbb{R}^m$.

Definition 2.6.10 (Tangent space) *Let $M \subset \mathbb{R}^m$ be a surface with boundary of C^1 class and dimension $m+1$. For each point $p \in M$, we may define the tangent space of dimension $m+1$, denoted by $T_p M \subset \mathbb{R}^n$, as*

$$T_p M = \mathbf{r}'(\mathbf{u})[\mathbb{R}^{m+1}],$$

where $\mathbf{r}(\mathbf{u}) = p$ and $\mathbf{r} : U_0 \to U$ is any parametrization of C^1 class of an open set in M, $U \subset M$ where $U_0 \subset \mathbb{R}^{m+1}$ is an open semi-space $H \subset \mathbb{R}^{m+1}$.

Remark 2.6.11 *Let $p \in \partial M$.*

Thus, for a parametrization $\mathbf{r} : U_0 \to U \subset M$, com $p = \mathbf{r}(\mathbf{u})$, we have that $\mathbf{u} \in \partial U_0$, so that $\mathbf{r}'(\mathbf{u})[\partial U_0] = T_p(\partial M)$ is the tangent sub-space to the boundary of M at $p \in \partial M$.

Observe that

$$T_p(\partial M) \subset T_p M,$$

where $T_p(\partial M)$ is a sub-space of $T_p(M)$ of dimension m (co-dimension 1).

Proposition 2.6.12 *Let $M \subset \mathbb{R}^n$ be a m-dimensional surface with a boundary. The definition of $T_p M$ does not depend on the parametrization used.*

Proof 2.10   Indeed, let $\mathbf{r} : U_0 \to U$ and $\mathbf{r} : V_0 \to V$ parameterizations in M where $p = \mathbf{r}(\mathbf{u}) = \mathbf{s}(\mathbf{v}) \in U \cap V$.

Observe that

$$\xi = (\mathbf{s}^{-1} \circ \mathbf{r}) : \mathbf{r}^{-1}(U \cap V) \to \mathbf{s}^{-1}(U \cap V)$$

is a diffeomorphism, where

$$\mathbf{s} \circ \xi = \mathbf{r}.$$

Thus

$$\mathbf{s}'(\mathbf{v})\xi'(\mathbf{u}) = \mathbf{r}'(\mathbf{u}).$$

Since $\xi' : \mathbb{R}^{m+1} \to \mathbb{R}^{m+1}$ is an isomorphism, we have that

$$\mathbf{r}'(\mathbf{u})[\mathbb{R}^{m+1}] = \mathbf{s}'(\mathbf{v})[\mathbb{R}^{m+1}].$$

The proof is complete.

Remark 2.6.13 *Let $p \in \partial M$. Thus, we have the tangent space $T_p M$ and its tangent sub-space $T_p(\partial M)$.*

Moreover, in $T_p M$, we have the semi-space comprised by its vectors outer directed concerning M and the vectors of $T_p(\partial M)$.

Let us see more rigorously such an intuitive idea.

Definition 2.6.14 *We say that a vector $\mathbf{w} \in \mathbb{R}^m$ is outer directed concerning the semi-space H, if $\mathbf{w} \notin H$.*

In such a case, if

$$H = \{\mathbf{u} \in \mathbb{R}^m : \alpha(\mathbf{u}) \le 0\},$$

we have $\alpha(\mathbf{w}) > 0$.

Let us also see the next theorem.

Theorem 2.6.15 *Let $\mathbf{f} : A \to B$ be a diffeomorphism between $A \subset H$ and $B \subset H_1$, where A is open in H and B is open in H_1 and, $H, H_1 \subset \mathbb{R}^m$ are semi-spaces.*

Suppose that $\mathbf{w}$ is outer directed concerning H. Under such hypotheses, for each $\mathbf{u} \in \partial A$ we have that $\mathbf{f}'(\mathbf{u})\mathbf{w}$ is outer directed relating H_1.

Proof 2.11   Observe that

$$H = \{\mathbf{u} \in \mathbb{R}^m : \alpha(\mathbf{u}) \le 0\},$$

and

$$H_1 = \{\mathbf{u} \in \mathbb{R}^m : \beta(\mathbf{u}) \le 0\},$$

where $\alpha, \beta : \mathbb{R}^m \to \mathbb{R}$ are appropriate linear functions.

Let $\mathbf{u} \in \partial A = A \cap \partial H$. From Theorem 2.6.4, $\mathbf{f}$ is a diffeomorphism between ∂A and ∂B, so that $\mathbf{f}'(\mathbf{u}) : \mathbb{R}^m \to \mathbb{R}^m$ isomorphically transforms ∂H in ∂H_1.

Thus, given $\mathbf{v} \in \mathbb{R}^m$, we have that

$$\beta[\mathbf{f}'(\mathbf{u})\mathbf{v}] = 0,$$

if, and only if,

$$\alpha(\mathbf{v}) = 0.$$

Since $\alpha(\mathbf{w}) > 0$, it suffices to show that

$$\beta[\mathbf{f}'(\mathbf{u})\mathbf{w}] \geq 0.$$

Observe that for $t < 0$ we have $\mathbf{u} + t\mathbf{w} \in H$, since $\alpha(\mathbf{u} + t\mathbf{w}) = 0 + t\alpha(\mathbf{w}) < 0$. Therefore, for a sufficiently small $t < 0$ in absolute value, we have

$$\mathbf{u} + t\mathbf{w} \in A \setminus \partial A,$$

so that

$$\mathbf{f}(\mathbf{u} + t\mathbf{w}) \in B \setminus \partial B$$

and hence

$$\beta(\mathbf{f}(\mathbf{u} + t\mathbf{w})) < 0.$$

Thus, for such values of t, we obtain

$$\frac{\beta(\mathbf{f}(\mathbf{u} + t\mathbf{w})) - \beta(\mathbf{f}(\mathbf{u}))}{t} = \frac{\beta(\mathbf{f}(\mathbf{u} + t\mathbf{w}))}{t} > 0.$$

Letting $t \to 0^-$, we get

$$\beta(\mathbf{f}'(\mathbf{u})\mathbf{w}) \geq 0.$$

The proof is complete.

Definition 2.6.16 *Let* $M \subset \mathbb{R}^n$ *be a* $m+1$*-dimensional surface of* C^1 *class with a boundary and let* $p \in \partial M$.

We say that a vector $\mathbf{w} \in T_pM$ *is outer directed concerning M, if there exists a parametrization* $\mathbf{r} : U_0 \to U$ *of* C^1 *class of an open set* U_0 *in H, where* $H \subset \mathbb{R}^{m+1}$ *is a semi-space,* $U \subset M$ *is open in M, and where* $p = \mathbf{r}(\mathbf{u}) \in U$, $\mathbf{u} \in \partial U_0$ *and* $\mathbf{w} = \mathbf{r}'(\mathbf{u})\mathbf{w}_0$, *for some* $\mathbf{w}_0 \in \mathbb{R}^{n+1}$ *which is outer directed concerning H.*

Observe that in such a case, from the last theorem, for each parametrization $\mathbf{s} : V_0 \to V$, com $p = \mathbf{s}(\mathbf{v}) \in V$, $\mathbf{v} \in \partial V_0$, we have that

$$\mathbf{w} = \mathbf{s}'(\mathbf{v})\mathbf{w}_1,$$

for some $\mathbf{w}_1 \in \mathbb{R}^{m+1}$, which is outer directed concerning the semi-space H_1, such that $\partial V_0 = V_0 \cap \partial H_1$ and in which V_0 is open.

Indeed, defining

$$\xi = \mathbf{s}^{-1}\mathbf{r},$$

and

$$\mathbf{w}_1 = \xi'(\mathbf{u})\mathbf{w}_0$$

we have that

$$
\begin{aligned}
\mathbf{s}'(\mathbf{v})\mathbf{w}_1 &= \mathbf{s}'(\mathbf{v})\xi'(\mathbf{u})\mathbf{w}_0 \\
&= \mathbf{r}'(\mathbf{u})\mathbf{w}_0 \\
&= \mathbf{w}.
\end{aligned}
\tag{2.9}
$$

From the last theorem, $\mathbf{w}_1$ is outer directed concerning H_1, since $\mathbf{w}_0$ is outer directed concerning H.

Remark 2.6.17 *At each point $p \in \partial M$, the tangent vectors to ∂M and the outer directed vectors concerning M form a semi-space of $T_p M$. Since $dim(T_p M) = m+1$, we have that $dim(T_p(\partial M)) = m$, so that there exists a unique vector $\mathbf{n}(p)$ which is orthogonal to $T_p(\partial M)$, it is in $T_p(M)$ and is also outer directed concerning M. O vector $\mathbf{n}(p)$ is said to be the outward normal vector to ∂M at $p \in \partial M$. (outward normal to the boundary of M at p).*

Let $p \in \partial M$, where $p = \mathbf{r}(\mathbf{u})$ and $\mathbf{u} = (0, u_1, \ldots, u_m) \in \partial U_0$ (for an appropriate standard parametrization).

Observe that $\mathbf{n}(p) \in T_p M$ has the expression

$$\mathbf{n}(p) = a_0 \frac{\partial \mathbf{r}(\mathbf{u})}{\partial u_0} + a_1 \frac{\partial \mathbf{r}(\mathbf{u})}{\partial u_1} + \ldots + a_m \frac{\partial \mathbf{r}(\mathbf{u})}{\partial u_m},$$

where a basis for $T_p(\partial M)$ is given by

$$\frac{\partial \mathbf{r}(\mathbf{u})}{\partial u_1}, \ldots, \frac{\partial \mathbf{r}(\mathbf{u})}{\partial u_m}.$$

Thus, we have

$$\mathbf{n}(p) \cdot \frac{\partial \mathbf{r}(\mathbf{u})}{\partial u_k} = 0, \ \forall k \in \{1, \ldots, m\}.
\tag{2.10}$$

We have then m linear equations and $m+1$ unknown variables, $a_0, a_1, \ldots, a_{m+1}$.

Observe that through (2.10) we may have $a_1(a_0), \ldots, a_m(a_0)$, where such relations are linear.

Through such a condition $\mathbf{n}(p) \cdot \mathbf{n}(p) = 1$, we have a_0^2, that is $|a_0|$.

Finally, the sign of a_0 must be such that $\mathbf{n}(p)$ is outer directed concerning M.

Now we shall show in a formal fashion that, if M is oriented, so is its boundary.

Let $M \subset \mathbb{R}^n$ be a $m+1$-dimensional surface of C^1 class, oriented and with a boundary.

Denote by $\mathscr{P}$ the set of parameterizations $\mathbf{r} : U_0 \to U \subset M$, of C^1 class, with the following properties:

1. U_0 *is connected.*

2. U_0 *is an open set in the semi-space*

$$H = \{(u_0, u_1, \ldots, u_m) \in \mathbb{R}^{m+1} : u_0 \leq 0\}.$$

3. $\mathbf{r}$ *is positive relating the orientation of M.*

We have already seen that the set of standard parameterizations $\mathbf{r} : U_0 \to U \subset M$ *which satisfies 1 and 2 (standard) comprises an atlas of M.*

If we include the condition 3 we have still an atlas.

Indeed, let $\mathbf{s} : V_0 \to V$ *be such that 1 and 2 hold. If* $\mathbf{s}$ *is negative we may compose it with the linear transformation*

$$T(u_0, u_1, \ldots, u_m) = (u_0, u_1, \ldots, -u_m)$$

and thus defining $U_0 = T^{-1}(V_0)$*, we have that*

$$\mathbf{r} = \mathbf{s} \circ T : U_0 \to V$$

satisfies the conditions 1, 2 and 3, since $\det T < 0$*.*

Therefore $\mathscr{P}$ *is an atlas.*

Let us now identify $\mathbb{R}^m$ *with*

$$\partial H_0 = \{\mathbf{u} = (u_0, u_1, \ldots, u_m) : u_0 = 0\}.$$

Let $\mathscr{P}_0$ *be the set of restrictions* $\mathbf{r}_0 = \mathbf{r}|_{\partial U_0}$ *of the parameterizations* $\mathbf{r}$ *of* $\mathscr{P}$ *such that* $\partial U_0 = U_0 \cap \partial H_0 \neq \emptyset$*.*

Thus $\mathscr{P}_0$ *is an atlas of* C^1 *class for* ∂M*.*

We shall now show that $\mathscr{P}_0$ *is coherent.*

Indeed, let $\mathbf{r}_0 : \partial U_0 \to \partial U \in \mathscr{P}_0$ *and* $\mathbf{s}_0 : \partial V_0 \to \partial V \in \mathscr{P}_0$ *be such that*

$$\partial U \cap \partial V \neq \emptyset.$$

Thus, the change of parametrization $\xi_0 = \mathbf{s}_0^{-1} \circ \mathbf{r}_0$ *is the restriction of diffeomorphism* $\xi = \mathbf{s}^{-1} \circ \mathbf{r}$ *to the boundary of the domain in question.*

Denote by $A = \xi'(\mathbf{u}) : \mathbb{R}^{m+1} \to \mathbb{R}^{m+1}$ *the derivative* ξ *at* $\mathbf{u}$ *such that*

$$\mathbf{r}(\mathbf{u}) \in \partial U \cap \partial V$$

(domain of ξ_0*).*

Since $\mathscr{P}$ *is coherent, we have that* $\det A > 0$*.*

On the other hand, ξ *is a diffeomorphism from the open set*

$$\mathbf{r}^{-1}(U \cap V) \subset H_0$$

to the open set

$$\mathbf{s}^{-1}(U \cap V) \subset H_0.$$

From Theorem 2.6.4, we have that

$$A(\partial H_0) = \partial H_0$$

that is,

$$Ae_i = (0, a_{1i}, \ldots, a_{mi}),$$

$\forall i \in \{1, \ldots, m\}$.

Finally, since $e_0 = (1, 0, \ldots, 0)$ is outer directed concerning H_0, we have, from Theorem 2.6.15, that $Ae_0 = (a_{00}, a_{10}, \ldots a_{m_0})$ is also outer directed.

Hence, we must have $a_{00} > 0$, where the matrix A has the following expression

$$A = \begin{bmatrix} a_{00} & 0 & \cdots & 0 \\ a_{10} & a_{11} & \cdots & a_{1m} \\ \vdots & \vdots & \ddots & \vdots \\ a_{m0} & a_{m1} & \cdots & a_{mm} \end{bmatrix}_{(m+1) \times (m+1)} . \tag{2.11}$$

Observe that $\det A = a_{00} \det A_0$ where

$$A_0 = \begin{bmatrix} a_{11} & a_{12} & \cdots & a_{1m} \\ a_{21} & a_{22} & \cdots & a_{2m} \\ \vdots & \vdots & \ddots & \vdots \\ a_{m1} & a_{m2} & \cdots & a_{mm} \end{bmatrix}_{m \times m} . \tag{2.12}$$

and where A_0 is the derivative of ξ_0 at $\mathbf{u}$.

Since $\det A > 0$ and $a_{00} > 0$, we have $\det A_0 > 0$.

From this, we may infer that the atlas $\mathscr{P}_0$ is coherent.

The orientation of ∂M is to be induced by the orientation of M.

2.7 The tangent space

Let $M \subset \mathbb{R}^n$ be a m-dimensional surface, where $1 \leq m \leq n$, of C^1 class, with a boundary ∂M.

Observe that for each $p \in M$, there exists a parametrization $\mathbf{r}^p : U_0^p \to U^p \subset M$, where $U_0^p \subset H_0$ is open in the semi-space H_0 (for standard parametrization), for a local system of coordinates. Here

$$H_0 = \{(u_1, \ldots, u_m) \in \mathbb{R}^m : u_1 \leq 0\}.$$

We denote $D = \cup_{p \in M} U_0^p$ and generically also denote $M = \mathbf{r}(D) = \cup_{p \in M} \mathbf{r}^p(U_0^p)$. where, to simplify the notation, we have written

$$\mathbf{r}^p(\mathbf{u}) \equiv \mathbf{r}(\mathbf{u}),$$

emphasizing we are in fact referring to a particular system of coordinates relating U_0^p and for a specific parametrization defined on U_0^p.

Definition 2.7.1 (The tangent space to M) *In the context of the last lines above, let $p = \mathbf{r}(\mathbf{u}) \in M$. We define the tangent space to M at p, denoted by $T_p(M)$, by*

$$T_p(M) = \left\{ \sum_{i=1}^{m} \alpha_i \frac{\partial \mathbf{r}(\mathbf{u})}{\partial u_i} \ : \ \alpha_1, \cdots, \alpha_m \in \mathbb{R} \right\}.$$

We recall that for a C^k class surface,

$$\left\{ \frac{\partial \mathbf{r}(\mathbf{u})}{\partial u_i} \right\}_{i=1}^{m}$$

is linearly independent $\forall p \in M$.

Remark 2.7.2 *We recall that the dual space of a real vector space V is formally defined as the set of all linear continuous functionals (in this context, in fact real functions) defined on V. From the Riesz representation theorem, considering the specific case where $V \subset \mathbb{R}^n$, given a linear continuous functional $F : V \to \mathbb{R}$, there exists $\alpha \in V$ such that*

$$F(\mathbf{u}) = \alpha \cdot \mathbf{u}, \ \forall \mathbf{u} \in V.$$

In such a case, as for any Hilbert space, we say that $V^* \approx V$, that is the dual space V^*, is indeed identified with V.

In the general case, from elementary linear algebra, given a basis $(\mathbf{e}_1, \cdots, \mathbf{e}_m)$ for V, we may obtain a corresponding dual basis for V^*, given by $F_1, \cdots, F_m \in V^*$, such that

$$F_i(\mathbf{e}_j) = \delta_{ij},$$

where

$$\delta_{ij} = \begin{cases} 1 & \text{if } i = j \\ 0 & \text{otherwise.} \end{cases} \tag{2.13}$$

In the next lines, we present the formal details on these last statements.

Theorem 2.7.3 *Let V be a real m-dimensional vector space and let $\{\mathbf{v}_1, \cdots, \mathbf{v}_m\}$ be a basis for V*

Under such hypotheses, there exists a unique basis $\{F_1, \cdots, F_m\}$ for V^* such that

$$F_i(v_j) = \delta_{ij},$$

where

$$\delta_{ij} = \begin{cases} 1 & \text{if } i = j \\ 0 & \text{if } i \neq j, \forall i, j \in \{1, \cdots, m\} \end{cases} \tag{2.14}$$

Proof 2.12   For each $i \in \{1, \cdots, m\}$, define $F_i : V \to \mathbb{R}$ by

$$F_i(\mathbf{v}) = a_i,$$

where,

$$\mathbf{v} = \sum_{j=1}^{m} a_j \mathbf{v}_j.$$

Therefore,

$$F_i(\mathbf{v}_j) = \delta_{ij}, \ \forall i, j \in \{1, \cdots, m\}.$$

Now we are going to show that such $\{F_i\}$ are unique.
Let $\tilde{F}_1, \cdots, \tilde{F}_m \in V^*$ be such that $\tilde{F}_i(\mathbf{v}_j) = \delta_{ij}, \ \forall i, j \in \{1, \cdots, m\}$.
Let $\mathbf{v} = \sum_{j=1}^{m} a_j \mathbf{v}_j \in V$.
Thus, fixing $i \in \{1, \cdots, m\}$, we have,

$$
\begin{aligned}
\tilde{F}_i(\mathbf{v}) &= \tilde{F}_i \left(\sum_{j=1}^{m} a_j \mathbf{v}_j \right) \\
&= \sum_{j=1}^{m} a_j \tilde{F}_i(\mathbf{v}_j) \\
&= \sum_{j=1}^{m} a_j \delta_{ij} \\
&= a_i \\
&= F_i(\mathbf{v}).
\end{aligned}
\tag{2.15}
$$

Therefore,

$$\tilde{F}_i(\mathbf{v}) = F_i(\mathbf{v}), \ \forall \mathbf{v} \in V,$$

so that,

$$\tilde{F}_i = F_i, \ \forall i \in \{1, \cdots, m\}.$$

We may conclude that the $F_1, \cdots, F_m \in V^*$ in question are unique.
At this point, we are going to show that $F_1, \cdots, F_m \in V^*$ are linearly independent.
Suppose that $b_1, \cdots, b_m \in \mathbb{R}$ are such that

$$F = \sum_{i=1}^{m} b_i F_i = \mathbf{0}.$$

Thus,

$$0 = F(\mathbf{v}_j) = \sum_{i=1}^{m} b_i F_i(\mathbf{v}_j) = \sum_{i=1}^{m} b_i \delta_{ij} = b_j, \ \forall j \in \{1, \cdots, m\}.$$

Hence, $\{F_1, \cdots, F_m\}$ is a linearly independent set.
To finish the proof, we are going to show that $\{F_1, \cdots, F_m\}$ spans V^*.
Let $S \in V^*$.
Denote $b_i = S(\mathbf{v}_i), \forall i \in \{1, \cdots, m\}$.

Define

$$F = \sum_{i=1}^{m} b_i F_i \in V^*.$$

Thus

$$F(\mathbf{v}_j) = \sum_{i=1}^{m} b_i F_i(\mathbf{v}_j) = \sum_{i=1}^{m} b_i \delta_{ij} = b_j.$$

Let $\mathbf{v} = \sum_{j=1}^{m} a_j \mathbf{v}_j$.
Thus,

$$S(\mathbf{v}) = S\left(\sum_{j=1}^{m} a_j \mathbf{v}_j\right) = \sum_{j=1}^{m} a_j S(\mathbf{v}_j) = \sum_{j=1}^{m} a_j b_j.$$

On the other hand,

$$F(\mathbf{v}) = F\left(\sum_{j=1}^{m} a_j \mathbf{v}_j\right) = \sum_{j=1}^{m} a_j F(\mathbf{v}_j) = \sum_{j=1}^{m} a_j b_j.$$

Hence,

$$F(\mathbf{v}) = S(\mathbf{v}), \forall \mathbf{v} \in V,$$

so that

$$S = F = \sum_{i=1}^{m} b_i F_i.$$

Thus, $\{F_1, \cdots, F_m\}$ spans V^*, so that it is a basis for V^*.
This completes the proof.

Theorem 2.7.4 *Let $M \subset \mathbb{R}^n$ be an m dimensional surface and let $f : M \to \mathbb{R}$ be C^1 class function.*

Let $p = \mathbf{r}(\mathbf{u}) \in M$. Under such hypotheses, we may associate to such a function f, a functional

$$F : T_p(M) \to \mathbb{R}$$

where, for each

$$\mathbf{v} = v_i \frac{\partial \mathbf{r}(\mathbf{u})}{\partial u_i},$$

we have

$$
\begin{aligned}
F(\mathbf{v}) &= F\left(v_i \frac{\partial \mathbf{r}(\mathbf{u})}{\partial u_i}\right) \\
&= \lim_{\varepsilon \to 0} \frac{(f \circ \mathbf{r})(\{u_i + \varepsilon v_i\}) - (f \circ \mathbf{r})(\{u_i\})}{\varepsilon} \\
&= df(\mathbf{v}). \tag{2.16}
\end{aligned}
$$

Conversely, if $F : T_p(M) \to \mathbb{R}$ is a continuous linear functional, that is, if there exists $\alpha \in \mathbb{R}^m$ such that

$$F(\mathbf{v}) = \alpha \cdot \left(v_i \frac{\partial \mathbf{r}(\mathbf{u})}{\partial u_i}\right), \forall \mathbf{v} \in T_p(M),$$

then there exists $f : M \to \mathbb{R}$ of C^1 class such that,

$$F(\mathbf{v}) \equiv df(\mathbf{v}), \ \forall \mathbf{v} \in T_p(M).$$

Proof 2.13   For $f \in C^1(M)$ and $p \in M$ in question, define $F : T_p(M) \to \mathbb{R}$ by

$$F(\mathbf{v}) = df(\mathbf{v}),$$

for all $\mathbf{v} = v_i \frac{\partial \mathbf{r}(\mathbf{u})}{\partial u_i} \in T_p(M)$.

Hence,

$$
\begin{aligned}
F(\mathbf{v}) &= df(\mathbf{v}) \\
&= \lim_{\varepsilon \to 0} \frac{(f \circ \mathbf{r})(\{u_i + \varepsilon v_i\}) - (f \circ \mathbf{r})(\{u_i\})}{\varepsilon} \\
&= \sum_{j=1}^{n} \frac{\partial (f \circ \mathbf{r})(u)}{\partial X_j} \frac{\partial X_j(\mathbf{u})}{\partial u_i} v_i \\
&= \alpha \cdot \left(v_i \frac{\partial \mathbf{r}(\mathbf{u})}{\partial u_i} \right),
\end{aligned}
\tag{2.17}
$$

where

$$\alpha_j = \frac{\partial (f \circ \mathbf{r})(\mathbf{u})}{\partial X_j}, \ \forall j \in \{1, \cdots, n\}.$$

We may conclude that F is continuous and linear on $T_p(M)$, that is,

$$F \in T_p(M)^*.$$

Conversely, assume that $F : T_p(M) \to \mathbb{R}$ is linear and continuous, so that by the Riesz representation theorem, there exists $\alpha \in \mathbb{R}^n$ such that

$$F(v) = \alpha \cdot \left(v_i \frac{\partial \mathbf{r}(\mathbf{u})}{\partial u_i} \right), \ \forall \mathbf{v} \in T_p(M).$$

Define $f : M \to \mathbb{R}$ by

$$f(\mathbf{w}) = \alpha \cdot \mathbf{w}, \forall \mathbf{w} \in M.$$

Thus, $f(\mathbf{r}(u)) = \alpha \cdot \mathbf{r}(\mathbf{u}) = \alpha_j X_j(\mathbf{u})$.

From this,

$$
\begin{aligned}
df(\mathbf{v}) &= \lim_{\varepsilon \to 0} \frac{\alpha_j X_j(\{u_i + \varepsilon v_i\}) - \alpha_j X_j(\{u_i\})}{\varepsilon} \\
&= \alpha_j \frac{\partial X_j(\mathbf{u})}{\partial u_i} v_i \\
&= \alpha \cdot \left(v_i \frac{\partial \mathbf{r}(\mathbf{u})}{\partial u_i} \right) \\
&= F(\mathbf{v}).
\end{aligned}
\tag{2.18}
$$

This completes the proof.

Corollary 2.7.5 *In an appropriate sense, the corresponding dual basis to $T_p(M)^*$, to the primal basis for $T_p(M)$*

$$\left\{ \frac{\partial \mathbf{r}(\mathbf{u})}{\partial u_i} \right\}_{i=1}^m,$$

is given by $\{dw_i(r(\mathbf{u}))\}$, where $\{w_i(\mathbf{r}(u)\} = \mathbf{r}^{-1}(\mathbf{r}(\mathbf{u})) = \{u_i\}$, so that we could denote such basis for $T_p(M)^*$ by

$$\{du_1, \cdots, du_m\}.$$

Proof 2.14   Let us denote

$$\left\{ \tilde{\mathbf{e}}_i = \frac{\partial \mathbf{r}(\mathbf{u})}{\partial u_i} \right\}_{i=1}^m,$$

and

$$\tilde{\mathbf{v}}_j = (0, 0, \cdots, 0, 1, 0, \cdots, 0),$$

that is, value 1 at the $j-th$ entry and value 0 at the remaining ones.

Hence,

$$
\begin{aligned}
du_i(\tilde{\mathbf{e}}_j) &= dw_i(\mathbf{r}(\mathbf{u}))(\tilde{\mathbf{e}}_j) \\
&= \lim_{\varepsilon \to 0} \frac{w_i(\mathbf{r}(\mathbf{u} + \varepsilon \tilde{\mathbf{v}}_j)) - w_i(\mathbf{r}(\mathbf{u}))}{\varepsilon} \\
&= \frac{\partial w_i(r(\mathbf{u}))}{\partial u_j} \\
&= \frac{\partial u_i}{\partial u_j} = \delta_{ij}.
\end{aligned}
\tag{2.19}
$$

Summarizing, we have obtained

$$du_i(\tilde{\mathbf{e}}_j) = \delta_{ij},$$

so that, in the context of the last theorem, $\{du_i, \cdots, du_m\}$ is a basis for $T_p(M)^*$.

The proof is complete.

2.8   Vector fields

Definition 2.8.1 *Let $M \subset \mathbb{R}^n$ be an m-dimensional surface that is, $1 \leq m < n$. Generically denoting*

$$M = \{\mathbf{r}(\mathbf{u}) \ : \ \mathbf{u} \in D\},$$

let $p = \mathbf{r}(\mathbf{u}) \in M$ and consider the set

$$\left\{ \frac{\partial \mathbf{r}(\mathbf{u})}{\partial u_i} \right\}_{i=1}^m,$$

be a basis for $T_p(M)$.

We define a vector field

$$X : M \rightarrow \{T_p(M), \; p \in M\},$$

by

$$X = \{X_p, \; p \in M\},$$

where point-wise

$$X_p(\mathbf{u}) = \sum_{i=1}^{m} X_{p_i}(\mathbf{u}) \frac{\partial \mathbf{r}(\mathbf{u})}{\partial u_i}.$$

Moreover, for each $p \in M$ we define the operator

$$\tilde{X}_p : C^1(M) \rightarrow C^1(D)$$

where

$$X_p = \sum_{i=1}^{m} X_{p_i} \frac{\partial \mathbf{r}(\mathbf{u})}{\partial u_i},$$

by

$$
\begin{aligned}
\tilde{X}_p(f) &= df(X_p) \\
&= \lim_{\varepsilon 0} \frac{(f \circ \mathbf{r})(\{u_i + \varepsilon X_{p_i}(\mathbf{u})\}) - (f \circ \mathbf{r})(\{u_i\})}{\varepsilon} \\
&= \frac{\partial(f \circ \mathbf{r})(\mathbf{u})}{\partial X_j} \frac{\partial X_j(\mathbf{u})}{\partial u_i} X_{p_i}(\mathbf{u}) \\
&= \nabla f(\mathbf{r}(u)) \cdot \left(X_{p_i}(\mathbf{u}) \frac{\partial \mathbf{r}(\mathbf{u})}{\partial u_i} \right).
\end{aligned}
\tag{2.20}
$$

Moreover, we denote by $\mathscr{X}(M)$ the set of C^1 class vector fields on M.

Definition 2.8.2 (The exterior product) *Let V be a real vector space and V^* its dual one. Let $(F_1, \cdots, F_k) \in V^* \times V^* \times \cdots \times V^*$ (k times) and $(\mathbf{v}_1, \mathbf{v}_2, \cdots, \mathbf{v}_k) \in V \times V \times \cdots \times V$ (k times). We define point-wise the exterior product $(F_1 \wedge F_2 \wedge \cdots \wedge F_k)(\mathbf{v}_1, \mathbf{v}_2, \cdots, \mathbf{v}_k)$, by*

$$(F_1 \wedge F_2 \wedge \cdots \wedge F_k)(\mathbf{v}_1, \mathbf{v}_2, \cdots, \mathbf{v}_k) = det\{F_i(\mathbf{v}_j)\}.$$

Remark 2.8.3 *Concerning a C^1 class m-dimensional surface $M \subset \mathbb{R}^n$, where $1 \leq m < n$ and*

$$M = \{\mathbf{r}(\mathbf{u}) \; : \; \mathbf{u} \in D\},$$

for a vector field $X \in \mathscr{X}(M)$ where

$$X = X_{p_i} \frac{\partial \mathbf{r}(u)}{\partial u_i}$$

and $f : M \to \mathbb{R}$ of C^1 class, we shall denote,

$$X \cdot f = df(X),$$

where point-wisely,

$$
\begin{aligned}
(X \cdot f)(\mathbf{r}(\mathbf{u})) &= df(X)[\mathbf{r}(\mathbf{u})] \\
&= \lim_{\varepsilon \to 0} \frac{(f \circ \mathbf{r})(\{u_i + \varepsilon X_{p_i}\}) - (f \circ \mathbf{r})(\mathbf{u})}{\varepsilon},
\end{aligned}
\tag{2.21}
$$

where $p = \mathbf{r}(\mathbf{u})$.

Definition 2.8.4 (Lie bracket) *Let $M \subset \mathbb{R}^n$ be a C^1 class m-dimensional surface where $1 \leq m < n$.*

Let $X, Y \in \tilde{\mathscr{X}}(M)$, where $\tilde{\mathscr{X}}(M)$ denotes the set of the $C^\infty(M) = \cap_{k \in \mathbb{N}} C^k(M)$ class vector fields. We define the Lie bracket of X and Y, denoted by $[X,Y] \in \tilde{\mathscr{X}}(M)$ by,

$$[X,Y] = (X \cdot Y_i - X \cdot Y_i)\frac{\partial \mathbf{r}(\mathbf{u})}{\partial u_i},$$

where

$$X = X_i \frac{\partial \mathbf{r}(\mathbf{u})}{\partial u_i}$$

and

$$Y = Y_i \frac{\partial \mathbf{r}(\mathbf{u})}{\partial u_i}.$$

Theorem 2.8.5 *Let $M \subset \mathbb{R}^n$ be a C^1 class m-dimensional surface where $1 \leq m < n$. Let $X, Y \in \tilde{\mathscr{X}}(M)$ and let $f \in C^2(M)$.*

Under such hypotheses,

$$[X,Y] \cdot f = X \cdot (Y \cdot f) - Y \cdot (X \cdot f).$$

Proof 2.15 Observe that

$$
\begin{aligned}
X \cdot (Y \cdot f) &= X \cdot (df(Y)) \\
&= X \cdot \left(\frac{\partial (f \circ r)(\mathbf{u})}{\partial u_i} Y_i \right) \\
&= \left[d\left(\frac{\partial (f \circ r)(\mathbf{u})}{\partial u_i} Y_i \right) \right](X) \\
&= \frac{\partial^2 (f \circ r)(\mathbf{u})}{\partial u_j \partial u_i} Y_i X_j + \frac{\partial (f \circ r)(\mathbf{u})}{\partial u_i} \frac{\partial Y_i}{\partial u_j} X_j.
\end{aligned}
\tag{2.22}
$$

Similarly,

$$Y \cdot (X \cdot f) = \frac{\partial^2 (f \circ \mathbf{r})(\mathbf{u})}{\partial u_j \partial u_i} X_i Y_j + \frac{\partial (f \circ \mathbf{r})(\mathbf{u})}{\partial u_i} \frac{\partial X_i}{\partial u_j} Y_j,$$

so that

$$X \cdot (Y \cdot f) - Y \cdot (X \cdot f)$$

$$= \frac{\partial (f \circ r)(\mathbf{u})}{\partial u_i} \frac{\partial Y_i}{\partial u_j} X_j - \frac{\partial (f \circ r)(\mathbf{u})}{\partial u_i} \frac{\partial X_i}{\partial u_j} Y_j. \qquad (2.23)$$

On the other hand,

$$[X, Y] \cdot f = \left[(X \cdot Y_i - Y \cdot X_i) \frac{\partial \mathbf{r}(u)}{\partial u_i} \right] \cdot f. \qquad (2.24)$$

Observe that

$$\left[(X \cdot Y_i) \frac{\partial \mathbf{r}(u)}{\partial u_i} \right] \cdot f = df \left[(X \cdot Y_i) \frac{\partial \mathbf{r}(u)}{\partial u_i} \right]$$

$$= \frac{\partial (f \circ \mathbf{r})(\mathbf{u})}{\partial u_i} (X \cdot Y_i)$$

$$= \frac{\partial (f \circ \mathbf{r})(\mathbf{u})}{\partial u_i} dY_i(X)$$

$$= \frac{\partial (f \circ \mathbf{r})(\mathbf{u})}{\partial u_i} \frac{\partial Y_i}{\partial u_j} X_j. \qquad (2.25)$$

Similarly,

$$\left[(Y \cdot X_i) \frac{\partial \mathbf{r}(u)}{\partial u_i} \right] \cdot f$$

$$= \frac{\partial (f \circ \mathbf{r})(\mathbf{u})}{\partial u_i} \frac{\partial X_i}{\partial u_j} Y_j. \qquad (2.26)$$

Therefore,

$$[X, Y] \cdot f = \left[(X \cdot Y_i - Y \cdot X_i) \frac{\partial \mathbf{r}(u)}{\partial u_i} \right] \cdot f$$

$$= \frac{\partial (f \circ \mathbf{r})(\mathbf{u})}{\partial u_i} \frac{\partial Y_i}{\partial u_j} X_j - \frac{\partial (f \circ \mathbf{r})(\mathbf{u})}{\partial u_i} \frac{\partial X_i}{\partial u_j} Y_j$$

$$= X \cdot (Y \cdot f) - Y \cdot (X \cdot f). \qquad (2.27)$$

The proof is complete.

Exercises 2.8.6

1. *Let $M \subset \mathbb{R}^n$ be a C^1 class m-dimensional surface where $1 \leq m \leq n$ and let $X, Y, Z \in \tilde{\mathcal{X}}(M)$. Show that*

   (a) $[\alpha X + \beta Y, Z] = \alpha[X, Z] + \beta[Y, Z], \ \forall \alpha, \beta \in \mathbb{R},$

   (b) $[X, \alpha Y + \beta Z] = \alpha[X, Y] + \beta[X, Z], \ \forall \alpha, \beta \in \mathbb{R}.$

(c) Antisymmetry:

$$[X,Y] = -[Y,X],$$

(d) Jacob Identity:

$$[[X,Y],Z] + [[Y,Z],X] + [[Z,X],Y] = 0.$$

(e) Leibnitz Rule:

$$[fX,gY] = fg[X,Y] + f(X \cdot g)Y - g(Y \cdot f)X, \; \forall f,g \in C^2(M).$$

At this point we introduce the definition of Lie Algebra.

Definition 2.8.7 (Lie Algebra) *A Lie Algebra is a vector space V for which an anti-symmetric bilinear form* $[\cdot,\cdot] : V \times V \to V$ *is defined, which satisfies the Jacob identity.*

Remark 2.8.8 *Observe that, from the exposed above,* $\tilde{\mathscr{X}}(M)$ *is a Lie Algebra with the Lie bracket*

$$[X,Y] = (X \cdot Y_i - Y \cdot X_i)\frac{\partial \mathbf{r}(\mathbf{u})}{\partial u_i}.$$

Definition 2.8.9 (Integral curve) *Let* $M \subset \mathbb{R}^n$ *be a* C^1 *class m-dimensional surface where* $1 \leq m < n$. *Let* $X \in \mathscr{X}(M)$.
Let A curve $(\mathbf{r} \circ \mathbf{u}) : I = (-\varepsilon,\varepsilon) \to M$ *is said to be an integral curve for X on I around* $p = \mathbf{r}(u(0))$, *if*

$$\frac{d\mathbf{r}(\mathbf{u}(t))}{dt} = X(\mathbf{u}(t)), \; \forall t \in I.$$

Observe that in such a case:

$$\begin{aligned}
\frac{d\mathbf{r}(\mathbf{u}(t))}{dt} &= \frac{\partial \mathbf{r}(\mathbf{u}(t))}{\partial u_i}\frac{du_i(t)}{dt} = X(\mathbf{u}(t)) \\
&= X_i(\mathbf{u}(t))\frac{\partial \mathbf{r}(\mathbf{u}(t))}{\partial u_i}, \quad (2.28)
\end{aligned}$$

so that

$$\frac{du_i(t)}{dt} = X_i(\mathbf{u}(t)). \quad (2.29)$$

2.9 The generalized derivative

At this section, we introduce the definition of Lie derivative. We start with a preliminary definition, namely, the generalized derivative.

Definition 2.9.1 (Generalized derivative) *Let $M \subset \mathbb{R}^n$ be a C^1 class m-dimensional surface where $1 \le m \le n$. Let $X \in \mathscr{X}(M)$. Let $\mathbf{r}(\mathbf{u}(t))$ be the point-wise representation of the integral curve of X around $p = \mathbf{r}(\mathbf{u}(0))$.*

Let $Y \in \tilde{\mathscr{X}}(M)$ and $f \in C^2(M)$. We define the generalized derivative of f relating Y, along the direction X, at the point $p = \mathbf{r}(\mathbf{u}(0))$, denoted by $D_X Y(f)(p)$, by

$$D_X Y(f) = \frac{d[Y(\mathbf{u}(t)) \cdot f(\mathbf{r}(\mathbf{u}(t)))]}{dt}\Big|_{t=0}.$$

Theorem 2.9.2 *Let $M \subset \mathbb{R}^n$ be a C^1 class m-dimensional surface where $1 \le m < n$. Let $X \in \mathscr{X}(M)$. Let $\mathbf{r}(\mathbf{u}(t))$ be the point-wise representation of the integral curve of X around $p = \mathbf{r}(\mathbf{u}(0))$.*

Let $Y \in \mathscr{X}(M)$ and $f \in C^2(M)$.

Under such assumptions,

$$D_X Y(f)(p) = X \cdot (Y \cdot f)(p).$$

Proof 2.16   Observe that,

$$
\begin{aligned}
D_X Y(f)(p) &= \frac{d[df(Y)]}{dt}\Big|_{t=0} \\
&= \frac{d}{dt}\left[\frac{\partial (f \circ \mathbf{r})(\mathbf{u}(t))}{\partial u_i} Y_i(\mathbf{u}(t))\right]_{t=0} \\
&= \left[\frac{\partial^2 (f \circ \mathbf{r})(\mathbf{u}(t))}{\partial u_j \partial u_i} \frac{du_j(t)}{dt} Y_i(\mathbf{u}(t))\right]_{t=0} \\
&\quad + \left[\frac{\partial (f \circ \mathbf{r})(\mathbf{u}(t))}{\partial u_i} \frac{\partial Y_i(\mathbf{u}(t))}{\partial u_j} \frac{du_j(t)}{dt}\right]_{t=0} \\
&= \left[\frac{\partial^2 (f \circ \mathbf{r})(\mathbf{u}(t))}{\partial u_j \partial u_i} X_j(\mathbf{u}(t)) Y_i(\mathbf{u}(t))\right]_{t=0} \\
&\quad + \left[\frac{\partial (f \circ \mathbf{r})(\mathbf{u}(t))}{\partial u_i} \frac{\partial Y_i(\mathbf{u}(t))}{\partial u_j} X_j(\mathbf{u}(t))\right]_{t=0} \\
&= X \cdot (Y \cdot f)(p). \quad\quad\quad\quad\quad\quad\quad\quad\quad (2.30)
\end{aligned}
$$

Definition 2.9.3 (Derivative of $Y \in \mathscr{X}(M)$ on the direction of $X \in \mathscr{X}(M)$) *Let $M \subset \mathbb{R}^m$ be a m-dimensional surface where $1 \le m < n$. Let $X, Y \in \mathscr{X}(M)$. Let $p = \mathbf{r}(\mathbf{u}(0))$ where $\mathbf{r}(\mathbf{u}(t))$ is the integral curve of X about $p = \mathbf{r}(\mathbf{u}_0)$, where $\mathbf{u}_0 = \mathbf{u}(0)$.*

We define the derivative of Y on the direction X at the point $p = \mathbf{r}(\mathbf{u}_0)$, denoted by

$$(D_X Y)(p),$$

by

$$(D_X Y)(p) = \lim_{t \to 0} \frac{Y(\mathbf{u}(t)) - Y(\mathbf{u}(0))}{t} = \frac{dY(\mathbf{u}(t))}{dt}\Big|_{t=0}.$$

Remark 2.9.4 *Observe that*

$$Y(\mathbf{u}(t)) = Y_i(\mathbf{u}(t))\frac{\partial \mathbf{r}(\mathbf{u}(t))}{\partial u_i},$$

so that

$$
\begin{aligned}
(D_X Y)(p) &= \frac{\partial Y_i(\mathbf{u}(0))}{\partial u_j}\frac{du_j(0)}{dt}\frac{\partial \mathbf{r}(\mathbf{u}(0))}{\partial u_i} + Y_i(\mathbf{u}(0))\frac{\partial^2 \mathbf{r}(\mathbf{u}(0))}{\partial u_i \partial u_j}\frac{du_j(0)}{dt} \\
&= \frac{\partial Y_i(\mathbf{u}(0))}{\partial u_j}X_j(\mathbf{u}(0)) + Y_i(\mathbf{u}(0))\frac{\partial^2 \mathbf{r}(\mathbf{u}(0))}{\partial u_i \partial u_j}X_j(\mathbf{u}(0)) \\
&= dY_i(X)(\mathbf{u}_0)\frac{\partial \mathbf{r}(\mathbf{u}_0)}{\partial u_i} + Y_i(\mathbf{u}_0)X_j(\mathbf{u}_0)\frac{\partial^2 \mathbf{r}(\mathbf{u}_0)}{\partial u_i \partial u_j}. \quad\quad (2.31)
\end{aligned}
$$

Summarizing,

$$(D_X Y)(p) = dY_i(X)(\mathbf{u}_0)\frac{\partial \mathbf{r}(\mathbf{u}_0)}{\partial u_i} + Y_i(\mathbf{u}_0)X_j(\mathbf{u}_0)\frac{\partial^2 \mathbf{r}(\mathbf{u}_0)}{\partial u_i \partial u_j}.$$

Definition 2.9.5 (Lie derivative of a vector field) *Let $M \subset \mathbb{R}^n$ be a C^1 class m-dimensional surface where $1 \le m < n$. Let $X \in \tilde{\mathscr{X}}(M)$. Let $\mathbf{r}(\mathbf{u}(t))$ be the pointwise representation of the integral curve of X around $p = \mathbf{r}(\mathbf{u}(0))$.*
Let $Y \in \tilde{\mathscr{X}}(M)$ and $f \in C^2(M)$. We define the Lie derivative of Y, along the direction X, at $f \in C^2(M)$ at the point $p = \mathbf{r}(\mathbf{u}(0))$, denoted by $L_X Y(f)(p)$, by

$$L_X Y(f)(p) = X \cdot (Y \cdot f)(p) - Y \cdot (X \cdot f)(p).$$

Observe that in such a case

$$L_X Y(f)(p) = ([X,Y] \cdot f)(p), \ \forall f \in C^2(M),$$

so that we denote simply,

$$L_X Y = [X,Y].$$

Relating the statements of this last definition, consider the following exercises.

Exercises 2.9.6

1. *Prove that*

$$L_X[Y,Z] = [L_X, Y, Z] + [Y, L_X Z], \ \forall X,Y,Z \in \tilde{\mathscr{X}}(M).$$

2. *Prove that*

$$L_X(L_Y)Z - L_Y(L_X)Z = L_{[X,Y]}Z, \ \forall X,Y,Z \in \tilde{\mathscr{X}}((M)).$$

2.9.1 On the integral curve existence

In this section, we present some well known results about the existence of solutions for a nonlinear system of ordinary differential equations. At this point we recall, in a more general fashion, the definition of an integral curve.

Let $M \subset \mathbb{R}^n$ be a C^1 class m-dimensional surface, where $1 \leq m \leq n$, where

$$M = \{\mathbf{r}(\mathbf{u}) : \mathbf{u} \in D\},$$

where $D \subset \mathbb{R}^m$ is a connected set such that ∂D is of C^1 class.

Let $X \in \mathscr{X}(M)$, that is

$$X = X_i(\mathbf{u}) \frac{\partial \mathbf{r}(\mathbf{u})}{\partial u_i}.$$

Let $\mathbf{u}_0 \in D^\circ$ and $p = \mathbf{r}(\mathbf{u}_0)$. An integral curve for X around p is a curve

$$(\mathbf{r} \circ u) : I \to M,$$

where I is a closed interval, such that $t_0 \in I^\circ$,

$$\mathbf{r}(\mathbf{u}(0)) = r(\mathbf{u}_0),$$

and

$$\frac{d\mathbf{r}(\mathbf{u}(t))}{dt} = X(\mathbf{u}(t)),$$

that is

$$\frac{\partial \mathbf{r}(\mathbf{u}(t))}{\partial u_i} \frac{du_i(t)}{\partial t} = X_i(\mathbf{u}(t)) \frac{\partial \mathbf{r}(\mathbf{u}(t))}{\partial u_i},$$

so that, in a component wise fashion,

$$\frac{du_i(t)}{\partial t} = X_i(\mathbf{u}(t)), \ \forall i \in \{1, ...m\}.$$

We present now the results concerning the existence of solutions for the ordinary differential equation systems in question.

Theorem 2.9.7 *Let $V \subset \mathbb{R}^m$ be an open set. Let $\hat{X} : V \to \mathbb{R}^m$ be a continuous function such that*

$$|\hat{X}(\mathbf{u}) - \hat{X}(\mathbf{v})| \leq K|\mathbf{u} - \mathbf{v}|, \ \forall \mathbf{u}, \mathbf{v} \in V,$$

for some $K > 0$.

Let $\mathbf{u}_0 \in V$ and let $r > 0$ be such that $B_r(\mathbf{u}_0) \in V$, where

$$B_r(\mathbf{u}_0) = \{\mathbf{u} \in \mathbb{R}^m : |\mathbf{u} - \mathbf{u}_0| < r\}.$$

Let $C > 0$ be such that

$$|\hat{X}(\mathbf{u})| < C, \ \forall \mathbf{u} \in B_r(\mathbf{u}_0).$$

Let $t_0 \in \mathbb{R}$ and let $\alpha = \min\{1/K, r/C\}$. Under such hypotheses there exists a curve $\mathbf{u} : I_\alpha \to B_r(\mathbf{u}_0)$ such that

$$\begin{cases} \frac{d\mathbf{u}(t)}{dt} = \hat{X}(\mathbf{u}(t)) \\ \mathbf{u}(t_0) = \mathbf{u}_0, \end{cases} \tag{2.32}$$

where,

$$I_\alpha = [t_0 - \alpha, t_0 + \alpha].$$

Proof 2.17 Observe that

$$\begin{cases} \frac{d\mathbf{u}(t)}{dt} = \hat{X}(\mathbf{u}(t)) \\ \mathbf{u}(t_0) = \mathbf{u}_0, \end{cases} \tag{2.33}$$

is equivalent to

$$\mathbf{u}(t) = \mathbf{u}_0 + \int_{t_0}^{t} \hat{X}(\mathbf{u}(s)) \, ds.$$

Define the sequence of functions $\{\mathbf{u}_n : I_\alpha \to \mathbb{R}^m\}$ by

$$\mathbf{u}_1(t) = \mathbf{u}_0,$$

$$\mathbf{u}_{n+1}(t) = \mathbf{u}_0 + \int_{t_0}^{t} \hat{X}(\mathbf{u}_n(s)) \, ds, \ \forall t \in I_\alpha, \ \forall n \in \mathbb{N}.$$

First, we shall prove by induction that

$$\mathbf{u}_n(t) \in B_r(\mathbf{u}_0), \ \forall n \in \mathbb{N}, \ t \in I_\alpha.$$

Clearly

$$\mathbf{u}_1(t) \in B_r(\mathbf{u}_0), \forall t \in I_\alpha.$$

Suppose

$$\mathbf{u}_n(t) \in B_r(\mathbf{u}_0), \ \forall t \in I_\alpha.$$

Thus,

$$|\hat{X}(\mathbf{u}_n(t))| < C, \ \forall t \in I_\alpha,$$

so that

$$\begin{aligned} |\mathbf{u}_{n+1}(t) - \mathbf{u}_0| \ &\leq \ \left| \int_{t_0}^{t} |\hat{X}(\mathbf{u}_n(s))| \, ds \right| \\ &< \ C|t - t_0| \\ &\leq \ C\alpha \\ &\leq \ C\frac{r}{C} \\ &= \ r, \end{aligned} \tag{2.34}$$

so that

$$\mathbf{u}_{n+1}(t) \in B_r(\mathbf{u}_0), \ \forall t \in I_\alpha.$$

The induction is complete, that is,

$$\mathbf{u}_n(t) \in B_r(\mathbf{u}_0), \ \forall t \in I_\alpha, \ \forall n \in \mathbb{N}.$$

Moreover, for $t \geq t_0$ we have

$$
\begin{aligned}
|\mathbf{u}_{n+1}(t) - \mathbf{u}_n(t)| &\leq \int_{t_0}^t |\hat{X}(\mathbf{u}_n(s)) - \hat{X}(\mathbf{u}_{n-1}(s))| \, ds \\
&\leq \int_{t_0}^t K|\mathbf{u}_n(s) - \mathbf{u}_{n-1}(s)| \, ds \\
&\leq K \int_{t_0}^t \int_{t_0}^s |\hat{X}(\mathbf{u}_{n-1}(s_1)) - \hat{X}(\mathbf{u}_{n-2}(s_1))| \, ds_1 \, ds \\
&\leq K^2 \int_{t_0}^t \int_{t_0}^s |\mathbf{u}_{n-1}(s_1) - \mathbf{u}_{n-2}(s_1)| \, ds_1 \, ds \\
&\leq K^2 \int_{t_0}^t \int_{t_0}^s \int_{t_0}^{s_1} |\hat{X}(\mathbf{u}_{n-2}(s_2)) - \hat{X}(\mathbf{u}_{n-3}(s_2))| \, ds_2 \, ds_1 \, ds \\
&\leq K^3 \int_{t_0}^t \int_{t_0}^s \int_{t_0}^{s_1} |\mathbf{u}_{n-2}(s_2) - \mathbf{u}_{n-3}(s_2)| \, ds_2 \, ds_1 \, ds. \quad (2.35)
\end{aligned}
$$

Proceeding inductively in this fashion, and recalling that point-wise $\mathbf{u}_1(t) = \mathbf{u}_0$, we finally would obtain

$$
\begin{aligned}
|\mathbf{u}_{n+1}(t) - \mathbf{u}_n(t)| &\leq K^n \int_{t_0}^t \int_{t_0}^s \int_{t_0}^{s_1} \cdots \int_{t_0}^{s_{n-1}} |\mathbf{u}_2(s_n) - \mathbf{u}_0| \, ds_n \, ds_{n_1} \cdots ds_1 \, ds \\
&\leq K^n r \int_{t_0}^t \int_{t_0}^s \int_{t_0}^{s_1} \cdots \int_{t_0}^{s_{n-1}} ds_n \, ds_{n-1} \cdots ds_1 \, ds \\
&\leq K^n r \frac{|t - t_0|^{n+1}}{(n+1)!}. \quad (2.36)
\end{aligned}
$$

The same estimate is valid for $t \leq t_0$.

Therefore,

$$
\begin{aligned}
|\mathbf{u}_{n+1}(t) - \mathbf{u}_n(t)| &\leq rK^n |t - t_0|^n \frac{|t - t_0|}{(n+1)!} \\
&\leq rK^n \alpha^n \frac{|t - t_0|}{(n+1)!} \\
&\leq r \frac{|t - t_0|}{(n+1)!} \\
&\to 0, \text{ uniformly as } n \to \infty. \quad (2.37)
\end{aligned}
$$

Fix $n, p \in \mathbb{N}$.

From the last inequality

$$|\mathbf{u}_{n+p}(t) - \mathbf{u}_n(t)|$$
$$= |\mathbf{u}_{n+p}(t) - \mathbf{u}_{n+p-1}(t) + \mathbf{u}_{n+p-1}(t)$$
$$-\mathbf{u}_{n+p-2}(t) + \mathbf{u}_{n+p-2}(t) + \cdots + \mathbf{u}_{n+1}(t) - \mathbf{u}_n(t)|$$
$$\leq |\mathbf{u}_{n+p}(t) - \mathbf{u}_{n+p-1}(t)| + |\mathbf{u}_{n+p-1}(t) - \mathbf{u}_{n+p-2}(t)| + \cdots + |\mathbf{u}_{n+1}(t) - \mathbf{u}_n(t)|$$
$$\leq r|t - t_0| \sum_{k=1}^{p} \frac{1}{(n+k)!}$$
$$\leq r|t - t_0| \sum_{k=(n+1)}^{\infty} \frac{1}{k!}. \tag{2.38}$$

Let us denote $\{a_k\} = \frac{1}{k!}$. Observe that

$$\lim_{k \to \infty} \frac{a_{k+1}}{a_k} = \lim_{k \to \infty} \frac{k!}{(k+1)!} = \lim_{k \to \infty} \frac{1}{k+1} = 0.$$

Hence,

$$\sum_{k=1}^{\infty} \frac{1}{k!}$$

is converging so that

$$\sum_{k=(n+1)}^{\infty} \frac{1}{k!} \to 0, \text{ as } n \to \infty.$$

From this and (2.38), $\{\mathbf{u}_n\}$ is a uniform Cauchy sequence of continuous functions, which uniformly converges to some continuous $\mathbf{u} : I_\alpha \to \overline{B}_r(\mathbf{u}_0)$, such that

$$\mathbf{u}(t) = \mathbf{u}_0 + \int_{t_0}^{t} \hat{X}(\mathbf{u}(s)) \, ds,$$

so that

$$\begin{cases} \frac{d\mathbf{u}(t)}{dt} = \hat{X}(\mathbf{u}(t)) \\ \mathbf{u}(t_0) = \mathbf{u}_0. \end{cases} \tag{2.39}$$

The proof of uniqueness of $\mathbf{u}$ is left as an exercise.

Theorem 2.9.8 (Gronwall's inequality) *Let* $f, g : [a, b] \to \mathbb{R}$ *be continuous and nonnegative functions. Suppose there exists* $A > 0$ *such that*

$$f(t) \leq A + \int_{a}^{t} f(s)g(s) \, ds, \ \forall t \in [a, b].$$

Under such hypotheses,

$$f(t) \leq A e^{\int_{a}^{t} g(s) \, ds}, \ \forall t \in [a, b].$$

Proof 2.18   Define $h : [a,b] \to \mathbb{R}$ by

$$h(t) = A + \int_a^t f(s)g(s)\, ds.$$

From the hypotheses, $h(t) > 0$ and

$$f(t) \le h(t), \ \forall t \in [a,b].$$

Moreover,

$$h'(t) = f(t)g(t) \le h(t)g(t), \ \forall t \in [a,b].$$

Therefore,

$$\frac{h'(t)}{h(t)} \le g(t), \ \forall t \in [a,b],$$

so that

$$\frac{d\ln(h(t))}{dt} \le g(t), \ \forall t \in [a,b],$$

and hence

$$\ln(h(t)) - \ln(A) \le \int_a^t g(s)\, ds, \ \forall t \in [a,b],$$

that is,

$$\ln(h(t)/A) \le \int_a^t g(s)\, ds,$$

so that

$$\frac{h(t)}{A} \le e^{\int_a^t g(s)\, ds},$$

that is,

$$f(t) \le h(t) \le A e^{\int_a^t g(s)\, ds}, \ \forall t \in [a,b].$$

The proof is complete.

Theorem 2.9.9 *Let $V \subset \mathbb{R}^m$ be an open set. Let $\hat{X} : V \to \mathbb{R}^m$ be a continuous function such that*

$$|\hat{X}(\mathbf{u}) - \hat{X}(\mathbf{v})| \le K|\mathbf{u} - \mathbf{v}|, \ \forall \mathbf{u}, \mathbf{v} \in V,$$

for some $K > 0$.

Let $\mathbf{u}_0 \in V$ and let $r > 0$ be such that $B_r(\mathbf{u}_0) \in V$, where

$$B_r(\mathbf{u}_0) = \{\mathbf{u} \in \mathbb{R}^m : |\mathbf{u} - \mathbf{u}_0| < r\}.$$

Let $C > 0$ be such that

$$|\hat{X}(\mathbf{u})| < C, \ \forall \mathbf{u} \in B_r(\mathbf{u}_0).$$

Let $F_t(\mathbf{u}_0)$ *denote the unique integral curve* $\mathbf{u} : I_\alpha \to \mathbb{R}$ *such that*

$$\begin{cases} \frac{d\mathbf{u}(t)}{dt} = \hat{X}(\mathbf{u}(t)) \\ \mathbf{u}(0) = \mathbf{u}_0, \end{cases} \tag{2.40}$$

where $\alpha = \min\{1/K, r/C\}$, *and* $I_\alpha = [-\alpha, \alpha]$.

Under such hypotheses there exists $r_1 > 0$ *and* $\varepsilon > 0$ *such that for each* $\mathbf{v}_0 \in B_r(\mathbf{u}_0)$ *there exists an integral* $\mathbf{v} : [-\varepsilon, \varepsilon] \to \mathbb{R}^m$ *such that*

$$\begin{cases} \frac{d\mathbf{v}(t)}{dt} = \hat{X}(\mathbf{v}(t)) \\ \mathbf{v}(0) = \mathbf{v}_0. \end{cases} \tag{2.41}$$

Moreover,

$$|\mathbf{u}(t) - \mathbf{v}(t)| \le e^{K|t|}|\mathbf{u}_0 - \mathbf{v}_0|, \forall t \in [-\varepsilon, \varepsilon].$$

Proof 2.19 Define $r_1 = r/2$ and $\varepsilon = \min\{1/K, r/(2C)\}$.

Let $\mathbf{v}_0 \in B_r(\mathbf{u}_0)$. Thus $B_{r_1}(\mathbf{v}_0) \subset B_r(\mathbf{u}_0)$, so that

$$|\hat{X}(\mathbf{v})| \le C, \ \forall \mathbf{v} \in B_{r_1}(\mathbf{u}_0).$$

From Theorem 2.9.7, with $\mathbf{v}_0$ in place of $\mathbf{u}_0$, r_1 in place of r, 0 in place of t_0 and ε in place of α, there exists a curve $\mathbf{v} : I_\varepsilon \to \mathbb{R}^m$, where $I_\varepsilon = [-\varepsilon, \varepsilon]$, such that

$$\begin{cases} \frac{d\mathbf{v}(t)}{dt} = \hat{X}(\mathbf{v}(t)) \\ \mathbf{v}(0) = \mathbf{v}_0. \end{cases} \tag{2.42}$$

Now define

$$f(t) = |\mathbf{u}(t) - \mathbf{v}(t)|.$$

Thus,

$$\begin{aligned} f(t) &= |\mathbf{u}(t) - \mathbf{v}(t)| \\ &= \left| \int_0^t (\hat{X}(\mathbf{u}(s)) - \hat{X}(\mathbf{v}(s))) \, ds + \mathbf{u}_0 - \mathbf{v}_0 \right| \\ &\le \left| \int_0^t K|\mathbf{u}(s) - \mathbf{v}(s)| \, ds \right| + |\mathbf{u}_0 - \mathbf{v}_0| \\ &\le |\mathbf{u}_0 - \mathbf{v}_0| + K \int_0^{|t|} f(s) \, ds. \end{aligned} \tag{2.43}$$

From the Gronwall's inequality with $A = |\mathbf{u}_0 - \mathbf{v}_0|$ and $g(t) = K$ we obtain,

$$\begin{aligned} |\mathbf{u}(t) - \mathbf{v}(t)| &= f(t) \\ &\le h(t) \\ &= |\mathbf{u}_0 - \mathbf{v}_0| + K \int_0^{|t|} f(s) \, ds \\ &\le A e^{\int_0^{|t|} K \, ds} \\ &= |\mathbf{u}_0 - \mathbf{v}_0| e^{K|t|}. \end{aligned} \tag{2.44}$$

The proof is complete.

2.10 Differential forms

We start with the following preliminary result.

Theorem 2.10.1 (Partition of unity) *Let $K \subset \mathbb{R}^n$ be a compact set such that*

$$K \subset \cup_{i=1}^m V_i,$$

where $V_i \subset \mathbb{R}^n$ is bounded and open, $\forall i \in \{1, \cdots m\}$.
Under such hypotheses, there exist functions $h_1, \cdots, h_m : \mathbb{R}^n \to \mathbb{R}$ such that

$$\sum_{i=1}^m h_i(u) = 1, \ \forall u \in K,$$

$$0 \leq h_i(u) \leq 1, \ \forall u \in \mathbb{R}^n$$

and

$$h_i \in C_c(V_i),$$

that is, h_i is continuous and with compact support contained in V_i, $\forall i \in \{1, \cdots, m\}$.
We recall that the support of h_i, denoted by supp h_i is defined by

$$supp\, h_i = \overline{\{u \in \mathbb{R}^n \ : \ h_i(u) \neq 0\}},$$

$\forall i \in \{1, \cdots, m\}$.

Proof 2.20 Let $u \in K \subset \cup_{i=1}^m V_i$.
Thus, there exists $j \in \{1, \cdots, m\}$ such that $u \in V_j$.
Since V_j is open, there exists $r_u > 0$ such that

$$\overline{B}_{r_u}(u) \subset V_j.$$

Observe that

$$K \subset \cup_{u \in K} B_{r_u}(u).$$

Since K is compact, there exists $u_1, \cdots u_N \in K$ such that

$$K \subset \cup_{j=1}^N B_{r_j}(u_j),$$

where we have denoted $r_{u_j} = r_j$, $\forall j \in \{1, \cdots, N\}$.
For each $i \in \{1, \cdots m\}$ define $\tilde{W}_i$ as the union of all $\overline{B}_{r_j}(u_j)$ which are contained in V_i.
For each $i \in \{1, \cdots, m\}$, select also an open set W_i such that

$$\overline{\tilde{W}}_i \subset W_i \subset \overline{W}_i \subset V_i.$$

At this point, define $g_i : \mathbb{R}^n \to \mathbb{R}$, by

$$g_i(u) = \frac{d(u, W_i^c)}{d(u, W_i^c) + d(u, \tilde{W}_i)},$$

where generically, for $B \subset \mathbb{R}^n$ we define

$$d(u,B) = \inf\{|u-v| : v \in B\}.$$

Observe that

$$g_i(u) = 1, \forall u \in \tilde{W}_i,$$
$$0 \le g_i(u) \le 1, \forall u \in \mathbb{R}^n$$

and

$$g_i(u) = 0, \forall u \in \overline{W_i^c},$$

so that

$$supp\, g_i \subset V_i, \forall i \in \{1, \cdots, m\}.$$

Define

$$h_1 = g_1,$$
$$h_2 = (1-g_1)g_2,$$
$$h_3 = (1-g_1)(1-g_2)g_3,$$
$$\vdots$$
$$h_m = (1-g_1)(1-g_2)\cdots(1-g_{m-1})g_m.$$

Hence

$$0 \le h(u) \le 1, \forall u \in \mathbb{R}^n,$$

and $h_i \in C_c(V_i), \forall i\{1, \cdots, m\}$.

Now we are going to show by induction that

$$h_1 + h_2 + \cdots + h_j = 1 - (1-g_1)(1-g_2)\cdots(1-g_j), \forall j \in \{1, \cdots m\}.$$

For $j = 1$, we have $h_1 = g_1 = 1 - (1-g_1)$.
Suppose that for $2 \le j < m$ we have

$$h_1 + h_2 + \cdots + h_j = 1 - (1-g_1)(1-g_2)\cdots(1-g_j).$$

Thus,

$$
\begin{aligned}
& h_1 + h_2 + \cdots + h_j + h_{j+1} \\
= & 1 - (1-g_1)(1-g_2)\cdots(1-g_j) + (1-g_1)(1-g_2)\cdots(1-g_j)g_{j+1} \\
= & 1 - (1-g_1)(1-g_2)\cdots(1-g_{j+1}).
\end{aligned}
\tag{2.45}
$$

The induction is complete so that, in particular, we have obtained

$$h_1 + h_2 + \cdots + h_m = 1 - (1-g_1)(1-g_2)\cdots(1-g_m). \tag{2.46}$$

Hence, if $u \in K$, then $u \in K \subset \cup_{j=1}^N B_{r_j}(u_j) = \cup_{i=1}^m \tilde{W}_i$ so that $u \in \tilde{W}_j$, for some $j \in \{1, \cdots, m\}$.

Thus, $g_j(u) = 0$, so that from this and (2.46), we obtain

$$h_1 + \cdots + h_m(u) = 1, \forall u \in K.$$

The proof is complete.

Remark 2.10.2 *Concerning this last theorem, the set $\{h_1, \cdots, h_m\}$ is said to be a partition of unity subordinate to $V_1, \cdots, V_m$ and related to K.*

At this point, we recall to have already established that $\{du_i\}_{i=1}^m$ is the dual basis for $T_p(M)^*$ corresponding to the primal basis

$$\left\{ \frac{\partial \mathbf{r}(\mathbf{u})}{\partial u_i} \right\}_{i=1}^m,$$

of $T_p(M)$.

Observe that, for $f \in C^1(M)$ and

$$\mathbf{v} = v_i \frac{\partial \mathbf{r}(\mathbf{u})}{\partial u_i} \in T_p(M),$$

we have,

$$df(\mathbf{v}) = \frac{\partial f(\mathbf{r}(\mathbf{u}))}{\partial X_j} \frac{\partial X_j(\mathbf{u})}{\partial u_i} v_i. \tag{2.47}$$

On the other hand, denoting

$$\tilde{\mathbf{e}}_j = \frac{\partial \mathbf{r}(\mathbf{u})}{\partial u_j},$$

$$
\begin{aligned}
du_i(\mathbf{v}) &= du_i\left(\sum_{j=1}^m v_j \tilde{\mathbf{e}}_j \right) \\
&= v_j du_i(\tilde{\mathbf{e}}_j) = v_j \delta_{ij} = v_i. \tag{2.48}
\end{aligned}
$$

From these last results we obtain,

$$
\begin{aligned}
df(\mathbf{v}) &= \frac{\partial f(\mathbf{r}(\mathbf{u}))}{\partial X_j} \frac{\partial X_j(\mathbf{u})}{\partial u_i} v_i \\
&= \frac{\partial f(\mathbf{r}(\mathbf{u}))}{\partial X_j} \frac{\partial X_j(\mathbf{u})}{\partial u_i} du_i(\mathbf{v}) \\
&= \frac{\partial f(\mathbf{r}(\mathbf{u}))}{\partial u_i} du_i(\mathbf{v}). \tag{2.49}
\end{aligned}
$$

We have just got the differential expression

$$df = \frac{\partial f(\mathbf{r}(\mathbf{u}))}{\partial u_i} du_i,$$

that is,

$$df = \frac{\partial f(\mathbf{r}(\mathbf{u}))}{\partial X_j} \frac{\partial X_j(\mathbf{u})}{\partial u_i} du_i,$$

so that

$$df = \frac{\partial f(\mathbf{r}(\mathbf{u}))}{\partial X_j} dX_j(\mathbf{u}).$$

Definition 2.10.3 *Let* $M \subset \mathbb{R}^n$ *be a* C^1 *class m dimensional surface, where* $1 \leq m \leq n$. *For* $p = \mathbf{r}(\mathbf{u}) \in M$, *we define a 1-form in M, locally as an element* ω *of* $(T_p M)^*$, *and globally as a field of elements in* $T(M)^* = \{(T_p M)^* : p \in M\}$, *expressed locally by*

$$\omega = \sum_{k=1}^{m} \omega_k(\mathbf{u}) du_k.$$

where $\omega_k : D \to \mathbb{R}$ *are functions of* C^1 *class,* $\forall k \in \{1, \cdots, m\}$.
   We define also a $k - form$ *in M, locally by*

$$\omega = \sum_{I} \omega_I(\mathbf{u}) du_I,$$

that is, locally ω *is an element of* $(T_p M)^* \times \cdots \times (T_p M)^* = [(T_p M)^*]^k$, *where* $1 \leq k \leq m$, ω_I *is a* C^1 *class function,* $du_I = du_{i_1} \wedge \cdots \wedge du_{i_k}$, *and*

$$I = (i_1, ..., i_k)$$

is any collection of k indices $i_j \in \{1, \cdots, m\}$.

Definition 2.10.4 (Differential of a form) *For a 1-form*

$$\omega = \sum_{k=1}^{m} \omega_k(\mathbf{u}) du_k,$$

we define its exterior differential, denoted by $d\omega$, *locally by*

$$d\omega = \sum_{k=1}^{m} d\omega_k(\mathbf{u}) \wedge du_k,$$

that is,

$$d\omega = \sum_{k=1}^{m} \sum_{j=1}^{m} \frac{\partial \omega_k(\mathbf{u})}{\partial u_j} du_j \wedge du_k.$$

   Similarly, we may define the exterior differential of a C^1 *class k-form, that is, point-wise for*

$$\omega = \sum_{I} \omega_I(\mathbf{u}) du_I,$$

we have

$$d\omega = \sum_{I} d\omega_I(\mathbf{u}) \wedge du_I,$$

that is

$$d\omega = \sum_{I} \sum_{j=1}^{m} \frac{\partial \omega_I(\mathbf{u})}{\partial u_j} du_j \wedge du_I.$$

2.11 Integration of differential forms

In this section, we first present a discussion about the subject, before presenting the main result, namely, the Stokes Theorem in its general form.

Let $M \subset \mathbb{R}^n$ be a m-dimensional C^1 class surface with a boundary ∂M, where generically,

$$M = \{\mathbf{r}(\mathbf{u}) \, : \, \mathbf{u} \in D\},$$

and

$$\partial M = \mathbf{r}(\partial D).$$

Here, we assume $\overline{D}$ to be compact and, therefore, $\overline{M}$ is compact. Let $\omega = \omega_I \, du_I$ be a k-form and consider the problem of calculating the integral:

$$I = \int_M \omega_I \, du_I.$$

At this point, we suppose the manifold M is oriented, in the specific sense that, considering a canonical system

$$\{\mathbf{e}_1, \cdots, \mathbf{e}_n\} \subset \mathbb{R}^n,$$

and

$$\left\{ \frac{\partial \mathbf{r}(\mathbf{u})}{\partial u_i} \right\}_{i=1}^m,$$

we may select normal continuous fields, point-wise denoted by

$$\{\mathbf{n}_1(\mathbf{r}(\mathbf{u})), \mathbf{n}_2(\mathbf{r}(\mathbf{u})), \ldots, \mathbf{n}_{n-m}(\mathbf{r}(\mathbf{u}))\},$$

where each $\mathbf{n}_j(\mathbf{r}(\mathbf{u}))$ is orthogonal to

$$\left\{ \frac{\partial \mathbf{r}(\mathbf{u})}{\partial u_i} \right\}_{i=1}^m,$$

and such that, for each $\mathbf{u} \in D$, $\mathbb{R}^n$ is spanned by

$$\left\{ \left\{ \frac{\partial \mathbf{r}(\mathbf{u})}{\partial u_i} \right\}_{i=1}^m, \mathbf{n}_1(\mathbf{r}(\mathbf{u})), \mathbf{n}_2(\mathbf{r}(\mathbf{u})), \ldots, \mathbf{n}_{n-m}(\mathbf{r}(\mathbf{u})) \right\}.$$

Moreover, we assume the matrix to change from this local basis

$$\left\{ \left\{ \frac{\partial \mathbf{r}(\mathbf{u})}{\partial u_i} \right\}_{i=1}^m, \mathbf{n}_1(\mathbf{r}(\mathbf{u})), \mathbf{n}_2(\mathbf{r}(\mathbf{u})), \ldots, \mathbf{n}_{n-m}(\mathbf{r}(\mathbf{u})) \right\}.$$

to the canonical one of $\mathbb{R}^n$ has always positive determinant, $\forall \mathbf{u} \in D$. In such a case, we say that M is positively oriented concerning the canonical base of $\mathbb{R}^n$ (the other possibility for a oriented surface, is that the mentioned determinant be negative $\forall p \in M$. In this case we say that M is negatively oriented related to the canonical base of $\mathbb{R}^n$).

Let $p \in \partial M$. Thus, $\mathbf{u}_p = \mathbf{r}^{-1}(p) \in \partial D$.

We also assume that, for an appropriate local system of coordinates compatible with the orientation, there exists a C^1 class function point-wise indicated by $g^p(u_1, \cdots, u_{m-1})$ such that for a rectangle $B_p = \prod_{k=1}^{m}[\alpha_k^p, \beta_k^p]$ we have,

$$
D \cap B_p = \begin{aligned} &\{(u_1, ..., u_m) \in \mathbb{R}^m \ : \ \alpha_k^p \leq u_k \leq \beta_k^p, \ \forall k \in \{1, .., m-1\}, \\ &\text{and } \alpha_m^p \leq u_m \leq g^p(u_1, \cdots, u_{m-1}) \leq \beta_m^p\}. \end{aligned} \tag{2.50}
$$

Thus,

$$
\cup_{p \in \partial M} B_p^{\circ} \supset \partial D,
$$

and since $\overline{\partial D}$ is compact, There exists, $p_1, \cdots, p_s \in M$ such that

$$
\overline{\partial D} \subset \cup_{l=1}^{s} B_{p_l}^{\circ}.
$$

Let B_{p_0} be an open set such that

$$
D^0 \supset B_{p_0} \supset D^{\circ} \setminus \cup_{l=1}^{s} B_{p_l}.
$$

Select a partition of unit

$$
\{\rho_{p_l}\}_{l=0}^{s}
$$

subordinate to

$$
\{B_{p_0}^{\circ}, B_{p_1}^{\circ}, \cdots, B_{p_s}^{\circ}\}.
$$

Observe that

$$
\mathrm{supp}\{\rho_{p_l}\} \subset B_{p_l}^{\circ}, \ \forall l \in \{0, \cdots, s\},
$$

$$
0 \leq \rho_{p_l}(\mathbf{u}) \leq 1, \ \forall l \in \{0, \cdots, s\}, \ \mathbf{u} \in D,
$$

$$
\sum_{l=0}^{s} \rho_{p_l} = 1, \forall \mathbf{u} \in D.
$$

Let us first consider the specific general case in which:

$$
\omega = \sum_{j=1}^{m} (-1)^{j+1} f_j(\mathbf{u}) \, du_1 \wedge du_2 \cdots \wedge \hat{du}_j \wedge ... \wedge du_m,
$$

where $f_j : D \to \mathbb{R}$ are C^1 class functions $\forall j \in \{1, \cdots, m\}$, and where $\hat{du}_j$ means that the term du_j is absent in the products in question.

Observe that

$$
d\omega = \sum_{j=1}^{m} \frac{\partial f_j(\mathbf{u})}{\partial u_j} du_1 \wedge \cdots \wedge du_m.
$$

Denote also,

$$
\mathbf{s}_k = \frac{\partial \mathbf{r}(\mathbf{u})}{\partial u_k} \, du_k,
$$

$$S_k = (s_1, s_2, \cdots, \hat{s}_k, \cdots, s_m),$$
$$S = (s_1, s_2, \cdots, s_m),$$

Thus,

$$\int_M d\omega = \int_D \sum_{j=1}^m \frac{\partial f_j(\mathbf{u})}{\partial u_j} (du_1 \wedge \cdots \wedge du_m)(S), \qquad (2.51)$$

and hence,

$$\int_M d\omega$$

$$= \int_D \sum_{j=1}^m (-1)^{j+1} \frac{\partial f_j(\mathbf{u})}{\partial u_j} (du_j \wedge (du_1 \wedge du_2 \cdots \wedge \hat{du}_j \wedge \cdots \wedge du_m))(S)$$

$$= \int_D \sum_{j=1}^m (-1)^{j+1} \frac{\partial [(\sum_{l=1}^s \rho_{p_l}) f_j(\mathbf{u})]}{\partial u_j} (du_j \wedge (du_1 \wedge du_2 \cdots \wedge \hat{du}_j \wedge \cdots \wedge du_m))(S)$$

$$= \sum_{l=1}^s \int_D \sum_{j=1}^m \left[(-1)^{j+1} \frac{\partial (\rho_{p_l}(\mathbf{u}) f_j(\mathbf{u}))}{\partial u_j} du_j \right] du_1 du_2 \cdots \hat{du}_j \cdots du_m$$

$$= \sum_{l=1}^s \int_{\partial B_l} \sum_{j=1}^{m-1} (-1)^{j+1} [(\rho_{p_l} f_j)(u_1, \cdots, \beta_j^l, \cdots, u_m)$$

$$\qquad - (\rho_{p_l} f_j)(u_1, \cdots, \alpha_j^l, \cdots, u_m)] du_1 du_2 \cdots \hat{du}_j \cdots du_m$$

$$\qquad + \int_{\partial D \cap B_l} (-1)^{m+1} (\rho_{p_l} f_m)(u_1, \cdots, u_{m-1}, g^l(u_1, \cdots, u_{m-1})) \, du_1 \cdots du_{m-1}$$

$$\qquad - \int_{\partial B_l} (-1)^{m+1} (\rho_{p_l} f_m)(u_1, \cdots, u_{m-1}, \alpha_m^l) \, du_1 \cdots du_{m-1}$$

$$= \sum_{l=1}^s \int_{\partial D \cap B_l} (-1)^{m+1} (\rho_{p_l} f_m)(u_1, \cdots, u_{m-1}, g^l(u_1, \cdots, u_{m-1})) \, du_1 \cdots du_{m-1}$$

$$= \int_{\partial D} (-1)^{m+1} \left(\sum_{l=1}^s \rho_{p_l} f_m \right) (u_1, \cdots, u_{m-1}, g^l(u_1, \cdots, u_{m-1})) \, du_1 \cdots du_{m-1}$$

$$= \int_{\partial D} \left(\sum_{l=1}^s \rho_{p_l}(\mathbf{u}) \right) \sum_{j=1}^m (-1)^{j+1} f_j(\mathbf{u}) du_1 du_2 \cdots \hat{du}_j \cdots du_m$$

$$= \int_{\partial D} \sum_{j=1}^m (-1)^{j+1} f_j(\mathbf{u}) du_1 du_2 \cdots \hat{du}_j \cdots du_m, \qquad (2.52)$$

so that

$$\int_M d\omega = \int_{\partial D} \sum_{j=1}^m (-1)^{j+1} f_j(\mathbf{u}) du_1 du_2 \cdots \hat{du}_j \cdots du_m$$

$$= \int_{\partial D} \sum_{j=1}^m (-1)^{j+1} f_j(\mathbf{u}) (du_1 \wedge du_2 \cdots \wedge \hat{du}_j \wedge \cdots \wedge du_m)(S_j)$$

$$= \int_{\partial M} \omega. \qquad (2.53)$$

2.12 A simple example to illustrate the integration process

We would emphasize, in the final of last section, when the wedge product is absent, the differential forms and integrations are relating the usual calculus sense.

In the next example, we see how to connect the general differential form concept to a more usual one, in the ordinary calculus sense.

So, let $M \subset \mathbb{R}^n$ be C^1 class surface given by:

$$M = \{\mathbf{r}(\mathbf{u}) : \mathbf{u} \in D\},$$

where, for simplicity $D = [a,b] \times [c,d]$.

Consider the integral

$$
\begin{aligned}
I &= \int_M f \, dX_1 \wedge dX_2 \\
&= \int_D f(\mathbf{u})(dX_1(\mathbf{u}) \wedge dX_2(\mathbf{u})) \left(\frac{\partial \mathbf{r}(\mathbf{u})}{\partial u_1} \, du_1, \frac{\partial \mathbf{r}(\mathbf{u})}{\partial u_2} \, du_2 \right) \quad (2.54)
\end{aligned}
$$

where, $X_1 : D \to \mathbb{R}$ and $X_2 : D \to \mathbb{R}$ are C^1 class functions.

Observe that

$$dX_1(\mathbf{u}) = \frac{\partial X_1(\mathbf{u})}{\partial u_1} \, du_1 + \frac{\partial X_1(\mathbf{u})}{\partial u_2} \, du_2$$

and

$$dX_2(\mathbf{u}) = \frac{\partial X_2(\mathbf{u})}{\partial u_1} \, du_1 + \frac{\partial X_2(\mathbf{u})}{\partial u_2} \, du_2$$

Consider the elementary rectangle

$$(u_1, u_1 + \Delta u_1) \times (u_2, u_2 + \Delta u_2),$$

which has as its sides the vectors

$$(\Delta u_1, 0) \in \mathbb{R}^2,$$

and

$$(0, \Delta u_2) \in \mathbb{R}^2.$$

Define

$$\tilde{\mathbf{s}}_1 = \frac{\partial \mathbf{r}(u)}{\partial u_1} \Delta u_1,$$

and

$$\tilde{\mathbf{s}}_2 = \frac{\partial \mathbf{r}(u)}{\partial u_2} \Delta u_2,$$

Considering the basis

$$\left\{ \frac{\partial \mathbf{r}(u)}{\partial u_1}, \frac{\partial \mathbf{r}(u)}{\partial u_2} \right\},$$

all to simplify the notation we shall denote

$$\tilde{\mathbf{s}}_1 = (\Delta u_1, 0),$$

and

$$\tilde{\mathbf{s}}_2 = (0, \Delta u_2).$$

Recall that, for the general case, for

$$\mathbf{v} = v_i \frac{\partial \mathbf{r}(u)}{\partial u_i}.$$

we have

$$du_j(\mathbf{v}) = v_j, \forall j \in \{1, \cdots, m\}.$$

Recall also that, from Definition 2.8.2,

$$(F_1 \wedge \cdots \wedge F_k)(\mathbf{v}_1, \cdots, \mathbf{v}_k) = \det\{F_i(\mathbf{v}_j)\}_{i,j=1}^k.$$

Let us evaluate

$$
\begin{aligned}
& (dX_1(\mathbf{u}) \wedge dX_2(\mathbf{u})) \left(\frac{\partial \mathbf{r}(u)}{\partial u_1} \Delta u_1, \frac{\partial \mathbf{r}(u)}{\partial u_2} \Delta u_2 \right) \\
=\ & (dX_1(\mathbf{u}) \wedge dX_2(\mathbf{u}))[(\tilde{\mathbf{s}}_1, \tilde{\mathbf{s}}_2)] \\
=\ & (dX_1(\mathbf{u}) \wedge dX_2(\mathbf{u}))[(\Delta u_1, 0), (0, \Delta u_2)] \\
=\ & \left[\left(\frac{\partial X_1(\mathbf{u})}{\partial u_1} du_1 + \frac{\partial X_1(\mathbf{u})}{\partial u_2} du_2 \right) \wedge \left(\frac{\partial X_2(\mathbf{u})}{\partial u_1} du_1 + \frac{\partial X_2(\mathbf{u})}{\partial u_2} du_2 \right) \right] \\
& ((\Delta u_1, 0), (0, \Delta u_2)) \\
=\ & \left[\left(\frac{\partial X_1(\mathbf{u})}{\partial u_1} \frac{\partial X_2(\mathbf{u})}{\partial u_2} du_1 \wedge du_2 \right) + \left(\frac{\partial X_1(\mathbf{u})}{\partial u_2} \frac{\partial X_2(\mathbf{u})}{\partial u_1} du_2 \wedge du_1 \right) \right] \\
& ((\Delta u_1, 0), (0, \Delta u_2)) \\
=\ & \left(\frac{\partial X_1(\mathbf{u})}{\partial u_1} \frac{\partial X_2(\mathbf{u})}{\partial u_2} du_1 \wedge du_2 \right) ((\Delta u_1, 0), (0, \Delta u_2)) \\
& + \left(\frac{\partial X_1(\mathbf{u})}{\partial u_2} \frac{\partial X_2(\mathbf{u})}{\partial u_1} du_2 \wedge du_1 \right) ((\Delta u_1, 0), (0, \Delta u_2)) \\
=\ & \frac{\partial X_1(\mathbf{u})}{\partial u_1} \frac{\partial X_2(\mathbf{u})}{\partial u_2} \begin{vmatrix} du_1(\Delta u_1, 0) & du_1(0, \Delta u_2) \\ du_2(\Delta u_1, 0) & du_2(0, \Delta u_2) \end{vmatrix} \\
& + \frac{\partial X_1(\mathbf{u})}{\partial u_2} \frac{\partial X_2(\mathbf{u})}{\partial u_1} \begin{vmatrix} du_2(\Delta u_1, 0) & du_2(0, \Delta u_2) \\ du_1(\Delta u_1, 0) & du_1(0, \Delta u_2) \end{vmatrix} \\
=\ & \frac{\partial X_1(\mathbf{u})}{\partial u_1} \frac{\partial X_2(\mathbf{u})}{\partial u_2} \begin{vmatrix} \Delta u_1 & 0 \\ 0 & \Delta u_2 \end{vmatrix} \\
& + \frac{\partial X_1(\mathbf{u})}{\partial u_2} \frac{\partial X_2(\mathbf{u})}{\partial u_1} \begin{vmatrix} 0 & \Delta u_2 \\ \Delta u_1 & 0 \end{vmatrix} \\
=\ & \left(\frac{\partial X_1(\mathbf{u})}{\partial u_1} \frac{\partial X_2(\mathbf{u})}{\partial u_2} - \frac{\partial X_1(\mathbf{u})}{\partial u_2} \frac{\partial X_2(\mathbf{u})}{\partial u_1} \right) \Delta u_1 \Delta u_2. \quad (2.55)
\end{aligned}
$$

So, to summarize,

$$
\begin{aligned}
&(dX_1(\mathbf{u}) \wedge dX_2(\mathbf{u}))[(\Delta u_1, 0), (0, \Delta u_2)] \\
&= \left(\frac{\partial X_1(\mathbf{u})}{\partial u_1} \frac{\partial X_2(\mathbf{u})}{\partial u_2} - \frac{\partial X_1(\mathbf{u})}{\partial u_2} \frac{\partial X_2(\mathbf{u})}{\partial u_1} \right) \Delta u_1 \Delta u_2,
\end{aligned}
\tag{2.56}
$$

or in its differential form:

$$
\begin{aligned}
&(dX_1(\mathbf{u}) \wedge dX_2(\mathbf{u}))[(du_1, 0), (0, du_2)] \\
&= \left(\frac{\partial X_1(\mathbf{u})}{\partial u_1} \frac{\partial X_2(\mathbf{u})}{\partial u_2} - \frac{\partial X_1(\mathbf{u})}{\partial u_2} \frac{\partial X_2(\mathbf{u})}{\partial u_1} \right) du_1 du_2.
\end{aligned}
\tag{2.57}
$$

Observe that this last differential form is one in the usual sense calculus, which is also used in the final usual integration process.

2.13   Volume (area) of a surface

Consider the problem of calculating the volume defined by the vectors

$$\mathbf{v}_1, \mathbf{v}_2, \cdots, \mathbf{v}_m \in \mathbb{R}^n$$

where $1 \leq m \leq n$.

Definition 2.13.1 *Given* $\mathbf{v}_1, \mathbf{v}_2 \in \mathbb{R}^n$, *we shall define the volume (in fact area) defined by* $\mathbf{v}_1$ *and* $\mathbf{v}_2$, *denoted by* V_2, *by*

$$V_2 = |\mathbf{v}_1||\mathbf{v}_2|| \sin \theta|,$$

where θ *is the angle between* $\mathbf{v}_1$ *and* $\mathbf{v}_2$, *which is given by:*

$$\cos \theta = \frac{\mathbf{v}_1 \cdot \mathbf{v}_2}{|\mathbf{v}_1||\mathbf{v}_2|},$$

so that

$$
\begin{aligned}
V_2 &= |\mathbf{v}_1||\mathbf{v}_2| \sqrt{1 - \cos^2 \theta} \\
&= |\mathbf{v}_1||\mathbf{v}_2| \sqrt{1 - \frac{(\mathbf{v}_1 \cdot \mathbf{v}_2)^2}{|\mathbf{v}_1|^2|\mathbf{v}_2|^2}} \\
&= \sqrt{|\mathbf{v}_1|^2|\mathbf{v}_2|^2 - (\mathbf{v}_1 \cdot \mathbf{v}_2)^2},
\end{aligned}
\tag{2.58}
$$

that is,

$$
\begin{aligned}
V_2^2 &= |\mathbf{v}_1|^2|\mathbf{v}_2|^2 - (\mathbf{v}_1 \cdot \mathbf{v}_2)^2 \\
&= \begin{vmatrix} \mathbf{v}_1 \cdot \mathbf{v}_1 & \mathbf{v}_1 \cdot \mathbf{v}_2 \\ \mathbf{v}_2 \cdot \mathbf{v}_1 & \mathbf{v}_2 \cdot \mathbf{v}_2 \end{vmatrix} \\
&\equiv g_2,
\end{aligned}
\tag{2.59}
$$

so that,

$$V_2 = \sqrt{g_2},$$

where

$$g_2 = \det\{\mathbf{v}_i \cdot \mathbf{v}_j\}_{i,j=1}^2.$$

Definition 2.13.2 *Let* $\mathbf{v}_1, \cdots, \mathbf{v}_m \subset \mathbb{R}^n$ *be non zero vectors, where* $2 \le m \le n$. *For* $2 \le s < m$, *we shall define inductively the volume defined by* $\mathbf{v}_1, \mathbf{v}_2, \cdots, \mathbf{v}_{s+1}$, *by*

$$V_2^2 = g_2,$$

and

$$V_{k+1}^2 = V_k^2 |\tilde{\mathbf{v}}_{k+1}|^2, \forall k \in \{2, \cdots, s\},$$

where

$$\tilde{\mathbf{v}}_{k+1} = \mathbf{v}_{k+1} - \sum_{j=1}^k \tilde{a}_j \mathbf{v}_j,$$

and where

$$\{\tilde{a}_j\} = \arg\min \left\{ \left| \mathbf{v}_{k+1} - \sum_{j=1}^k a_j \mathbf{v}_j \right|^2 \; : \; a_1, \cdots, a_k \in \mathbb{R} \right\}$$

Theorem 2.13.3 *Under the statements of the last definition, we have:*

$$V_k^2 = g_k, \forall k \in \{2, \cdots, s+1\},$$

where

$$g_k = \det\{\mathbf{v}_i \cdot \mathbf{v}_j\}_{i,j=1}^k.$$

Proof 2.21 We prove the result by induction.
For $k = 2$ we have already:

$$V_2^2 = g_2.$$

For $2 \ge k < m$ assume

$$V_k^2 = g_k.$$

It suffices, to complete the induction, to show that

$$V_{k+1}^2 = g_{k+1}.$$

By definition,

$$V_{k+1}^2 = V_k^2 |\tilde{\mathbf{v}}_{k+1}|^2,$$

where

$$\tilde{\mathbf{v}}_{k+1} = \mathbf{v}_{k+1} - \sum_{j=1}^k \tilde{a}_j \mathbf{v}_j,$$

and where

$$\{\tilde{a}_j\} = \arg\min\left\{\left|\mathbf{v}_{k+1} - \sum_{j=1}^{k} a_j \mathbf{v}_j\right|^2 \; : \; a_1, \cdots, a_k \in \mathbb{R}\right\}$$

For this last optimization problem, the extremal necessary conditions are given by:

$$\left(\mathbf{v}_{k+1} - \sum_{j=1}^{k} \tilde{a}_j \mathbf{v}_j\right) \cdot \mathbf{v}_s = 0, \forall s \in \{1, \cdots, k\},$$

so that

$$\{\mathbf{v}_s \cdot \mathbf{v}_j\}\{\tilde{a}_j\} = \{\mathbf{v}_{k+1} \cdot \mathbf{v}_s\},$$

that is,

$$\{\tilde{a}_j\} = \{\mathbf{v}_s \cdot \mathbf{v}_j\}^{-1}\{\mathbf{v}_{k+1} \cdot \mathbf{v}_s\}.$$

Observe that

$$g_{k+1} = \{\mathbf{v}_i \cdot \mathbf{v}_j\}_{i,j=1}^{k+1}$$

$$= \begin{vmatrix} \mathbf{v}_1 \cdot \mathbf{v}_1 & \mathbf{v}_1 \cdot \mathbf{v}_2 & \cdots & \mathbf{v}_1 \cdot \mathbf{v}_{k+1} \\ \mathbf{v}_2 \cdot \mathbf{v}_1 & \mathbf{v}_2 \cdot \mathbf{v}_2 & \cdots & \mathbf{v}_2 \cdot \mathbf{v}_{k+1} \\ \vdots & \vdots & \ddots & \vdots \\ \mathbf{v}_{k+1} \cdot \mathbf{v}_1 & \mathbf{v}_{k+1} \cdot \mathbf{v}_2 & \cdots & \mathbf{v}_{k+1} \cdot \mathbf{v}_{k+1} \end{vmatrix} \qquad (2.60)$$

so that,

$$g_{k+1} = \begin{vmatrix} \mathbf{v}_1 \cdot \mathbf{v}_1 & \mathbf{v}_1 \cdot \mathbf{v}_2 & \cdots & \mathbf{v}_1 \cdot \left(\mathbf{v}_{k+1} - \sum_{j=1}^{k} \tilde{a}_j \mathbf{v}_j\right) \\ \mathbf{v}_2 \cdot \mathbf{v}_1 & \mathbf{v}_2 \cdot \mathbf{v}_2 & \cdots & \mathbf{v}_2 \cdot \left(\mathbf{v}_{k+1} - \sum_{j=1}^{k} \tilde{a}_j \mathbf{v}_j\right) \\ \vdots & \vdots & \ddots & \vdots \\ \mathbf{v}_{k+1} \cdot \mathbf{v}_1 & \mathbf{v}_{k+1} \cdot \mathbf{v}_2 & \cdots & \mathbf{v}_{k+1} \cdot \left(\mathbf{v}_{k+1} - \sum_{j=1}^{k} \tilde{a}_j \mathbf{v}_j\right) \end{vmatrix} \qquad (2.61)$$

that is,

$$g_{k+1} = \begin{vmatrix} \mathbf{v}_1 \cdot \mathbf{v}_1 & \mathbf{v}_1 \cdot \mathbf{v}_2 & \cdots & 0 \\ \mathbf{v}_2 \cdot \mathbf{v}_1 & \mathbf{v}_2 \cdot \mathbf{v}_2 & \cdots & 0 \\ \vdots & \vdots & \ddots & \vdots \\ \mathbf{v}_{k+1} \cdot \mathbf{v}_1 & \mathbf{v}_{k+1} \cdot \mathbf{v}_2 & \cdots & \mathbf{v}_{k+1} \cdot \left(\mathbf{v}_{k+1} - \sum_{j=1}^{k} \tilde{a}_j \mathbf{v}_j\right) \end{vmatrix}. \qquad (2.62)$$

From this, since

$$\mathbf{v}_s \cdot \left(\mathbf{v}_{k+1} - \sum_{j=1}^{k} \tilde{a}_j \mathbf{v}_j\right) = 0, \forall s \in \{1, .., k\},$$

we obtain,

$$g_{k+1} = \begin{vmatrix} \mathbf{v}_1 \cdot \mathbf{v}_1 & \mathbf{v}_1 \cdot \mathbf{v}_2 & \cdots & 0 \\ \mathbf{v}_2 \cdot \mathbf{v}_1 & \mathbf{v}_2 \cdot \mathbf{v}_2 & \cdots & 0 \\ \vdots & \vdots & \ddots & \vdots \\ \mathbf{v}_{k+1} \cdot \mathbf{v}_1 & \mathbf{v}_{k+1} \cdot \mathbf{v}_2 & \cdots & \left(\mathbf{v}_{k+1} - \sum_{j=1}^{k} \tilde{a}_j \mathbf{v}_j\right) \cdot \left(\mathbf{v}_{k+1} - \sum_{j=1}^{k} \tilde{a}_j \mathbf{v}_j\right) \end{vmatrix}$$

so that

$$g_{k+1} = \begin{vmatrix} \mathbf{v}_1 \cdot \mathbf{v}_1 & \mathbf{v}_1 \cdot \mathbf{v}_2 & \cdots & 0 \\ \mathbf{v}_2 \cdot \mathbf{v}_1 & \mathbf{v}_2 \cdot \mathbf{v}_2 & \cdots & 0 \\ \vdots & \vdots & \ddots & \vdots \\ \mathbf{v}_{k+1} \cdot \mathbf{v}_1 & \mathbf{v}_{k+1} \cdot \mathbf{v}_2 & \cdots & \tilde{\mathbf{v}}_{k+1} \cdot \tilde{\mathbf{v}}_{k+1} \end{vmatrix} \tag{2.63}$$

from which, calculating the determinant through the last column, we have,

$$g_{k+1} = g_k |\tilde{\mathbf{v}}_{k+1}|^2.$$

This completes the induction.

The proof is complete.

Let us turn our attention to the problem of calculating the volume of a manifold.

Consider an m-dimensional C^1 class surface $M \subset \mathbb{R}^n$, where, $1 \leq m \leq n$ and

$$M = \{\mathbf{r}(\mathbf{u}) : \mathbf{u} \in D\}.$$

We define, the volume of M, denoted by V, by:

$$V = \int_M dM,$$

where dM will be specified in the next lines.

We may also denote,

$$V = \int_D dM(\mathbf{u}).$$

At this point we address the problem of finding $dM(\mathbf{u})$.

Let $\mathbf{u} = (u_1, \cdots, u_m) \in D$, and let $p = \mathbf{r}(\mathbf{u})..$

Consider the $m-$ dimensional elementary rectangle

$$(u_1, u_1 + \Delta u_1) \times (u_2, u_2 + \Delta u_2) \times \cdots \times (u_m, u_m + \Delta u_m).$$

We are going to obtain the corresponding approximate volume in the surface, which we denote by

$$\Delta M(u).$$

Observe that,

$$\begin{aligned} \Delta \mathbf{r}_{u_j} &= \mathbf{r}(u_1, \cdots, u_j + \Delta u_j, \cdots, u_m) - \mathbf{r}(u_1, \cdots, u_j, \cdots, u_m) \\ &= \frac{\partial \mathbf{r}(\mathbf{u})}{\partial u_j} \Delta u_j + o(\Delta u_j) \\ &\approx \frac{\partial \mathbf{r}(\mathbf{u})}{\partial u_j} \Delta u_j, \end{aligned} \tag{2.64}$$

for Δu_j sufficiently small.

Hence, $\Delta M(\mathbf{u})$ is approximately defined by the set of vectors

$$\{\Delta \mathbf{r}_{u_1}, \cdots, \Delta \mathbf{r}_{u_m}\} \subset \mathbb{R}^n,$$

where, $\Delta \mathbf{r}_{u_j} \approx \frac{\partial \mathbf{r}(\mathbf{u})}{\partial u_j} \Delta u_j$, from the last theorem, we obtain,

$$\Delta M(\mathbf{u}) \approx \sqrt{g} \Delta u_1 \Delta u_2 \cdots \Delta u_m,$$

where

$$g = \det\{g_{ij}\}_{i,j=1}^m,$$

and

$$\mathbf{g}_i \equiv \frac{\partial \mathbf{r}(\mathbf{u})}{\partial u_j}, \forall i \in \{1, \cdots, m\},$$

and finally

$$g_{ij} = \mathbf{g}_i \cdot \mathbf{g}_j, \ \forall i, j \in \{1, \cdots, m\}.$$

So, in its differential form, we could write,

$$dM(\mathbf{u}) = \sqrt{g} \, du_1 \cdots du_m,$$

so that up to local coordinates and concerning partition of unity, we have

$$
\begin{aligned}
\int_M dM &= \int_D dM(\mathbf{u}) \\
&= \int_D \sqrt{g} \, du_1 \cdots du_m. \quad (2.65)
\end{aligned}
$$

2.14 Change of variables, the general case

Let $D \subset \mathbb{R}^n$ be a compact block (or even a more general compact region analogous to a simple one in $\mathbb{R}^3$). Let $f : D \to \mathbb{R}$ be a continuous function.

Consider the integral I, where

$$I = \int_D f(\mathbf{x}) \, d\mathbf{x} = \int_D f(x_1, \cdots, x_n) \, dx_1 \cdots dx_n.$$

Consider also the change in variables, given by the C^1 class functions $X_1, \cdots, X_n : D_0 \to \mathbb{R}$, where we denote

$$\mathbf{r}(\mathbf{u}) = X_1(\mathbf{u})\mathbf{e}_1 + \cdots + X_n(\mathbf{u})\mathbf{e}_n.$$

We assume $\mathbf{r} : D_0 \to D$ to be a bijection, where $\{\mathbf{e}_1, \cdots, \mathbf{e}_n\}$ denotes the canonical basis for $\mathbb{R}^n$ and

$$\mathbf{u} = (u_1, \cdots, u_n) \in D_0 \subset \mathbb{R}^n.$$

More specifically, the change of variables is given by,

$$x_1 = X_1(\mathbf{u}), \cdots, x_n = X_n(\mathbf{u}).$$

At this point we shall show that

$$\left| \det\left\{ \frac{\partial X_i(\mathbf{u})}{\partial u_j} \right\} \right|^2 = g,$$

where

$$g = \det\{g_{ij}\}$$

and

$$g_{ij} = \frac{\partial \mathbf{r}(\mathbf{u})}{\partial u_i} \cdot \frac{\partial \mathbf{r}(\mathbf{u})}{\partial u_j},$$

so that

$$
\begin{aligned}
g_{ij} &= \frac{\partial \mathbf{r}(\mathbf{u})}{\partial u_i} \cdot \frac{\partial \mathbf{r}(\mathbf{u})}{\partial u_j} \\
&= \sum_{k=1}^{n} \frac{\partial X_k}{\partial u_i} \frac{\partial X_k}{\partial u_j} \\
&= \sum_{k=1}^{n} (X_k)_{u_i} (X_k)_{u_j}.
\end{aligned}
\tag{2.66}
$$

Observe that

$$
\begin{aligned}
&\left\{ \frac{\partial X_i(\mathbf{u})}{\partial u_j} \right\}^T \left\{ \frac{\partial X_i(\mathbf{u})}{\partial u_j} \right\} \\
&= \left\{ \begin{array}{cccc} (X_1)_{u_1} & (X_2)_{u_1} & \cdots & (X_n)_{u_1} \\ (X_1)_{u_2} & (X_2)_{u_2} & \cdots & (X_n)_{u_2} \\ \vdots & \vdots & \ddots & \vdots \\ (X_1)_{u_n} & (X_2)_{u_n} & \cdots & (X_n)_{u_n} \end{array} \right\} \cdot \left\{ \begin{array}{cccc} (X_1)_{u_1} & (X_1)_{u_2} & \cdots & (X_1)_{u_n} \\ (X_2)_{u_1} & (X_2)_{u_2} & \cdots & (X_2)_{u_n} \\ \vdots & \vdots & \ddots & \vdots \\ (X_n)_{u_1} & (X_n)_{u_2} & \cdots & (X_n)_{u_n} \end{array} \right\}
\end{aligned}
$$

so that

$$
\begin{aligned}
&\left\{ \frac{\partial X_i(\mathbf{u})}{\partial u_j} \right\}^T \left\{ \frac{\partial X_i(\mathbf{u})}{\partial u_j} \right\} \\
&= \left\{ \sum_{k=1}^{n} (X_k)_{u_i} (X_k)_{u_j} \right\} \\
&= \left\{ \frac{\partial \mathbf{r}(\mathbf{u})}{\partial u_i} \cdot \frac{\partial \mathbf{r}(\mathbf{u})}{\partial u_j} \right\} \\
&= \{g_{ij}\}.
\end{aligned}
\tag{2.67}
$$

Hence,

$$\left| \det \left\{ \frac{\partial X_i(\mathbf{u})}{\partial u_j} \right\} \right|^2 = \det\{g_{ij}\} = g.$$

Thus,

$$\left| \det \left\{ \frac{\partial X_i(\mathbf{u})}{\partial u_j} \right\} \right| = \sqrt{g},$$

so that, from the exposed in the last two sections, we have,

$$
\begin{aligned}
I &= \int_D f(x_1, \cdots, x_n) \, dx_1 \cdots dx_n \\
&= \int_{D_0} f(X_1(\mathbf{u}), \cdots, X_n(\mathbf{u})) \sqrt{g} \, du_1 \cdots du_n \\
&= \int_{D_0} f(X_1(\mathbf{u}), \cdots, X_n(\mathbf{u})) \left| \det \left\{ \frac{\partial X_i(\mathbf{u})}{\partial u_j} \right\} \right| du_1 \cdots du_n. \quad (2.68)
\end{aligned}
$$

2.15 The Stokes Theorem

In this subsection, we present the Stokes theorem:

Theorem 2.15.1 *Let $M \subset \mathbb{R}^n$ be a C^1 class oriented compact m-dimensional surface with a boundary ∂M, where $1 \leq m \leq n$.*
Let $\omega = \sum_I \omega_I \, du_I$ be a (m-1)-form. Under such hypotheses,

$$\int_M d\omega = \int_{\partial M} \omega.$$

Proof 2.22 The proof follows from the discussion in section 2.11.
Indeed, denoting

$$\mathbf{s}_k = \frac{\partial \mathbf{r}(\mathbf{u})}{\partial u_k} \, du_k, \; \forall k \in \{1, \cdots, m\},$$

and

$$\mathbf{S}_j = (\mathbf{s}_1, \cdots, \hat{\mathbf{s}}_j, \cdots, \mathbf{s}_m),$$

where $\hat{\mathbf{s}}_j$ means the absence of such a term in the concerning list of vectors, we may infer that the general form

$$\sum_{j=1}^{m} \sum_I (\omega_I(\mathbf{u}) du_I(\mathbf{u}))(\mathbf{S}_j)$$

would stand for:

$$
\begin{aligned}
& \sum_{j=1}^{m} (-1)^{j+1} h_j(\mathbf{u}) \, (du_1 \wedge \cdots \wedge \hat{du}_j \wedge \cdots \wedge du_m)(\mathbf{S}_j), \\
&= \sum_{j=1}^{m} (-1)^{j+1} h_j(\mathbf{u}) \, du_1 \cdots \hat{du}_j \cdots du_m, \quad (2.69)
\end{aligned}
$$

for appropriate C^1 class functions $h_j, \forall j \in \{1, \cdots, m\}$.

2.15.1 *Recovering the classical results on vector calculus in $\mathbb{R}^3$ from the general Stokes Theorem*

■ Recovering the standard stokes Theorem in $\mathbb{R}^3$. Let $S \subset \mathbb{R}^3$ be a C^1 class surface with a boundary $C = \partial S$, where

$$S = \{(x,y,z) \in \mathbb{R}^3 \,:\, z = z(x,y) \text{ and } (x,y) \in D\},$$

where $D \subset \mathbb{R}^2$ is a simple region.

Consider the form $\omega = P\,dx + Q\,dy + R\,dz$, where we denote $\mathbf{F} = P\mathbf{i} + Q\mathbf{j} + R\mathbf{z}$, and where $P, Q, R : V \supset S \to \mathbb{R}$ are C^1 class scalar functions.

From the Stokes Theorem 2.15.1, we have

$$\int_C \omega = \int_S d\omega,$$

that is,

$$
\begin{aligned}
\int_C P\,dx + Q\,dy + R\,dz &= \int_S d\omega \\
&= \int_S (dP \wedge dx + dQ \wedge dy + dR \wedge dz) \\
&= \int_S \left(\frac{\partial P}{\partial dx}\,dx + \frac{\partial P}{\partial y}\,dy + \frac{\partial P}{\partial z}\,dz \right) \wedge dx \\
&\quad + \left(\frac{\partial Q}{\partial x}\,dx + \frac{\partial Q}{\partial y}\,dy + \frac{\partial Q}{\partial z}\,dz \right) \wedge dy \\
&\quad + \left(\frac{\partial R}{\partial x}\,dx + \frac{\partial R}{\partial y}\,dy + \frac{\partial R}{\partial z}\,dz \right) \wedge dz \\
&= \int_S (Q_x - P_y)\,dx \wedge dy + \int_S (R_y - Q_z)\,dy \wedge dz \\
&\quad + (P_z - R_x)\,dz \wedge dx. \qquad (2.70)
\end{aligned}
$$

We recall that in S, $z = z(x,y)$, so that, in the context of informal calculus language, we have

$$dz = z_x\,dx + z_y\,dy.$$

Also,

$$dx \wedge dy = dxdy,$$

$$dy \wedge dz = dy \wedge (z_x dx + z_y dy) = -z_x dxdy,$$

$$dz \wedge dx = (z_x dx + z_y dy) \wedge dx = -z_y dxdy,$$

so that from (2.71), we obtain,

$$
\begin{aligned}
&\int_C P\,dx + Q\,dy + R\,dz \\
=\;&\int_D (Q_x - P_y)\,dxdy + \int_D (R_y - Q_z)\,(-z_x)\,dxdy \wedge dz \\
&+ (P_z - R_x)\,(-z_y)\,dxdy \\
=\;&\int_D (R_y - Q_z)\,(-z_x)\,dxdy + (P_z - R_x)\,(-z_y)\,dxdy + \int_D (Q_x - P_y)\,dxdy \\
=\;&\int_D curl(\mathbf{F}) \cdot [-z_x\mathbf{i} - z_y\mathbf{j} + k]\,dxdy \\
=\;&\int_D curl(\mathbf{F}) \cdot \frac{[-z_x\mathbf{i} - z_y\mathbf{j} + k]}{\sqrt{z_x^2 + z_y^2 + 1}}\sqrt{z_x^2 + z_y^2 + 1}\,dxdy \\
=\;&\int_D curl(\mathbf{F}) \cdot \mathbf{n}\sqrt{z_x^2 + z_y^2 + 1}\,dxdy \\
=\;&\int_S curl(\mathbf{F}) \cdot \mathbf{n}\,dS,
\end{aligned}
\tag{2.71}
$$

where $\mathbf{n}$ is unit outward normal relating S.

So, to summarize, we have obtained,

$$
\int_C \mathbf{F} \cdot d\mathbf{r} = \int_S curl(\mathbf{F}) \cdot \mathbf{n}\,dS,
$$

which is the standard Stokes Theorem result in $\mathbb{R}^3$.

Remark 2.15.2 *We have used some informality here. In fact, in a more rigorous fashion, we should have:*

$$
\mathbf{r}(x,y) = (x, y, z(x,y)),
$$

and

$$
\begin{aligned}
&(dx \wedge dy)(\mathbf{r}_x(x,y)\,dx, \mathbf{r}_y(x,y)\,dy) \\
=\;&(dx \wedge dy)((1,0,z_x)\,dx, (0,1,z_y)\,dy) \\
=\;&dx\,dy,
\end{aligned}
\tag{2.72}
$$

and with more details,

$$(dy \wedge dz)(\mathbf{r}_x(x,y)\, dx,\, \mathbf{r}_y(x,y)\, dy)$$
$$= (dy \wedge dz)((1,0,z_x)\, dx,\, (0,1,z_y)\, dy)$$
$$= (dy \wedge (z_x dx + z_y dy))((1,0,z_x)\, dx,\, (0,1,z_y)\, dy)$$
$$= z_x(dy \wedge dx)((1,0,z_x)\, dx,\, (0,1,z_y)\, dy)$$
$$+ z_y(dy \wedge dy)((1,0,z_x)\, dx,\, (0,1,z_y)\, dy)$$
$$= z_x \begin{vmatrix} dy(dx,0,z_x dx) & dy(0,dy,z_y\, dy) \\ dx(dx,0,z_x dx) & dx(0,dy,z_y\, dy) \end{vmatrix}$$
$$= z_x \begin{vmatrix} 0 & dy \\ dx & 0 \end{vmatrix}$$
$$= -z_x dxdy. \tag{2.73}$$

A similar remark is valid for

$$dz \wedge dx.$$

■ Recovering the Divergence Theorem in $\mathbb{R}^3$:

Let $V \subset \mathbb{R}^3$ be a simple volume with a boundary $S = \partial V$. Let $\mathbf{F} = P\mathbf{i} + Q\mathbf{j} + R\mathbf{k}$ where $P, Q, R : V \to \mathbb{R}$ are C^1 class scalar functions.

From the last item, we have

$$(dx \wedge dy)((1,0,z_x)\, dx,\, (0,1,z_y)\, dy) = dxdy,$$

$$(dy \wedge dz)(((1,0,z_x)\, dx,\, (0,1,z_y)\, dy)) = -z_x dxdy,$$

$$(dz \wedge dx)(((1,0,z_x)\, dx,\, (0,1,z_y)\, dy)) = -z_y dxdy,$$

so that,

$$(P\, dy \wedge dz + Q\, dz \wedge dx + R\, dx \wedge dy)((dx,0),(0,dy))$$
$$= P(-z_x)\, dxdy + Q(-z_y)\, dxdy + R\, dxdy$$
$$= (P\mathbf{i} + Q\mathbf{j} + R\mathbf{k}) \cdot (-z_x\mathbf{i} - z_y\mathbf{j} + \mathbf{k})\, dxdy$$
$$= \mathbf{F} \cdot \frac{(-z_x\mathbf{i} - z_y\mathbf{j} + \mathbf{k})}{\sqrt{z_x^2 + z_y^2 + 1}} \sqrt{z_x^2 + z_y^2 + 1}\, dxdy$$
$$= \mathbf{F} \cdot \mathbf{n}\sqrt{z_x^2 + z_y^2 + 1}\, dxdy$$
$$= \mathbf{F} \cdot \mathbf{n}\, dS, \tag{2.74}$$

where

$$\mathbf{n} = \frac{(-z_x\mathbf{i} - z_y\mathbf{j} + \mathbf{k})}{\sqrt{z_x^2 + z_y^2 + 1}},$$

is the unit outward normal relating S and

$$dS = \sqrt{z_x^2 + z_y^2 + 1}\, dxdy.$$

Consider, with some informality here, the form

$$\omega = P\,dy \wedge dz + Q\,dz \wedge dx + R\,dx \wedge dy = \mathbf{F} \cdot \mathbf{n}\,dS.$$

From the Stokes Theorem 2.15.1, we have

$$\int_S \omega = \int_V d\omega.$$

Observe that

$$
\begin{aligned}
\int_V d\omega &= \int_V (dP \wedge (dy \wedge dz)) + (dQ \wedge (dz \wedge dx)) + (dR \wedge (dx \wedge dy)) \\
&= \int_V \left(\frac{\partial P}{\partial x}\,dx + \frac{\partial P}{\partial y}\,dy + \frac{\partial P}{\partial z}\,dz \right) \wedge (dy \wedge dz) \\
&\quad + \left(\frac{\partial Q}{\partial x}\,dx + \frac{\partial Q}{\partial y}\,dy + \frac{\partial Q}{\partial z}\,dz \right) \wedge (dz \wedge dx) \\
&\quad + \left(\frac{\partial R}{\partial x}\,dx + \frac{\partial R}{\partial y}\,dy + \frac{\partial R}{\partial z}\,dz \right) \wedge (dx \wedge dy) \\
&= \int_V \left(\frac{\partial P}{\partial x} + \frac{\partial Q}{\partial y} + \frac{\partial R}{\partial z} \right) dx \wedge dy \wedge dz \\
&= \int_V div(\mathbf{F})\,dx\,dy\,dz. \qquad\qquad (2.75)
\end{aligned}
$$

Joining the pieces we obtain

$$
\begin{aligned}
\int_S \omega &= \int_S (\mathbf{F} \cdot \mathbf{n})\,dS \\
&= \int_S P\,dy \wedge dz + Q\,dz \wedge dx + R\,dx \wedge dy \\
&= \int_V div(\mathbf{F})\,dx\,dy\,dz, \qquad\qquad (2.76)
\end{aligned}
$$

which is the standard Gauss Divergence Theorem.

Exercises 2.15.3

1. *Find the domain of the one variable vectorial functions* $\mathbf{r}$ *below indicated.*

   (a)
   $$\mathbf{r}(t) = \frac{1}{t^2 + 1}\mathbf{i} + \sqrt{(t-1)(t+3)}\mathbf{j},$$

   (b)
   $$\mathbf{r}(t) = \ln(t^2 - 16)\mathbf{i} + \sqrt{t^2 + 2t - 15}\mathbf{j} + \tan(t+1)\mathbf{k},$$

(c)
$$\mathbf{r}(t) = \sqrt{25 - t^2}\mathbf{i} + \sqrt{t^2 + 2t - 8}\mathbf{j}.$$

2. *Through the formula*
$$\frac{dy}{dx} = \frac{dy/dt}{dx/dt},$$

calculate the derivatives of the functions defined by the parametric equations indicated,

(a)
$$\mathbf{r}(t) = \frac{e^t}{1 + e^t}\mathbf{i} + t^2 \ln(t)\mathbf{j},$$

(b)
$$\mathbf{r}(t) = \frac{\cos(t)}{5 + \sin(t)}\mathbf{i} + \ln(\sqrt{t^4 + t^2})\mathbf{j}.$$

3. *Let* $\mathbf{r} : \mathbb{R} \setminus \{-1\} \to \mathbb{R}^2$ *be defined by*
$$\mathbf{r}(t) = \frac{t}{t + 1}\mathbf{i} + \ln(t^2 + 1)\mathbf{j}.$$

Find the equation of the tangent line to the graph of the curve defined by $\mathbf{r}$ *at the point corresponding to* $t = 1$.

4. *Let* $\mathbf{r}, \mathbf{s} : \mathbb{R} \to \mathbb{R}^2$ *be defined by*
$$\mathbf{r}(t) = t\mathbf{i} + t^2\mathbf{j}$$

and
$$\mathbf{s}(t) = (t^2 + t)\mathbf{i} + t^3\mathbf{j}.$$

Calculate the angle between $\mathbf{r}'(t)$ *and* $\mathbf{s}'(t)$ *at the point corresponding to* $t = 1$.

5. *Let* $\mathbf{r} : \mathbb{R} \to \mathbb{R}^3$ *be defined by*
$$\mathbf{r}(t) = \frac{2t}{1 + t^2}\mathbf{i} + \frac{1 - t^2}{1 + t^2}\mathbf{j} + \mathbf{k}.$$

Show that the angle between $\mathbf{r}(t)$ *and* $\mathbf{r}'(t)$ *is constant.*

6. *A vectorial function* $\mathbf{r}$ *satisfies the equation,*
$$t\mathbf{r}'(t) = \mathbf{r}(t) + t\mathbf{A}, \ \forall t > 0$$

where
$$\mathbf{A} \in \mathbb{R}^3.$$

Suppose that $\mathbf{r}(1) = 2\mathbf{A}$. *Calculate* $\mathbf{r}''(1)$ *and* $\mathbf{r}(3)$ *as functions of* $\mathbf{A}$.

7. *Find a function* $\mathbf{r} : (0, +\infty) \to \mathbb{R}^3$ *such that*

$$\mathbf{r}(x) = xe^x \mathbf{A} + \frac{1}{x} \int_1^x \mathbf{r}(t)\, dt.$$

where $\mathbf{A} \in \mathbb{R}^3$, $\mathbf{A} \neq \mathbf{0}$.

8. *Using the Green Theorem, calculate the areas of the regions D, where,*

   (a)
   $$D = \{(x,y) \in \mathbb{R}^2 \ : \ x^2 + y^2 \leq 1 \text{ and } y \geq 1/2\}.$$

   (b)
   $$D = \{(x,y) \in \mathbb{R}^2 \ : \ x^2 + y^2 \leq 1 \text{ and } -1/2 \leq y \leq \sqrt{3}/2\}.$$

   (c)
   $$D = \{(x,y) \in \mathbb{R}^2 \ : \ x^2 + y^2 \leq 1 \text{ and } 0 \leq x \leq 1/2\}.$$

9. *Calculate the area of surface S, where*

$$S = \left\{ (x,y,z) \in \mathbb{R}^3 \ : \ x^2 + y^2 + z^2 = 1 \text{ and } \frac{1}{2} \leq z \leq \frac{\sqrt{3}}{2} \right\}.$$

10. *Calculate the area of surface S, where*

$$S = \left\{ (x,y,z) \in \mathbb{R}^3 \ : \ x^2 + y^2 + z^2 = 1 \text{ and } \frac{-\sqrt{3}}{2} \leq z \leq \frac{1}{2} \right\}.$$

11. *Calculate the area of surface S, where*

$$S = \left\{ (x,y,z) \in \mathbb{R}^3 \ : \ z^2 = x^2 + y^2 \text{ and } x^2 + y^2 \leq 2ax \right\},$$

   where $a \in \mathbb{R}$.

12. *Calculate* $I = \int \int_S x\, dS$, *where*

$$S = \left\{ (x,y,z) \in \mathbb{R}^3 \ : \ x^2 + y^2 = R^2 \text{ and } |z| \leq 1 \right\}.$$

13. *Using the Divergence Theorem, calculate* $I = \int \int_S (y\mathbf{j} + z\mathbf{k}) \cdot \mathbf{n}\, dS$, *where*

$$S = \left\{ (x,y,z) \in \mathbb{R}^3 \ : \ x = \sqrt{R^2 - y^2 - z^2} \text{ and } x \geq \frac{\sqrt{3}R}{2} \right\},$$

   where $R > 0$.

14. *Using the Divergence Theorem, calculate* $I = \int \int_S \mathbf{F} \cdot \mathbf{n} \, dS$ *where*

$$S = \{(x,y,z) \in \mathbb{R}^3 \ : \ x^2 + y^2 + z^2 = 2R_0 x \text{ and } z \geq 0\}$$

and where $\mathbf{F} = x^2 \mathbf{i} + y^2 \mathbf{j} + z^2 \mathbf{k}$ *e* $R_0 > 0$.

15. *Let* $u : V \to \mathbb{R}$ *be a scalar field and let* $\mathbf{F} : V \to \mathbb{R}^3$ *be a vectorial one, where* $V \subset \mathbb{R}^3$ *is open* $u, \mathbf{F}$ *are of* C^1 *class. Show that*

$$div(u\mathbf{F}) = (\nabla u) \cdot \mathbf{F} + u \, (div\mathbf{F}).$$

16. *Let* $u, v : V \to \mathbb{R}$ *be* C^2 *class scalar fields, where* $V \subset \mathbb{R}^3$ *is open and its closure is simple. Defining*

$$\nabla^2 u = \frac{\partial^2 u}{\partial x^2} + \frac{\partial^2 u}{\partial y^2} + \frac{\partial^2 u}{\partial z^2}$$

show that $div(\nabla u) = \nabla^2 u$ *and prove the Green identities,*

(a)

$$\int \int \int_V (v\nabla^2 u + \nabla v \cdot \nabla u) \, dV = \int \int_S v(\nabla u \cdot \mathbf{n}) \, dS$$

where $S = \partial V$ *(that is, S is the boundary of V.)*

(b)

$$\int \int \int_V (v\nabla^2 u - u\nabla^2 v) \, dV = \int \int_S \left(v\frac{\partial u}{\partial \mathbf{n}} - u\frac{\partial v}{\partial \mathbf{n}} \right) dS,$$

where $S = \partial V$ *and* $\frac{\partial u}{\partial \mathbf{n}} = \nabla u \cdot \mathbf{n}$.

17. *Let* $u : V \to \mathbb{R}$, $\mathbf{F} : V \to \mathbb{R}^3$ *be* C^2 *class fields on the open set* $V \subset \mathbb{R}^3$. *Prove that* $curl(\nabla u) = \mathbf{0}$ *and* $div(curl(\mathbf{F})) = 0$, *on V.*

18. *Let* $D \subset \mathbb{R}^2$ *be a simple region. Let* $u, v \in C(\overline{D}) \cap C^1(D)$. *Prove that*

$$\int \int_D uv_x \, dxdy = \int_{\partial D} uv \, dy - \int \int_D u_x v \, dxdy,$$

and

$$\int \int_D uv_y \, dxdy = \int_{\partial D} uv \, dx - \int \int_D u_y v \, dxdy,$$

19. *Let* $V \subset \mathbb{R}^3$ *be an open region bounded by a closed surface of* C^1 *class. Using the first Green identity, prove the uniqueness of solution of the Dirichlet problem,*

$$\begin{cases} \nabla^2 u = f, & in V \\ \\ u = u_0, & on \ \partial V, \end{cases} \tag{2.77}$$

where $f : V \to \mathbb{R}$ *is continuous and* $u_0 : \partial V \to \mathbb{R}$ *is also continuous. Also, using the first Green identity, prove that*

(a) *for the Neumann problem*

$$\begin{cases} \nabla^2 u = f, & in \ V \\ \\ \frac{\partial u}{\partial \mathbf{n}} = u_0, & on \ \partial V, \end{cases} \tag{2.78}$$

to have a solution, it is necessary that

$$\int\int\int_V f \, dxdydz = \int\int_{\partial V} u_0 \, dS.$$

Hint: Consider $v \equiv 1$ in the first Green identity.

(b) *Prove that any two solutions of the Neumann problem differ by a constant.*

20. *Let $V \subset \mathbb{R}^3$ be a simple region. Let $\mathbf{F} : V \to \mathbb{R}^3$ be a vectorial field of C^1 class.*

Let $\mathbf{x}_0 \in V^\circ$. Show that

$$div(\mathbf{F}(\mathbf{x}_0)) = \lim_{r \to 0} \frac{\int\int_{\partial B_r(\mathbf{x}_0)} \mathbf{F} \cdot \mathbf{n} \, dS}{Vol(B_r(\mathbf{x}_0))},$$

where $\mathbf{n}$ denotes unit outward normal field to $B_r(\mathbf{x}_0)$.

21. *Let $V \subset \mathbb{R}^3$ be a simple region. Let $f : V \to \mathbb{R}$ be a scalar field of C^2 class.*

Let $\mathbf{x}_0 \in V^\circ$. Through the first Green identity, show that

$$\nabla^2 f(\mathbf{x}_0) = \lim_{r \to 0} \frac{\int\int_{\partial B_r(\mathbf{x}_0)} \frac{\partial f}{\partial \mathbf{n}} \, dS}{Vol(B_r(\mathbf{x}_0))},$$

where $\mathbf{n}$ denotes the unit outward normal field to $B_r(\mathbf{x}_0)$.

22. *Let $M \subset \mathbb{R}^n$ be a m-dimensional surface of C^2 class, where $1 \leq m < n$.*

Let $X, Y, Z \in \tilde{\mathscr{X}}(M)$, where $\tilde{\mathscr{X}}(M)$ denotes the set of tangential vector fields of C^∞ class defined on M.

Show that

(a) $[\alpha X + \beta Y, Z] = \alpha[X, Z] + \beta[Y, Z], \ \forall \alpha, \beta \in \mathbb{R},$

(b) $[X, \alpha Y + \beta Z] = \alpha[X, Y] + \beta[X, Z], \ \forall \alpha, \beta \in \mathbb{R}.$

(c) *Anti-symmetry:*

$$[X, Y] = -[Y, X],$$

(d) *Jacob Identity:*

$$[[X, Y], Z] + [[Y, Z], X] + [[Z, X], Y] = 0.$$

(e) Leibnitz rule:

$$[fX, gY] = fg[X, Y] + f(X \cdot g)Y - g(Y \cdot f)X, \ \forall f, g \in C^2(M).$$

(f) Recalling that $L_X Y = [X, Y]$, show that

    i.

$$L_X[Y, Z] = [L_X Y, Z] + [Y, L_X Z],$$

    ii.

$$L_X(L_Y)Z - L_Y(L_X)Z = L_{[X,Y]}Z.$$

23. *Consider a 3-dimensional surface $M \subset \mathbb{R}^4$ defined by*

$$M = \{(x, y, z, w) \in \mathbb{R}^4 \ : \ x^2 + y^2 + z^2 + \ln(w) = 1\}.$$

(a) Defining $\mathbf{u} = (u_1, u_2, u_3) = (x, y, z)$, write M in the form,

$$M = \{\mathbf{r}(\mathbf{u}) \in \mathbb{R}^4 \ : \ \mathbf{u} \in \mathbb{R}^3\}.$$

(b) Let $p = \mathbf{r}(\mathbf{u})$. Obtain the tangent space and equation of the hyperplan tangent to M at p.

(c) For $f : M \to \mathbb{R}$, $X, Y \in \mathscr{X}(M)$ such that

$$f(\mathbf{r}(\mathbf{u})) = x^2 + e^{x^2 y} + (\sin((x^2 + z^2)))^3 + w(x, y, z),$$

$$X(x, y, z) = e^x \frac{\partial \mathbf{r}(x, y, z)}{\partial x} + y \frac{\partial \mathbf{r}(x, y, z)}{\partial y} + (x + z^2)^2 \frac{\partial \mathbf{r}(x, y, z)}{\partial z},$$

and

$$Y(x, y, z) = (\sin(xy))^2 \frac{\partial \mathbf{r}(x, y, z)}{\partial x} + (\cos(x^2 + y^2))^3 \frac{\partial \mathbf{r}(x, y, z)}{\partial y} + e^{x + z^2} \frac{\partial \mathbf{r}(x, y, z)}{\partial z},$$

for $p = \mathbf{r}(x_0, y_0, z_0)$, calculate

    i. $(X \cdot f)(p)$,

    ii. $(D_X Y)(p)$,

    iii. $[X, Y](p)$,

    iv. $([X, Y] \cdot f)(p)$

    v. Compute numerically the results obtained the 4 last items at the point $p_0 = \mathbf{r}(x_0, y_0, z_0) = \mathbf{r}(\pi, 0, 1)$.

(d) Let $Z \in \mathscr{X}(M)$, where

$$Z(x, y, z) = (x + y + z) \frac{\partial \mathbf{r}(x, y, z)}{\partial x} + (2x + y + z) \frac{\partial \mathbf{r}(x, y, z)}{\partial y} + (-y + z) \frac{\partial \mathbf{r}(x, y, z)}{\partial z}.$$

Obtain the integral curve $r(\mathbf{u}(t))$ of Z, such that $\mathbf{u}(0) = (1, -1, 0)$.

24. *Obtain the differential $dM(x,y,z)$ to calculate the area of the surface $M \subset \mathbb{R}^4$, where*

$$M = \{(x,y,z,w) \in \mathbb{R}^4 : e^w = [\sin(x^2+y)]^3 + z^2 + 5 \text{ and } x^2+y^2+z^2 \le 1\}.$$

25. *Consider the surface $M \subset \mathbb{R}^4$ defined by*

$$M = \{(x,y,z) \in \mathbb{R}^4 : e^w - x^2 - y^2 - z^2 = 1\}.$$

Write its equation in the form,

$$M = \{\mathbf{r}(x,y,z) : (x,y,z) \in \mathbb{R}^3\},$$

where

$$\mathbf{r}(x,y,z) = X_1(x,y,z)\mathbf{e}_1 + \cdots + X_4(x,y,z)\mathbf{e}_4.$$

Let

$$dX_1 = \frac{\partial X_1}{\partial x}\,dx + \frac{\partial X_1}{\partial y}\,dy + \frac{\partial X_1}{\partial z}\,dz,$$

and

$$dX_4 = \frac{\partial X_4}{\partial x}\,dx + \frac{\partial X_4}{\partial y}\,dy + \frac{\partial X_4}{\partial z}\,dz.$$

(a) *Calculate*

$$(dX_1 \wedge dX_4)(\mathbf{s}_1,\mathbf{s}_2),$$

where

$$\mathbf{s}_1 = \frac{\partial \mathbf{r}(x,y,z)}{\partial x}\,\Delta x,$$

and

$$\mathbf{s}_2 = \frac{\partial \mathbf{r}(x,y,z)}{\partial y}\,\Delta y,$$

and where $\Delta x, \Delta y \in \mathbb{R}$.

(b) *Consider the differential form*

$$\omega = (w(x,y,z)+x^2y+z)dX_1 \wedge dX_4 + (w(x,y,z)^2+\sin(x^2+y)-z^2)dX_1 \wedge dX_2,$$

where

$$dX_2 = \frac{\partial X_2}{\partial x}\,dx + \frac{\partial X_2}{\partial y}\,dy + \frac{\partial X_2}{\partial z}\,dz.$$

Obtain the exterior differential $d\omega$ of ω at $(\mathbf{s}_1,\mathbf{s}_2,\mathbf{s}_3)$, where

$$\mathbf{s}_1 = \frac{\partial \mathbf{r}(x,y,z)}{\partial x}\,\Delta x,$$

$$\mathbf{s}_2 = \frac{\partial \mathbf{r}(x,y,z)}{\partial y}\,\Delta y,$$

and

$$\mathbf{s}_3 = \frac{\partial \mathbf{r}(x, y, z)}{\partial z} \Delta z,$$

and where Δx, Δy, $\Delta z \in \mathbb{R}$.

26. *Let $M \subset \mathbb{R}^n$ be a 3-dimensional C^1 class surface, where $n \geq 4$,*

$$M = \{\mathbf{r}(\mathbf{u}) = X_i(\mathbf{u})\mathbf{e}_i : \mathbf{u} \in D\},$$

$D \subset \mathbb{R}^3$ and $\{\mathbf{e}_1, ..., \mathbf{e}_n\}$ is the canonical basis for $\mathbb{R}^n$,

Let $\omega = dX_1 \wedge dX_4 \wedge dX_3$ be a 3-form on M, where,

$$dX_1(\mathbf{u}) = \frac{\partial X_1(\mathbf{u})}{\partial u_1} du_1 + \frac{\partial X_1(\mathbf{u})}{\partial u_2} du_2 + \frac{\partial X_1(\mathbf{u})}{\partial u_3} du_3,$$

$$dX_4(\mathbf{u}) = \frac{\partial X_4(\mathbf{u})}{\partial u_1} du_1 + \frac{\partial X_4(\mathbf{u})}{\partial u_2} du_2 + \frac{\partial X_4(\mathbf{u})}{\partial u_3} du_3,$$

and

$$dX_3(\mathbf{u}) = \frac{\partial X_3(\mathbf{u})}{\partial u_1} du_1 + \frac{\partial X_3(\mathbf{u})}{\partial u_2} du_2 + \frac{\partial X_3(\mathbf{u})}{\partial u_3} du_3.$$

Compute

$$(dX_1(\mathbf{u}) \wedge dX_4(\mathbf{u}) \wedge dX_3(\mathbf{u}))(\mathbf{s}_1, \mathbf{s}_2, \mathbf{s}_3),$$

where

$$\mathbf{s}_1 = \frac{\partial \mathbf{r}(\mathbf{u})}{\partial u_1} \Delta u_1,$$

$$\mathbf{s}_2 = \frac{\partial \mathbf{r}(\mathbf{u})}{\partial u_2} \Delta u_2$$

and

$$\mathbf{s}_3 = \frac{\partial \mathbf{r}(\mathbf{u})}{\partial u_3} \Delta u_3.$$

27. *Consider the vectorial field $F : \mathbb{R}^3 \to \mathbb{R}^3$ where $\mathbf{F} = x^2\mathbf{i} + y^2\mathbf{j} + (z - x^2)\mathbf{k}$. Using the Stokes Theorem, calculate*

$$I = \int\int_S curl(\mathbf{F}) \cdot \mathbf{n}\, dS$$

where

$$S = \{(x, y, z) \in \mathbb{R}^3 : z = 8 - x^2 - 2y^2 \text{ and } 2 \leq z \leq 4\}.$$

28. *Consider the vectorial field* $F : \mathbb{R}^3 \to \mathbb{R}^3$ *where* $\mathbf{F} = y\mathbf{i} + y\mathbf{j} + 5\mathbf{k}$.

    Using the Stokes theorem, calculate

    $$I = \int\int_S curl(\mathbf{F}) \cdot \mathbf{n}\, dS$$

    where

    $$S = \{(x,y,z) \in \mathbb{R}^3 \: : \: z = 16 - x^2 - 3y^2 \text{ and } z \geq y^2 + 2x + y\}.$$

29. *Let* $D \subset \mathbb{R}^n$ *be an open set and let* $f : D \to \mathbb{R}$ *be a function of* C^1 *class (therefore differentiable on D).*

    Let $\mathbf{x}_0 \in D$ *and* $\varepsilon > 0$. *Prove that there exists* $\delta > 0$ *such that if* $\mathbf{x} \in D$ *and* $|\mathbf{x} - \mathbf{x}_0| < \delta$, *then*

    (a)
    $$f(\mathbf{x}) - f(\mathbf{x}_0) = f'(\mathbf{x}_0) \cdot (\mathbf{x} - \mathbf{x}_0) + r(\mathbf{x}),$$

    where
    $$|r(\mathbf{x})| \leq \varepsilon |\mathbf{x} - \mathbf{x}_0|.$$

    (b) *Use the previous item to show that if* $\mathbf{x}, \mathbf{y} \in D$, $|\mathbf{x} - \mathbf{x}_0| < \delta$ *and* $|\mathbf{y} - \mathbf{x}_0| < \delta$ *then*

    $$f(\mathbf{y}) - f(\mathbf{x}) = f'(\mathbf{x}_0) \cdot (\mathbf{y} - \mathbf{x}) + r_1(\mathbf{x},\mathbf{y}),$$

    where
    $$|r_1(\mathbf{x},\mathbf{y})| \leq 2\varepsilon\delta.$$

30. *Let* $M \subset \mathbb{R}^n$ *be a m-dimensional surface of* C^1 *class, where* $1 \leq m < n$, *where*
    $$M = \{\mathbf{r}(\mathbf{u}) \: : \: \mathbf{u} \in D \subset \mathbb{R}^m\}.$$

    Let $f \in C^2(M)$ *and* $X, Y \in \mathscr{X}(M)$.

    In this chapter, we have denoted

    $$X \cdot f = df(X) = \frac{\partial(f \circ \mathbf{r})(\mathbf{u})}{\partial u_i} X_i(\mathbf{u}).$$

    (a) *Calculate*
    $$X \cdot (Y \cdot f).$$

    (b) *Show that*
    $$X \cdot (Y \cdot f) - Y \cdot (X \cdot f) = [X,Y] \cdot f,$$

    where
    $$[X,Y] = (dY_i(X) - dX_i(Y))\frac{\partial \mathbf{r}(\mathbf{u})}{\partial u_i}.$$

(c) *Consider the 3-dimensional manifold $M \subset \mathbb{R}^4$ defined by*

$$M = \{(x, y, z, w) \in \mathbb{R}^4 \ : \ e^w - x^6 - y^2 - z^4 = 5\}.$$

   i. *Defining $\mathbf{u} = (u_1, u_2, u_3) = (x, y, z)$, write M in the form,*

$$M = \{\mathbf{r}(\mathbf{u}) \in \mathbb{R}^4 \ : \ \mathbf{u} \in \mathbb{R}^3\}.$$

  ii. *For $f : M \to \mathbb{R}$, $X, Y \in \mathscr{X}(M)$ such that*

$$f(\mathbf{r}(\mathbf{u})) = 5y^2 + w(x, y, z),$$

$$X(x, y, z) = \cos(x - y) \frac{\partial \mathbf{r}(x, y, z)}{\partial x} + y^3 \frac{\partial \mathbf{r}(x, y, z)}{\partial y} + (x + z^2)^3 \frac{\partial \mathbf{r}(x, y, z)}{\partial z},$$

and

$$Y(x, y, z) = e^x \frac{\partial \mathbf{r}(x, y, z)}{\partial x} + (\sin(x^2 + y^3))^4 \frac{\partial \mathbf{r}(x, y, z)}{\partial y} + e^{x^3 z} \frac{\partial \mathbf{r}(x, y, z)}{\partial z},$$

for $p = \mathbf{r}(x, y, z)$, calculate

$$([X, Y] \cdot f)(p)$$

31. *Consider a 3-dimensional surface $M \subset \mathbb{R}^4$ defined by*

$$M = \{(x, y, z) \in \mathbb{R}^4 \ : \ \ln(w) - x + 2y^2 - z^3 = 1\}.$$

Write the equation of M in the form,

$$M = \{\mathbf{r}(x, y, z) \ : \ (x, y, z) \in \mathbb{R}^3\},$$

where

$$\mathbf{r}(x, y, z) = X_1(x, y, z)\mathbf{e}_1 + \cdots + X_4(x, y, z)\mathbf{e}_4,$$

  (a) *Obtain $(dX_1 \wedge dX_2 \wedge dX_4)(\mathbf{s}_1, \mathbf{s}_2, \mathbf{s}_3)$.*

  (b) *Consider the differential form*

$$\omega = e^{x^2 + y^5} \, dX_2 + \sin(xy^2) \, dX_1.$$

Obtain the exterior differential $d\omega$ of ω at $(\mathbf{s}_1, \mathbf{s}_2)$, where for the last two sub-items,

$$\mathbf{s}_1 = \frac{\partial \mathbf{r}(x, y, z)}{\partial x} \Delta x,$$

$$\mathbf{s}_2 = \frac{\partial \mathbf{r}(x, y, z)}{\partial y} \Delta y,$$

and

$$\mathbf{s}_3 = \frac{\partial \mathbf{r}(x, y, z)}{\partial z} \Delta z,$$

and where Δx, Δy, $\Delta z \in \mathbb{R}$.

APPLICATIONS TO VARIATIONAL QUANTUM MECHANICS AND RELATIVITY THEORY

II

Chapter 3

A Variational Formulation for the Relativistic Klein-Gordon Equation

3.1 Introduction

In this chapter we propose a variational formulation for the Klein-Gordon relativistic equation obtained through an extension of the classical mechanics approach to a more general context, which may include, in some sense, the quantum mechanics one. The main results are developed through the introduction of the normal field definition and concerning wave function concept.

The aim of introducing the energy part related to the normal field is to minimize and control, in a specific appropriate sense to be described in the next sections, the curvature field distribution along the concerned mechanical system.

Regarding the references, this work is based on the book "A Classical Description of Variational Quantum Mechanics and Related Models" [22], published by Nova Science Publishers. In the first two sections, we present a summary of the main introductory results presented in [22]. In the final section we develop in details the main result, namely, the establishment of the Klein-Gordon relativistic equation obtained from the respective variational formulation.

At this point, we remark that details on the Sobolev Spaces involved may be found in [1, 13]. For standard references in quantum mechanics, we refer to [40, 37, 12] and the non-standard [11].

Finally, we emphasize this article is not about Bohmian mechanics, even though the David Bohm's work has been always inspiring.

3.2 The Newtonian approach

In this section, specifically for a free particle context, we shall obtain a close relationship between classical and quantum mechanics.

Let $\Omega \subset \mathbb{R}^3$ be an open, bounded and connected set with a regular (Lipschitzian) boundary denoted by $\partial\Omega$, on which we define a position field, in a free volume context, denoted by $\mathbf{r} : \Omega \times [0,T] \to \mathbb{R}^3$, where $[0,T]$ is a time interval.

Suppose also an associated density distribution scalar field is given by $(\rho \circ \mathbf{r})$: $\Omega \times [0,T] \to [0,+\infty)$, so that the kinetics energy for such a system, denoted by $J : U \times V \to \mathbb{R}$, is defined as

$$J(\mathbf{r},\rho) = \frac{1}{2} \int_0^T \int_\Omega \rho(\mathbf{r}(\mathbf{x},t)) \frac{\partial \mathbf{r}(\mathbf{x},t)}{\partial t} \cdot \frac{\partial \mathbf{r}(\mathbf{x},t)}{\partial t} \sqrt{g} \, d\mathbf{x} dt,$$

subject to

$$\int_\Omega \rho(\mathbf{r}(\mathbf{x},t)) \sqrt{g} \, d\mathbf{x} = m, \text{ on } [0,T],$$

where m is the total system mass, t denotes time and $d\mathbf{x} = dx_1 \, dx_2 \, dx_3$.

Here,

$$\begin{aligned} U &= \{\mathbf{r} \in W^{1,2}(\Omega \times [0,T]) : \mathbf{r}(\mathbf{x},0) = \mathbf{r}_0(\mathbf{x}) \\ &\quad \text{and } \mathbf{r}(\mathbf{x},T) = \mathbf{r}_1(\mathbf{x}), \text{ in } \Omega\}, \end{aligned} \tag{3.1}$$

and

$$V = \{\rho(\mathbf{r}) \in L^2([0,T]; W^{1,2}(\Omega)) : \mathbf{r} \in U\}.$$

Also

$$\mathbf{g}_k = \frac{\partial \mathbf{r}(\mathbf{x},t)}{\partial x_k},$$

$$g_{jk} = \mathbf{g}_j \cdot \mathbf{g}_k,$$

and

$$g = \det\{g_{jk}\}.$$

For such a standard Newtonian formulation, the kinetics energy takes into account just the tangential field given by the time derivative

$$\frac{\partial \mathbf{r}(\mathbf{x},t)}{\partial t}.$$

At this point, the idea is to complement such an energy with a new term which would also consider the variation of a normal field $\mathbf{n}$ and concerning distribution of curvature, such that

$$\mathbf{n} \cdot \frac{\partial \mathbf{r}(\mathbf{x},t)}{\partial t} = 0, \text{ in } \Omega \times [0,T].$$

So, with such statements in mind, we redefine the concerning energy, denoting it again by $J : U \times V \times V_1 \to \mathbb{R}$, as

$$
\begin{aligned}
J(\mathbf{r},\mathbf{n},\rho) \;=\; & -\frac{1}{2}\int_0^T \int_\Omega \rho(\mathbf{r}(\mathbf{x},t)) \frac{\partial \mathbf{r}(\mathbf{x},t)}{\partial t} \cdot \frac{\partial \mathbf{r}(\mathbf{x},t)}{\partial t} \sqrt{g}\, d\mathbf{x}dt \\
& +\frac{\gamma}{2}\int_0^T \int_\Omega \hat{R}\sqrt{g}\, d\mathbf{x}dt,
\end{aligned}
\tag{3.2}
$$

where $\gamma > 0$ is an appropriate constant,

$$\hat{R} = g^{ij}\hat{R}_{ij},$$

$$\hat{R}_{jk} = \hat{R}^i_{jik},$$

$$\hat{R}^i_{jkl} = b^l_i\, b_{jk},$$

$$b_{ij} = -\frac{1}{\sqrt{m}}\frac{\partial\left(\sqrt{\rho(\mathbf{r})}\mathbf{n}(\mathbf{r})\right)}{\partial x_j} \cdot \mathbf{g}_i,$$

$$b^i_j = g^{il}b_{lj},$$

and,

$$\{g^{ij}\} = \{g_{ij}\}^{-1},$$

$\forall i,j,k,l \in \{1,2,3\}.$
subject to

$$\mathbf{n}(\mathbf{r}) \cdot \mathbf{n}(\mathbf{r}) = 1, \text{ in } \Omega \times [0,T],$$

$$\mathbf{n}(\mathbf{r}) \cdot \frac{\partial \mathbf{r}}{\partial t} = 0, \text{ in } \Omega \times [0,T],$$

and

$$\int_\Omega \rho(\mathbf{r}(\mathbf{x},t))\sqrt{g}\, d\mathbf{x} = m, \text{ on } [0,T].$$

Here

$$V_1 = \{\mathbf{n}(\mathbf{r}) \in L^2(\Omega \times [0,T]) \; : \; \mathbf{r} \in U\}.$$

Thus defining ϕ such that

$$|\phi| = \sqrt{\frac{\rho}{m}}$$

and already including the Lagrange multipliers concerning the restrictions, the final expression for the energy, denoted by $J : U \times V \times V_1 \times V_2 \times [V_3]^2 \to \mathbb{R}$, would be given by

$$
\begin{aligned}
J(\mathbf{r}, \mathbf{n}, \phi, E, \lambda_1, \lambda_2) \;=\; & -\frac{1}{2} \int_0^T \int_\Omega m |\phi(\mathbf{r}(\mathbf{x},t))|^2 \frac{\partial \mathbf{r}(\mathbf{x},t)}{\partial t} \cdot \frac{\partial \mathbf{r}(\mathbf{x},t)}{\partial t} \sqrt{g} \, d\mathbf{x}dt \\
& + \frac{\gamma}{2} \int_0^T \int_\Omega \hat{R} \sqrt{g} \, d\mathbf{x}dt \\
& - m \int_0^T E(t) \left(\int_\Omega |\phi(\mathbf{r})|^2 \sqrt{g} \, d\mathbf{x} - 1 \right) dt \\
& + \langle \lambda_1, \mathbf{n} \cdot \mathbf{n} - 1 \rangle_{L^2} \\
& + \left\langle \lambda_2, \mathbf{n} \cdot \frac{\partial \mathbf{r}}{\partial t} \right\rangle_{L^2},
\end{aligned}
\tag{3.3}
$$

where,

$$
\begin{aligned}
U \;=\; & \{ \mathbf{r} \in W^{1,2}(\Omega \times [0,T]) \;:\; \mathbf{r}(\mathbf{x},0) = \mathbf{r}_0(\mathbf{x}) \\
& \text{and } \mathbf{r}(\mathbf{x},T) = \mathbf{r}_1(\mathbf{x}), \text{ in } \Omega \},
\end{aligned}
\tag{3.4}
$$

$$
\begin{aligned}
V &= \{ \phi(\mathbf{r}) \in L^2([0,T]; W^{1,2}(\Omega; \mathbb{C})) \;:\; \mathbf{r} \in U \}, \\
V_1 &= \{ \mathbf{n}(\mathbf{r}) \in L^2(\Omega \times [0,T]) \;:\; \mathbf{r} \in U \}, \\
V_2 &= L^2([0,T]), \\
V_3 &= L^2(\Omega \times [0,T]),
\end{aligned}
$$

and generically

$$
\langle f, h \rangle_{L^2} = \int_0^T \int_\Omega f h \sqrt{g} \, d\mathbf{x} \, dt, \forall f, h \in L^2(\Omega \times [0,T]).
$$

Moreover,

$$
\hat{R} = g^{ij} \hat{R}_{ij},
$$

$$
\hat{R}_{jk} = \hat{R}^i_{jik},
$$

$$
\hat{R}^i_{jkl} = b^l_i \, b^*_{jk},
$$

$$
b_{ij} = -\frac{\partial \left(\phi(\mathbf{r}) \mathbf{n}(\mathbf{r}) \right)}{\partial x_j} \cdot \mathbf{g}_i,
$$

$$
b^i_j = g^{il} b_{lj},
$$

$\forall i, j, k, l \in \{1,2,3\}$.

Finally, in particular for the special case in which

$$
\mathbf{r}(\mathbf{x},t) \approx \mathbf{x},
$$

so that

$$\frac{\partial \mathbf{r}(\mathbf{x},t)}{\partial t} \approx 0,$$

and

$$\mathbf{n} \cdot \frac{\partial \mathbf{r}}{\partial t} \approx 0,$$

we may set

$$\mathbf{n} = \mathbf{c},$$

where $\mathbf{c} \in \mathbb{R}^3$ is a constant such that

$$\mathbf{c} \cdot \mathbf{c} = 1,$$

and obtain

$$\mathbf{g}_k \approx \mathbf{e}_k,$$

where

$$\{\mathbf{e}_1, \mathbf{e}_2, \mathbf{e}_3\}$$

is the canonical basis of $\mathbb{R}^3$.

Therefore, in such a case,

$$\frac{\gamma}{2} \int_0^T \int_\Omega \hat{R} \sqrt{g} \, d\mathbf{x}dt \approx \frac{\gamma T}{2} \sum_{k=1}^3 \int_\Omega \frac{\partial \phi}{\partial x_k} \frac{\partial \phi^*}{\partial x_k} \, d\mathbf{x}.$$

Hence, we would also obtain

$$\begin{aligned}
J(\mathbf{r}, \mathbf{n}, \phi, E, \lambda_1, \lambda_2)/T &\approx \tilde{J}(\phi, E) \\
&= \frac{\gamma}{2} \sum_{k=1}^3 \int_\Omega \frac{\partial \phi}{\partial x_k} \frac{\partial \phi^*}{\partial x_k} \, d\mathbf{x} \\
&\quad -E\left(\int_\Omega |\phi|^2 d\mathbf{x} - 1 \right).
\end{aligned} \tag{3.5}$$

This last energy is just the standard Schrödinger one in a free particle context.

3.3 A brief note on the relativistic context, the Klein-Gordon equation

Denoting by c the speed of light and

$$d\bar{t}^2 = c^2 dt^2 - dX_1^2 - dX_2^2 - dX_3^2,$$

in a relativistic free particle context, the Hilbert variational formulation could be extended, for a motion in a pseudo Riemannian relativistic C^1 class manifold M, where locally

$$M = \{\mathbf{r}(\mathbf{u}) \; : \; \mathbf{u} \in \Omega\},$$

$$\mathbf{u} = (u_1, u_2, u_3, u_4) \in \mathbb{R}^4,$$

and
$$\mathbf{r} : \Omega \subset \mathbb{R}^4 \to \mathbb{R}^4$$
point-wise stands for,
$$\mathbf{r}(\mathbf{u}) = (ct(\mathbf{u}), X_1(\mathbf{u}), X_2(\mathbf{u}), X_3(\mathbf{u})),$$
to a functional J_1 where denoting $\rho(\mathbf{r}) = |R(\mathbf{r})|^2$, the mass differential is given by
$$dm = \frac{\rho(\mathbf{r})}{\sqrt{1 - v^2/c^2}} \sqrt{|g|}\, d\mathbf{u} = \frac{|R(\mathbf{r})|^2}{\sqrt{1 - v^2/c^2}} \sqrt{|g|}\, d\mathbf{u},$$
the semi-classical kinetics energy differential is given by
$$
\begin{aligned}
dE_c &= \frac{\partial \mathbf{r}(\mathbf{u})}{\partial t} \cdot \frac{\partial \mathbf{r}(\mathbf{u})}{\partial t}\, dm \\
&= -\left(\frac{d\bar{t}}{dt}\right)^2 dm \\
&= -(c^2 - v^2)\, dm, \qquad\qquad (3.6)
\end{aligned}
$$
so that
$$dE_c = -c^2 \left(\sqrt{1 - v^2/c^2}\right) |R(\mathbf{r})|^2 \sqrt{|g|}\, d\mathbf{u},$$
and
$$
\begin{aligned}
J_1(\mathbf{r}, R, \mathbf{n}) &= -\int_\Omega dE_c + \frac{\gamma}{2} \int_\Omega \hat{R}\sqrt{|g|}\, d\mathbf{u} \\
&= c^2 \int_\Omega |R(\mathbf{r})|^2 \sqrt{1 - v^2/c^2}\, \sqrt{|g|}\, d\mathbf{u} \\
&\quad + \frac{\gamma}{2} \int_\Omega \hat{R}\sqrt{|g|}\, d\mathbf{u}, \qquad\qquad (3.7)
\end{aligned}
$$
subject to
$$\int_\Omega |R(\mathbf{r})|^2 \sqrt{|g|}\, d\mathbf{u} = m,$$
where m is the particle mass at rest.

Moreover,
$$\mathbf{n}(\mathbf{r}) \cdot \frac{\partial \mathbf{r}}{\partial \bar{t}} = 0, \quad \text{in } \Omega,$$
where
$$
\begin{aligned}
\frac{\partial \mathbf{r}}{\partial \bar{t}} &= \frac{\partial \mathbf{r}}{\partial t} \frac{\partial t}{\partial \bar{t}} \\
&= \frac{\frac{\partial \mathbf{r}}{\partial t}}{\frac{\partial \bar{t}}{\partial t}} \\
&= \frac{\partial \mathbf{r}}{c \partial t} \frac{1}{\sqrt{1 - v^2/c^2}}, \qquad\qquad (3.8)
\end{aligned}
$$

and

$$\mathbf{n}(\mathbf{r}) \cdot \mathbf{n}(\mathbf{r}) = 1, \text{ in } \Omega.$$

Where γ is an appropriate positive constant to be specified.
Also,

$$\mathbf{g}_k = \frac{\partial \mathbf{r}(\mathbf{u})}{\partial u_k},$$

$$g = det\{g_{ij}\},$$

$$g_{ij} = \mathbf{g}_i \cdot \mathbf{g}_j,$$

where here, in this subsection, such a product is given by

$$\mathbf{y} \cdot \mathbf{z} = -y_0 z_0 + \sum_{i=1}^{3} y_i z_i, \ \forall \mathbf{y} = (y_0, y_1, y_2, y_3), \ \mathbf{z} = (z_0, z_1, z_2, z_3) \in \mathbb{R}^4,$$

$$\hat{R} = g^{ij} \hat{R}_{ij},$$

$$\hat{R}_{jk} = \hat{R}^i_{jik},$$

$$\hat{R}^i_{jkl} = b^l_i \, b^*_{jk},$$

$$b_{ij} = -\frac{1}{\sqrt{m}} \frac{\partial (R(\mathbf{r})\mathbf{n}(\mathbf{r}))}{\partial u_j} \cdot \mathbf{g}_i,$$

$$b^i_j = g^{il} b_{lj},$$

and,

$$\{g^{ij}\} = \{g_{ij}\}^{-1},$$

$\forall i, j, k, l \in \{1, 2, 3, 4\}$.
Finally,

$$v = \sqrt{\left(\frac{\partial X_1}{\partial t}\right)^2 + \left(\frac{\partial X_2}{\partial t}\right)^2 + \left(\frac{\partial X_3}{\partial t}\right)^2},$$

where,

$$\frac{\partial X_k(\mathbf{u})}{\partial t} = \frac{\partial X_k(\mathbf{u})}{\partial u_j} \frac{\partial u_j}{\partial t}$$

$$= \sum_{j=1}^{4} \frac{\frac{\partial X_k(\mathbf{u})}{\partial u_j}}{\frac{\partial t(\mathbf{u})}{\partial u_j}}, \ \forall k \in \{1, 2, 3\}. \tag{3.9}$$

Here, the Einstein sum convention holds.

Remark 3.1 The role of the variable $\mathbf{u}$ concerns the idea of establishing a relation between t, X_1, X_2 and X_3. The dimension of M may vary with the problem in question.
■

3.3.1 Obtaining the Klein-Gordon equation

Of particular interest is the case in which

$$\mathbf{u} = (t, x_1, x_2, x_3) = (t, \mathbf{x}) \in \mathbb{R}^4,$$

where $\mathbf{x} = (x_1, x_2, x_3) \in \mathbb{R}^3$.

In such a case we could have, point-wise,

$$\mathbf{r}(\mathbf{x}, t) = (ct, X_1(t, \mathbf{x}), X_2(t, \mathbf{x}), X_3(t, \mathbf{x})),$$

and

$$M = \{\mathbf{r}(\mathbf{x}, t) \ : \ (\mathbf{x}, t) \in \Omega \times [0, T]\},$$

for an appropriate $\Omega \subset \mathbb{R}^3$.

Also, denoting $d\mathbf{x} = dx_1 dx_2 dx_3$, the mass differential would be given by

$$dm = \frac{\rho(\mathbf{r})}{\sqrt{1 - v^2/c^2}} \sqrt{-g} \, d\mathbf{x} = \frac{|R(\mathbf{r})|^2}{\sqrt{1 - v^2/c^2}} \sqrt{-g} \, d\mathbf{x},$$

the semi-classical kinetics energy differential would be expressed by

$$
\begin{aligned}
dE_c &= \frac{\partial \mathbf{r}(t, \mathbf{x})}{\partial t} \cdot \frac{\partial \mathbf{r}(t, \mathbf{x})}{\partial t} \, dm \\
&= -\left(\frac{d\bar{t}}{dt}\right)^2 dm \\
&= -(c^2 - v^2) \, dm,
\end{aligned}
\tag{3.10}
$$

so that

$$dE_c = -c^2 (\sqrt{1 - v^2/c^2}) |R(\mathbf{r})|^2 \sqrt{-g} \, d\mathbf{x},$$

where

$$d\bar{t}^2 = c^2 dt^2 - dX_1(t, \mathbf{x})^2 - dX_2(t, \mathbf{x})^2 - dX_3(t, \mathbf{x})^2,$$

and

$$
\begin{aligned}
J_1(\mathbf{r}, R, \mathbf{n}) &= -\int_0^T \int_\Omega dE_c \, dt + \frac{\gamma}{2} \int_0^T \int_\Omega \hat{R} \sqrt{-g} \, d\mathbf{x} \, dt \\
&= c^2 \int_0^T \int_\Omega |R(\mathbf{r})|^2 \sqrt{1 - v^2/c^2} \, \sqrt{-g} \, d\mathbf{x} \, dt \\
&\quad + \frac{\gamma}{2} \int_0^T \int_\Omega \hat{R} \sqrt{-g} \, d\mathbf{x} \, dt,
\end{aligned}
\tag{3.11}
$$

subject to

$$R(\mathbf{r}(\mathbf{x}, 0)) = R_0(\mathbf{x})$$
$$R(\mathbf{r}(\mathbf{x}, T)) = R_1(\mathbf{x})$$

and

$$R(\mathbf{r}(\mathbf{x},t)) = 0, \text{ on } \partial\Omega \times [0,T],$$

$$\int_{\Omega} |R(\mathbf{r})|^2 \sqrt{-g}\, d\mathbf{x} = m, \text{ on } [0,T],$$

$$\mathbf{n}(\mathbf{r}) \cdot \frac{\partial \mathbf{r}}{\partial \bar{t}} = 0, \text{ in } \Omega \times [0,T],$$

where

$$
\begin{aligned}
\frac{\partial \mathbf{r}}{\partial \bar{t}} &= \frac{\partial \mathbf{r}}{\partial t} \frac{\partial t}{\partial \bar{t}} \\
&= \frac{\frac{\partial \mathbf{r}}{\partial t}}{\frac{\partial \bar{t}}{\partial t}} \\
&= \frac{\partial \mathbf{r}}{c \partial t} \frac{1}{\sqrt{1 - v^2/c^2}},
\end{aligned}
\tag{3.12}
$$

and

$$\mathbf{n}(\mathbf{r}) \cdot \mathbf{n}(\mathbf{r}) = 1, \text{ in } \Omega \times [0,T].$$

Also, we have denoted

$$x_0 = ct,$$

$$(x_0, \mathbf{x}) = (x_0, x_1, x_2, x_3),$$

$$\mathbf{g}_k = \frac{\partial \mathbf{r}(t, \mathbf{x})}{\partial x_k},$$

$$g = det\{g_{ij}\},$$

$$g_{ij} = \mathbf{g}_i \cdot \mathbf{g}_j,$$

where here again, such a product is given by

$$\mathbf{y} \cdot \mathbf{z} = -y_0 z_0 + \sum_{i=1}^{3} y_i z_i, \ \forall \mathbf{y} = (y_0, y_1, y_2, y_3), \ \mathbf{z} = (z_0, z_1, z_2, z_3) \in \mathbb{R}^4,$$

$$\hat{R} = g^{ij} \hat{R}_{ij},$$

$$\hat{R}_{jk} = \hat{R}^i_{jik},$$

$$\hat{R}^i_{jkl} = b^l_i b^*_{jk},$$

$$b_{ij} = -\frac{1}{\sqrt{m}} \frac{\partial (R(\mathbf{r})\mathbf{n}(\mathbf{r}))}{\partial x_j} \cdot \mathbf{g}_i,$$

$$b^i_j = g^{il} b_{lj},$$

and,

$$\{g^{ij}\} = \{g_{ij}\}^{-1},$$

$\forall i,j,k,l \in \{0,1,2,3\}$.

Finally, we would also have

$$v = \sqrt{\left(\frac{\partial X_1}{\partial t}\right)^2 + \left(\frac{\partial X_2}{\partial t}\right)^2 + \left(\frac{\partial X_3}{\partial t}\right)^2}.$$

In particular for the special case in which

$$\mathbf{r}(\mathbf{x},t) \approx (ct,\mathbf{x}),$$

so that

$$\frac{\partial \mathbf{r}(\mathbf{x},t)}{\partial t} \approx (c,0,0,0),$$

and

$$\mathbf{n} \cdot \frac{\partial \mathbf{r}}{\partial t} \approx 0,$$

where we have set

$$\mathbf{n} = \mathbf{c} = (0,c_1,c_2,c_3).$$

Here, $\mathbf{c} \in \mathbb{R}^4$ is a constant such that

$$\mathbf{c} \cdot \mathbf{c} = 1,$$

and thus we would obtain

$$\mathbf{g}_0 \approx (1,0,0,0),\ \mathbf{g}_1 \approx (0,1,0,0),\ \mathbf{g}_2 \approx (0,0,1,0) \text{ and } \mathbf{g}_3 \approx (0,0,0,1) \in \mathbb{R}^4.$$

Therefore, defining $\phi \in W^{1,2}(\Omega \times [0,T];\mathbb{C})$ as

$$\phi(\mathbf{x},t) = \frac{R(ct,\mathbf{x})}{\sqrt{m}},$$

we have

$$\frac{\gamma}{2}\int_0^T \int_\Omega \hat{R}\sqrt{g}\,dxdt \approx \frac{\gamma}{2}\int_0^T \int_\Omega \left(-\frac{1}{c^2}\frac{\partial \phi(\mathbf{x},t)}{\partial t}\frac{\partial \phi^*(\mathbf{x},t)}{\partial t}\right.$$
$$\left. + \sum_{k=1}^3 \frac{\partial \phi(\mathbf{x},t)}{\partial x_k}\frac{\partial \phi^*(\mathbf{x},t)}{\partial x_k}\right)dxdt, \qquad (3.13)$$

and

$$c^2\int_0^T \int_\Omega |R(\mathbf{r})|^2\sqrt{1-v^2/c^2}\,\sqrt{-g}\,d\mathbf{x}\,dt \approx mc^2\int_0^T \int_\Omega |\phi(\mathbf{x},t)|^2\,dxdt.$$

Hence, we would also obtain

$$
\begin{aligned}
J(\mathbf{r}, \mathbf{n}, \phi, E, \lambda_1, \lambda_2) \approx \frac{\gamma}{2} &\left(\int_0^T \int_\Omega -\frac{1}{c^2} \frac{\partial \phi(\mathbf{x},t)}{\partial t} \frac{\partial \phi^*(\mathbf{x},t)}{\partial t} \, d\mathbf{x} dt \right. \\
&\left. + \sum_{k=1}^3 \int_\Omega \int_0^T \frac{\partial \phi(\mathbf{x},t)}{\partial x_k} \frac{\partial \phi^*(\mathbf{x},t)}{\partial x_k} \, d\mathbf{x} dt \right) \\
&+ mc^2 \int_0^T \int_\Omega |\phi(\mathbf{x},t)|^2 \, d\mathbf{x} dt \\
&- m \int_0^T E(t) \left(\int_\Omega |\phi(\mathbf{x},t)|^2 d\mathbf{x} - 1 \right) dt. \quad (3.14)
\end{aligned}
$$

The Euler Lagrange equations for such an energy are given by

$$
\begin{aligned}
\frac{\gamma}{2} &\left(\frac{1}{c^2} \frac{\partial^2 \phi(\mathbf{x},t)}{\partial t^2} - \sum_{k=1}^3 \frac{\partial^2 \phi(\mathbf{x},t)}{\partial x_k^2} \right) \\
&+ mc^2 \phi(\mathbf{x},t) - E_1(t)\phi(\mathbf{x},t) = 0, \text{ in } \Omega, \quad (3.15)
\end{aligned}
$$

where,

$$
\phi(\mathbf{x},0) = \phi_0(\mathbf{x}), \text{ in } \Omega,
$$
$$
\phi(\mathbf{x},T) = \phi_1(\mathbf{x}), \text{ in } \Omega,
$$
$$
\phi(\mathbf{x},t) = 0, \text{ on } \partial\Omega \times [0,T]
$$

and $E_1(t) = mE(t)$.

Equation (3.15) is the relativistic Klein-Gordon one.

For $E_1(t) = E_1 \in \mathbb{R}$ (not time dependent), at this point we suggest a solution (and implicitly related time boundary conditions) $\phi(\mathbf{x},t) = e^{-\frac{iE_1 t}{\hbar}} \phi_2(\mathbf{x})$, where

$$
\phi_2(\mathbf{x}) = 0, \text{ on } \partial\Omega.
$$

Therefore, replacing this solution into equation (3.15), we would obtain

$$
\left(\frac{\gamma}{2} \left(-\frac{E_1^2}{c^2\hbar^2} \phi_2(\mathbf{x}) - \sum_{k=1}^3 \frac{\partial^2 \phi_2(\mathbf{x})}{\partial x_k^2} \right) + mc^2 \phi_2(\mathbf{x}) - E_1 \phi_2(\mathbf{x}) \right) e^{-\frac{iE_1 t}{\hbar}} = 0,
$$

in Ω.

Denoting

$$
E_2 = -\frac{\gamma E_1^2}{2c^2\hbar^2} + mc^2 - E_1,
$$

the final eigenvalue problem would stand for

$$
-\frac{\gamma}{2} \sum_{k=1}^3 \frac{\partial^2 \phi_2(\mathbf{x})}{\partial x_k^2} + E_2 \phi_2(\mathbf{x}) = 0, \text{ in } \Omega
$$

where E_1 is such that

$$\int_{\Omega} |\phi_2(\mathbf{x})|^2 \, d\mathbf{x} = 1.$$

Moreover, from (3.15), such a solution $\phi(\mathbf{x},t) = e^{-\frac{iE_1 t}{\hbar}} \phi_2(\mathbf{x})$ is also such that

$$\frac{\gamma}{2} \left(\frac{1}{c^2} \frac{\partial^2 \phi(\mathbf{x},t)}{\partial t^2} - \sum_{k=1}^{3} \frac{\partial^2 \phi(\mathbf{x},t)}{\partial x_k^2} \right)$$

$$+ mc^2 \phi(\mathbf{x},t) = i\hbar \frac{\partial \phi(\mathbf{x},t)}{\partial t}, \text{ in } \Omega. \tag{3.16}$$

At this point, we recall that in quantum mechanics

$$\gamma = \hbar^2/m.$$

Finally, we remark that this last equation (3.16) is a kind of relativistic Schrödinger-Klein-Gordon equation.

3.4 About the role of normal field as the wave function in quantum mechanics

In this section, we develop a general approach which includes both classical and quantum mechanics in a unified framework. We would emphasize that models in standard quantum mechanics and related subjects are addressed in [40, 37, 3, 12, 12]. It is also worth mentioning the non-standard approach developed in [11].

Moreover, details on the Sobolev spaces involved may be found in [1].

The first important concept to be introduced refers to the idea of a normal field to a given manifold. At this point, we state such a definition.

Definition 3.1 Normal field   Let $D \subset \mathbb{R}^m$ be an open, bounded, simply connected set with a C^1 class boundary.

Consider a C^1 class manifold M where, $1 \le m < n$, so that locally

$$M = \{ \mathbf{r}(u_1, \cdots, u_m) \; : \; (u_1, \cdots, u_m) \in D \}$$

where $\mathbf{r} : D \to \mathbb{R}^n$ is given by

$$\mathbf{r}(u_1, \cdots, u_m) = X_1(u_1, \cdots, u_m)\mathbf{e}_1 + \cdots + X_n(u_1, \cdots, u_m)\mathbf{e}_n$$

and $\{\mathbf{e}_1, \cdots, \mathbf{e}_n\}$ is the canonical basis for $\mathbb{R}^n$.

We assume

$$\left\{ \frac{\partial \mathbf{r}(u_1, \cdots, u_m)}{\partial u_k} \right\},$$

is a linearly independent set, $\forall \mathbf{u} = (u_1, \cdots, u_m) \in D$.

Hence, at a specific point, such a set is a basis for the corresponding tangent space to M.

So, we define the normal field to M, denoted by $\mathbf{N} : D \to \mathbb{R}^n$ as the one which minimizes the functional $J : W^{1,2}(D; \mathbb{R}^n) \to \mathbb{R}$, which for an appropriate positive definite tensor $\{\gamma_{jk}\}$ is given by,

$$J(\mathbf{N}) = \frac{1}{2} \int_D \gamma_{jk}(\mathbf{u}) \frac{\partial \mathbf{N}(\mathbf{u})}{\partial u_j} \cdot \frac{\partial \mathbf{N}(\mathbf{u})}{\partial u_k} \sqrt{g} \, du_1 \cdots du_m,$$

subject to

$$\int_D \mathbf{N}(\mathbf{u}) \cdot \mathbf{N}(\mathbf{u}) \sqrt{g} \, d\mathbf{u} = V(M),$$

and

$$\mathbf{N}(\mathbf{u}) \cdot \frac{\partial \mathbf{r}(\mathbf{u})}{\partial u_k} = 0, \quad \text{a.e. in } D, \ \forall k \in \{1, \cdots, m\}.$$

Where the volume $V(M)$ of M is given by

$$V(M) = \int_D \sqrt{g} \, d\mathbf{u}$$

and

$$d\mathbf{u} = du_1 \cdots du_m.$$

Finally, unless otherwise indicated, generically the inner product in question is given by

$$\mathbf{y} \cdot \mathbf{z} = \sum_{i=1}^{n} y_i z_i, \ \forall \mathbf{y} = (y_1, \cdots, y_n), \ \mathbf{z} = (z_1, \cdots, z_n) \in \mathbb{R}^n.$$

Remark 3.2 Such a definition will make possible the connection between classical and quantum mechanics. The wave function will be closely related to the normal field for a concerned motion on a manifold. ∎

Remark 3.3 In the above lines and throughout this text, the Einstein convention of sum of repeated indices holds, as the meaning is clear. ∎

3.5 A more general proposal for the energy, one more example

In this section and in the next ones, we intend to introduce the normal field to the motion/displacements directions in question as a variable of interest. In a more specific context to be defined in the subsequent sections, the normal field will be closely related to the wave function of quantum mechanics.

So, the idea here is to define a initial path which will lead us to a connection between classical and quantum mechanics in an appropriate sense.

Consider first a Euler beam model, for which we define the problem of finding a critical point for the functional $J : U \times B \to \mathbb{R}$ given by

$$J(w, \rho) = \frac{1}{2} \int_0^T \int_\Omega E(\rho(x)) I(x) w_{xx}(x,t)^2 \, dx \, dt$$

$$-\frac{1}{2} \int_0^T \int_\Omega \rho(x) w_t(x,t)^2 \, dx \, dt$$

$$-\int_0^T \int_\Omega f(x,t) w(x,t) \, dx \, dt, \qquad (3.17)$$

subject to

$$\int_\Omega \rho(x) \, dx = m.$$

Here, $\rho \in B$ denotes the beam density, where

$$B = \{\rho \in W^{1,2}(\Omega) : \rho_{min} \le \rho(x) \le \rho_{max}, \text{ in } \Omega\}.$$

Also $\Omega = [0, l]$ denotes the beam straight axis, $[0, T]$ is a time interval,

$$w \in U = \{w \in W^{2,2}(\Omega \times [0,T]) :$$
$$w(0,t) = w(l,t) = w_x(0,t) = w_x(l,t) = 0, \text{ on } [0,T],$$
$$\text{and } w(x,0) = w_0(x), \ w(x,T) = w_1(x), \text{ in } \Omega\} \qquad (3.18)$$

stands for the field of vertical displacements, $E(\rho)$ is the generalized Young modulus which varies linearly with ρ, m the total beam mass, I the cross section moment of inertia, and $f \in L^2(\Omega \times [0,T])$ denotes an external vertical load applied along the beam.

We consider an extension for this conventional energy functional, by including a new term concerning the variation of the normal direction $\mathbf{n}$ to the field of displacements. So, $\mathbf{n}$ would be such that

$$\mathbf{n}(x,t) \cdot (1, w_x(x,t)) = 0, \text{ in } \Omega \times [0,T].$$

For an appropriate constant $\gamma > 0$, we would denote such a new term by E_q, where

$$E_q = \frac{\gamma}{2} \int_0^T \int_\Omega \frac{\partial[\sqrt{\rho(x)}\mathbf{n}(x,t)]}{\partial x} \cdot \frac{\partial[\sqrt{\rho(x)}\mathbf{n}(x,t)]}{\partial x} \, dx \, dt.$$

This last term, in some sense, overlaps with the elastic energy one, since both are acting to minimize the beam distribution of curvature.

Thus, denoting for convenience $\rho(x) = R(x)^2$, the final optimization problem would be represented by a not relabeled functional $J : U \times V \times V_1 \to \mathbb{R}$, given by

$$J(w, R, \mathbf{n}) = \frac{1}{2} \int_0^T \int_\Omega E(R(x)^2) I(x) w_{xx}(x,t)^2 \, dx \, dt$$

$$-\frac{1}{2} \int_0^T \int_\Omega R(x)^2 w_t(x,t)^2 \, dx \, dt$$

$$+\frac{\gamma}{2} \int_0^T \int_\Omega \frac{\partial[R(x)\mathbf{n}(x,t)]}{\partial x} \cdot \frac{\partial[R(x)\mathbf{n}(x,t)]}{\partial x} \, dx \, dt$$

$$-\int_0^T \int_\Omega f(x,t) w(x,t) \, dx \, dt, \qquad (3.19)$$

subject to

$$\int_{\Omega} R(x)^2 \, dx = m, \text{ on } [0,T],$$

$$\mathbf{n}(x,t) \cdot (1, w_x(x,t)) = 0, \text{ in } \Omega \times [0,T],$$

$$\mathbf{n}(x,t) \cdot \mathbf{n}(x,t) = 1, \text{ in } \Omega \times [0,T],$$

and,

$$\rho_{min} \leq R(x)^2 \leq \rho_{max}, \text{ in } \Omega.$$

Finally,

$$V = \{R \in L^2([0,T]; W^{1,2}(\Omega)) \ : \ R(0) = R_0, \ R(l) = R_1\},$$

and

$$
\begin{aligned}
V_1 \ = \ & \{\mathbf{n} \in L^2([0,T]; W^{1,2}(\Omega; \mathbb{R}^2)) \ : \\
& \mathbf{n}(0,t) = -\mathbf{j}, \ \mathbf{n}(l,t) = -\mathbf{j}, \text{ on } [0,T]\},
\end{aligned}
$$

where $\mathbf{j} = (0,1) \in \mathbb{R}^2$.

Hence, we would be looking for critical points of the extended functional $J_\lambda : U \times V \times V_1 \times V_2 \times [V_3]^2 \times [V_4]^2 \to \mathbb{R}$, where

$$
\begin{aligned}
& J_\lambda(w, R, \mathbf{n}, E, \lambda_1, \lambda_2, \lambda_3, \lambda_4) \\
= \ & \frac{1}{2} \int_0^T \int_\Omega E(R(x)^2) I(x) w_{xx}(x,t)^2 \, dx \, dt \\
& - \frac{1}{2} \int_0^T \int_\Omega R(x)^2 w_t(x,t)^2 \, dx \, dt \\
& + \frac{\gamma}{2} \int_0^T \int_\Omega \frac{\partial [R(x)\mathbf{n}(x,t)]}{\partial x} \cdot \frac{\partial [R(x)\mathbf{n}(x,t)]}{\partial x} \, dx \, dt \\
& - \int_0^T \int_\Omega f(x,t) w(x,t) \, dx \, dt \\
& - \frac{1}{2} ET \left(\int_\Omega R(x)^2 \, dx - m \right) \\
& + \int_0^T \int_\Omega \lambda_1(x,t) \mathbf{n}(x,t) \cdot (1, w_x(x,t)) \, dx \, dt \\
& + \frac{1}{2} \int_0^T \int_\Omega \lambda_2(x,t) (\mathbf{n}(x,t) \cdot \mathbf{n}(x,t) - 1) \, dx \, dt \\
& + \int_\Omega \lambda_3(x)^2 (-R(x)^2 + \rho_{min}) \, dx T \\
& + \int_\Omega \lambda_4(x)^2 (R(x)^2 - \rho_{max}) \, dx \, T, \qquad (3.20)
\end{aligned}
$$

where $V_2 = \mathbb{R}$, $V_3 = L^2(\Omega \times [0,T])$, $V_4 = L^4(\Omega)$ and $E, \lambda_1, \lambda_2, \lambda_3, \lambda_4$ are appropriate Lagrange multipliers concerning the constraints.

So, the corresponding Euler Lagrange equations would be given by,

1. variation in w,

$$(E(R(x)^2)I(x)w_{xx}(x,t))_{xx} + R(x)^2 w_{tt}(x,t)$$
$$-f(x,t) - (\lambda_1(x,t)\mathbf{n}_2(x,t))_x = 0, \text{ in } \Omega \times [0,T], \qquad (3.21)$$

where we have denoted

$$\mathbf{n}(x,t) = (\mathbf{n}_1(x,t), \mathbf{n}_2(x,t)), \text{ in } \Omega \times [0,T].$$

2. Variation in R,

$$\frac{1}{2}\frac{\partial E(R(x)^2)}{\partial R}I(x)\int_0^T w_{xx}(x,t)^2 \, dt - R(x)\int_0^T w_t(x,t)^2 \, dt$$
$$-\gamma\int_0^T \frac{\partial^2(R(x)\mathbf{n}(x,t))}{\partial x^2}\cdot\mathbf{n}(x,t) \, dt$$
$$-ETR(x) - 2\lambda_3(x)^2 TR(x) + 2\lambda_4(x)^2 TR(x) = 0, \text{ in } \Omega \times [0,T]. \quad (3.22)$$

3. Variation in $\mathbf{n}$,

$$-\gamma R(x)\frac{\partial^2(R(x)\mathbf{n}(x,t))}{\partial x^2}$$
$$+\lambda_1(x,t)(1,w_x(x,t)) + \lambda_2(x,t)\mathbf{n}(x,t) = 0, \text{ in } \Omega \times [0,T]. \quad (3.23)$$

4. Variation in E,

$$\int_\Omega R(x)^2 \, dx = m, \text{ on } [0,T].$$

5. Variation in λ_1,

$$\mathbf{n}(x,t)\cdot(1,w_x(x,t)) = 0, \text{ in } \Omega \times [0,T].$$

6. Variation in λ_2,

$$\mathbf{n}(x)\cdot\mathbf{n}(x,t) = 1, \text{ in } \Omega.$$

7. Variation in λ_3

$$\lambda_3(x)(-R(x)^2 + \rho_{min}) = 0, \text{ in } \Omega.$$

8. Variation in λ_4,

$$\lambda_4(x)(R(x)^2 - \rho_{max}) = 0, \text{ in } \Omega.$$

Remark 3.4   The only difference between equation (3.21) and the standard elasticity model is the final term

$$-(\lambda_1(x,t)\mathbf{n}_2(x,t))_x.$$

However, considering that

$$\mathbf{n}(x,t) \cdot (1, w_x(x,t)) = 0, \text{ in } \Omega \times [0,T],$$

from equation (3.23) we obtain

$$\lambda_1 \approx \mathscr{O}(\gamma).$$

So, for the case $\gamma \ll m$, we recover the classical mechanics approach in equation (3.21).

On the other hand, since

$$\mathbf{n} \cdot \mathbf{n} = 1, \text{ in } \Omega \times [0,T],$$

we obtain

$$\frac{\partial \mathbf{n}}{\partial x} \cdot \mathbf{n} = 0, \text{ in } \Omega \times [0,T],$$

so that

$$\int_0^T \int_\Omega \frac{\partial R(x)\mathbf{n}(x,t)}{\partial x} \cdot \frac{\partial R(x)\mathbf{n}(x,t)}{\partial x} \, dx \, dt$$

$$= \int_0^T \int_\Omega \frac{\partial R(x)}{\partial x} \cdot \frac{\partial R(x)}{\partial x} \, dx \, dt$$

$$+ \int_0^T \int_\Omega R(x)^2 \frac{\partial \mathbf{n}(x,t)}{\partial x} \cdot \frac{\partial \mathbf{n}(x,t)}{\partial x} \, dx \, dt. \qquad (3.24)$$

Thus, for the case in which $\|f\|_\infty$ is small, neglecting the classical part in equation (3.22) and for

$$\rho_{min} < R(x)^2 < \rho_{max}, \text{ in } \Omega,$$

we would obtain,

$$-\gamma \frac{\partial^2 R(x)}{\partial x^2} + \frac{\gamma}{T} R(x) \int_0^T \frac{\partial \mathbf{n}(x,t)}{\partial x} \cdot \frac{\partial \mathbf{n}(x,t)}{\partial x} \, dt$$

$$\approx E R(x), \text{ in } \Omega. \qquad (3.25)$$

This last equation is similar to the standard Schrödinger one of quantum mechanics, with R playing the wave function role. ∎

3.6 Another example, generalizing the standard quantum approach

Consider an electronic field for which the local position is defined mathematically by a function $\mathbf{r} : \Omega \times [0,T] \to \mathbb{R}^3$, with associated mass/charge density $\rho \circ \mathbf{r} : \Omega \times [0,T] \to \mathbb{R}$, where we shall denote, as above indicated,

$$\rho(\mathbf{r}) = |R(\mathbf{r})|^2,$$

and where, from now on, we assume $R(\mathbf{r})$ to be complex.

Here $\Omega \subset \mathbb{R}^3$ is a open, bounded, connected set with a Lipschitzian boundary, eventually we could have $\Omega = \mathbb{R}^3$.

Let us define here,

$$K = \frac{1}{4\pi\varepsilon_0}$$

and

$$K_1 = \left| \frac{\text{charge}}{\text{mass}} \right|$$

for an electron at rest.

Our proposal, in some sense, generalizes the approach in the Bohm's book (see [12] for details) of quantum mechanics.

The idea is to add a new energy term relating the variation of the normal field aiming to minimize the distribution of curvature for the motion in question.

So, let us examine the different parts of energy and constraints relating such a proposal.

1. Kinetics energy E_C (classical),

$$E_C = \frac{1}{2} \int_0^T \int_\Omega |R(\mathbf{r})|^2 \frac{\partial \mathbf{r}}{\partial t} \cdot \frac{\partial \mathbf{r}}{\partial t} \sqrt{g} \, dx \, dt,$$

where $x = (x_1, x_2, x_3) \in \mathbb{R}^3$ and $dx = dx_1 \, dx_2 \, dx_3$,

$$\mathbf{g}_k = \frac{\partial \mathbf{r}(x,t)}{\partial x_k},$$

$$g_{ij} = \mathbf{g}_i \cdot \mathbf{g}_j, \ \forall i, j, k \in \{1, 2, 3\},$$

and

$$g = \det\{g_{ij}\}.$$

2. Energy relating the presence of an electric potential V, E_p (classical),

$$E_p = \frac{1}{2} \int_0^T \int_\Omega K_1 V(\mathbf{r}) |R(\mathbf{r})|^2 \sqrt{g} \, dx \, dt.$$

3. Self-interacting Coulomb energy, E_{in} (classical),

$$
\begin{aligned}
&E_{in} \\
&= \int_0^T \int_\Omega \int_\Omega \frac{K K_1^2 |R(\mathbf{r}(x,t))|^2 |R(\mathbf{r}(\tilde{x},t))|^2}{|\mathbf{r}(x,t) - \mathbf{r}(\tilde{x},t)|} \\
&\quad \times \sqrt{g(\mathbf{r}(x,t))} \, dx \, \sqrt{g(\mathbf{r}(\tilde{x},t))} \, d\tilde{x} \, dt,
\end{aligned}
$$

4. Energy associated with the normal field relating the motion, E_q (quantum part),

$$E_q = \frac{\gamma}{2} \int_0^T \int_\Omega \frac{\partial [R(\mathbf{r})\mathbf{n}(\mathbf{r})]}{\partial x_k} \cdot \frac{\partial [R(\mathbf{r})^*\mathbf{n}(\mathbf{r})]}{\partial x_k} \sqrt{g} \, dx \, dt, \qquad (3.26)$$

Generically, for $a, b \in \mathbb{R}$ and $z = a + i_m b \in \mathbb{C}$ we have denoted $z^* = a - i_m b$, and where i_m denotes the imaginary unit.

5. Constraints: The system will be subject to following constraints:

   (a) $\int_\Omega |R(\mathbf{r})|^2 \sqrt{g}\, dx = m_e$, on $[0,T]$,

   (b) $\mathbf{n}(\mathbf{r}) \cdot \frac{\partial \mathbf{r}}{\partial t} = 0$, in $\Omega \times [0,T]$,

   (c) $\mathbf{n}(\mathbf{r}) \cdot \mathbf{n}(\mathbf{r}) = 1$, in $\Omega \times [0,T]$.

Hence, we will be looking for critical points of the functional $J_\lambda : \tilde{U} \to \mathbb{R}$, where $\tilde{U} = U \times U_1 \times U_2 \times V_1 \times V_2 \times V_2$,

$$
\begin{aligned}
U \;=\; & \{\mathbf{r} \in W^{1,2}(\Omega \times [0,T]; \mathbb{R}^3) \,:\, \mathbf{r}(x,t) = \mathbf{r}_0(x,t), \text{ in } \partial\Omega \times [0,T], \\
& \mathbf{r}(x,0) = \mathbf{r}_1(x),\ \mathbf{r}(x,T) = \mathbf{r}_2(x),\ \text{ in } \Omega\},
\end{aligned}
\tag{3.27}
$$

$$
\begin{aligned}
U_1 \;=\; & \{R(\mathbf{r}) \in L^2([0,T]; W^{1,2}(\Omega; \mathbb{C})) \,:\, \mathbf{r} \in U, \\
& R(\mathbf{r}(x,t)) = R_0(x), \text{ on } \partial\Omega \times [0,T]\},
\end{aligned}
\tag{3.28}
$$

$$
U_2 = \{\mathbf{n}(\mathbf{r}) \in L^2([0,T]; W^{1,2}(\Omega; \mathbb{R}^3)) \,:\, \mathbf{r} \in U\},
$$

$$
V_1 = L^2([0,T]),
$$

$$
V_2 = L^2(\Omega \times [0,T]),
$$

and,

$$
\begin{aligned}
& J_\lambda(\mathbf{r},R,\mathbf{n},E,\lambda_1,\lambda_2) \\
=\; & -\frac{1}{2}\int_0^T \int_\Omega R(\mathbf{r})^2 \frac{\partial \mathbf{r}}{\partial t} \cdot \frac{\partial \mathbf{r}}{\partial t} \sqrt{g}\, dx\, dt + \frac{1}{2}\int_0^T \int_\Omega K_1 V(\mathbf{r}) |R(\mathbf{r})|^2 \sqrt{g}\, dx\, dt \\
& + \int_0^T \int_\Omega \int_\Omega \frac{K K_1^2 |R(\mathbf{r}(x,t))|^2 |R(\mathbf{r}(\tilde{x},t))|^2}{|\mathbf{r}(x,t) - \mathbf{r}(\tilde{x},t)|} \sqrt{g(\mathbf{r}(x,t))}\, dx\, \sqrt{g(\mathbf{r}(\tilde{x},t))}\, d\tilde{x}\, dt \\
& \frac{\gamma}{2}\int_0^T \int_\Omega \frac{\partial [R(\mathbf{r})\mathbf{n}(\mathbf{r})]}{\partial x_k} \cdot \frac{\partial [R(\mathbf{r})^*\mathbf{n}(\mathbf{r})]}{\partial x_k} \sqrt{g}\, dx\, dt \\
& -\frac{1}{2}\int_0^T E(t)\left(\int_\Omega R(\mathbf{r})^2 \sqrt{g}\, dx - m_e\right) dt + \left\langle \lambda_1, \mathbf{n}(\mathbf{r}) \cdot \frac{\partial \mathbf{r}}{\partial t}\right\rangle \\
& +\frac{1}{2}\langle \lambda_2, \mathbf{n}(\mathbf{r}) \cdot \mathbf{n}(\mathbf{r}) - 1\rangle
\end{aligned}
\tag{3.29}
$$

where generically again

$$
\langle f, h\rangle = \int_0^T \int_\Omega f h \sqrt{g}\, dx\, dt,
$$

m_e denotes the system mass and E, λ_1, λ_2 are appropriate Lagrange multipliers for the concerning restrictions.

3.7 Recovering the standard quantum approach

Considering the expressions in (3.29), similarly as in the last section, considering that

$$
\mathbf{n} \cdot \mathbf{n} = 1, \text{ in } \Omega \times [0,T],
$$

we obtain

$$\frac{\partial \mathbf{n}}{\partial x_k} \cdot \mathbf{n} = 0, \text{ in } \Omega \times [0,T], \ \forall k \in \{1,2,3\},$$

so that

$$
\begin{aligned}
E_q &= \frac{\gamma}{2} \int_0^T \int_\Omega \frac{\partial [R(\mathbf{r})\mathbf{n}(\mathbf{r})]}{\partial x_k} \cdot \frac{\partial [R(\mathbf{r})^*\mathbf{n}(\mathbf{r})]}{\partial x_k} \sqrt{g} \, dx \, dt \\
&= \frac{\gamma}{2} \int_0^T \int_\Omega \frac{\partial R(\mathbf{r})}{\partial x_k} \cdot \frac{\partial R(\mathbf{r})^*}{\partial x_k} \sqrt{g} \, dx \, dt \\
&\quad + \frac{\gamma}{2} \int_0^T \int_\Omega |R(x)|^2 \frac{\partial \mathbf{n}(\mathbf{r})}{\partial x_k} \cdot \frac{\partial \mathbf{n}(\mathbf{r})}{\partial x_k} \sqrt{g} \, dx \, dt
\end{aligned}
\tag{3.30}
$$

Also, for the interacting energy, we have

$$
\begin{aligned}
E_{in} &= \int_\Omega \int_\Omega K K_1^2 \frac{|R(\mathbf{r}(x))|^2 |R(\mathbf{r}(\tilde{x}))|^2}{|\mathbf{r}(x) - \mathbf{r}(\tilde{x})|} \sqrt{g(\mathbf{r}(x))} \, dx \, \sqrt{g(\mathbf{r}(\tilde{x}))} \, d\tilde{x} \\
&= \int_\Omega \int_\Omega K K_1^2 \frac{|R(\mathbf{r})|^2 |R(\tilde{\mathbf{r}})|^2}{|\mathbf{r} - \tilde{\mathbf{r}}|} \, d\mathbf{r} \, d\tilde{\mathbf{r}}.
\end{aligned}
\tag{3.31}
$$

Moreover,

$$
\begin{aligned}
E_p &= \frac{1}{2} \int_0^T \int_\Omega K_1 V(\mathbf{r}) |R(\mathbf{r})|^2 \sqrt{g} \, dx \, dt \\
&= \frac{1}{2} \int_0^T \int_\Omega K_1 V(\mathbf{r}) |R(\mathbf{r})|^2 \, d\mathbf{r} \, dt.
\end{aligned}
\tag{3.32}
$$

From the results above and from (3.29) we would obtain the following energy expression

$$
\begin{aligned}
J_\lambda(\mathbf{r}, R, \mathbf{n}, E, \lambda_1, \lambda_2) &\approx J(R, \mathbf{r}, E) \\
&= E_q + E_{in} + E_p - \frac{1}{2} E \left(\int_\Omega |R(\mathbf{r})|^2 \, d\mathbf{r} - m_e \right) \\
&= \frac{\gamma}{2} \int_0^T \int_\Omega \frac{\partial R(\mathbf{r})}{\partial x_k} \frac{\partial R(\mathbf{r})^*}{\partial x_k} \sqrt{g} \, dx \, dt \\
&\quad + \frac{\gamma}{2} \int_0^T \int_\Omega |R(x)|^2 \frac{\partial \mathbf{n}(\mathbf{r})}{\partial x_k} \cdot \frac{\partial \mathbf{n}(\mathbf{r})}{\partial x_k} \sqrt{g} \, dx \, dt \\
&\quad + \int_0^T \int_\Omega \int_\Omega K K_1^2 \frac{|R(\mathbf{r})|^2 |R(\tilde{\mathbf{r}})|^2}{|\mathbf{r} - \tilde{\mathbf{r}}|} \, d\mathbf{r} \, d\tilde{\mathbf{r}} \, dt \\
&\quad + \frac{1}{2} \int_0^T \int_\Omega K_1 V(\mathbf{r}) |R(\mathbf{r})|^2 \, d\mathbf{r} \, dt \\
&\quad - \frac{1}{2} \int_0^T E(t) \left(\int_\Omega |R(\mathbf{r})|^2 \, d\mathbf{r} - m_e \right) dt
\end{aligned}
\tag{3.33}
$$

This last functional is just an analogous to the standard Schrödinger one with interaction terms.

3.7.1 About the role of the angular momentum in quantum mechanics

Neglecting the self-interaction term in (3.33), for the case in which

$$\mathbf{r}(x) \approx x,$$

$$\{g_{ij}\} \approx \{\delta_{ij}\}$$

and denoting

$$\phi(x) = R(\mathbf{r}(x))$$

we would obtain $J : W_0^{1,2}(\Omega; \mathbb{C}) \times \mathbb{R} \to \mathbb{R}$ given by

$$
\begin{aligned}
J(\phi, E) \quad \approx \quad & \frac{\gamma}{2} \int_\Omega \nabla \phi \cdot \nabla \phi^* \, dx \\
& + \frac{1}{2} \int_\Omega K_1 V(x) |\phi(x)|^2 \, dx \\
& - \frac{1}{2} E \left(\int_\Omega |\phi|^2 \, dx - m_e \right).
\end{aligned}
\tag{3.34}
$$

Remark 3.5 At this point, we remark that the case in which

$$V(x) = \frac{K_2}{|x|},$$

for $\gamma = \hbar^2 / m_e^2$ and for an appropriate constant K_2 refers to the standard variational formulation for an electron in a hydrogen atom. ■

The corresponding Euler-Lagrange equations for (3.34) for the specific case $\gamma = \frac{\hbar^2}{m_e^2}$ are given by

$$\nabla^2 \phi + \frac{m_e^2}{\hbar^2} (E - V) \phi = 0,
\tag{3.35}$$

and

$$\int_\Omega |\phi|^2 \, dx = m_e.$$

In spherical coordinates (r, θ, φ), the Laplace operator may be expressed by,

$$\nabla^2 \phi \quad = \quad \frac{1}{r^2} \frac{\partial}{\partial r} \left[r^2 \frac{\partial \phi}{\partial r} \right] + \frac{1}{r^2} \left(\frac{1}{\sin \theta} \frac{\partial}{\partial \theta} \left[\sin(\theta) \frac{\partial \phi}{\partial \theta} \right] + \frac{1}{\sin^2 \theta} \frac{\partial^2 \phi}{\partial \varphi^2} \right). \tag{3.36}$$

Assuming, for a not relabeled V, $V(x) \approx V(r)$, through the standard separation of variables procedure, we shall look for a solution of the form:

$$\phi(r, \theta, \varphi) = f(r) g(\theta, \varphi).$$

Replacing such a solution in (3.35) we obtain

$$\left(\frac{d^2 f(r)}{dr^2} + \frac{2}{r}\frac{df(r)}{dr} + \frac{m_e^2}{\hbar^2}[E - V(r)]f(r)\right)g(\theta,\varphi) + \frac{f(r)}{r^2}\tilde{L}g(\theta,\varphi) = 0 \quad (3.37)$$

Dividing the last equation by $f(r)g(\theta,\varphi)$, we have that ϕ will be a solution of (3.35) if f and g are such that

$$\frac{d^2 f(r)}{dr^2} + \frac{2}{r}\frac{df(r)}{dr} + \frac{m_e^2}{\hbar^2}[E - V(r)]f(r) = -c\frac{f(r)}{r^2},$$

and

$$\tilde{L}g(\theta,\varphi) = cg(\theta,\varphi),$$

$\forall c \in \mathbb{R}$.

Defining

$$L_z = \frac{\hbar}{i_m}\left(x\frac{\partial}{\partial y} - y\frac{\partial}{\partial x}\right),$$

$$L_x = \frac{\hbar}{i_m}\left(y\frac{\partial}{\partial z} - z\frac{\partial}{\partial y}\right),$$

$$L_y = \frac{\hbar}{i_m}\left(z\frac{\partial}{\partial x} - x\frac{\partial}{\partial z}\right),$$

we have that

$$[L_y, L_z] = i_m\hbar L_x,$$

$$[L_z, L_x] = i_m\hbar L_y,$$

$$[L_x, L_y] = i_m\hbar L_z,$$

where generically,

$$[A, B] = AB - BA.$$

Defining the angular momentum L by

$$L = (L_x, L_y, L_z),$$

we may obtain

$$L^2 = L_x^2 + L_y^2 + L_z^2,$$

where, in spherical coordinates,

$$L_z = \frac{\hbar}{i_m}\frac{\partial}{\partial\varphi},$$

$$L_x = -\frac{\hbar}{i_m}\left(\sin\varphi\frac{\partial}{\partial\theta} + \cot\theta\cos\varphi\frac{\partial}{\partial\varphi}\right),$$

$$L_y = \frac{\hbar}{i_m} \left(\cos\varphi \frac{\partial}{\partial\theta} - \cot\theta \sin\varphi \frac{\partial}{\partial\varphi} \right),$$

so that

$$L^2 = -\hbar^2 \tilde{L},$$

that is,

$$\tilde{L} = -\frac{L^2}{\hbar^2}.$$

Observe that, in the analogous classical mechanics system, we have

$$H = \frac{1}{2m_e} \left(p_r^2 + \frac{\hat{L}^2}{r^2} \right) + V(r),$$

where

$$\hat{L} = \mathbf{r} \times \mathbf{p}$$

and

$$\mathbf{p} = m_e \frac{\partial \mathbf{r}}{\partial t},$$

so that

$$\hat{L}_z = xp_y - yp_x,$$
$$\hat{L}_y = zp_x - xp_z,$$
$$\hat{L}_x = yp_z - zp_y.$$

Therefore, the analogous in quantum mechanics to $\hat{L}_x, \hat{L}_y, \hat{L}_z$ are L_x, L_y, L_z as defined above, so that the classical Hamiltonian,

$$H = \frac{1}{2m_e} \left(p_r^2 + \frac{\hat{L}^2}{r^2} \right) + V(r),$$

translates in a quantum mechanics sense, with a corresponding eigenvalue, into

$$H(\phi) = \left[-\frac{\hbar^2}{2m_e} \left(\frac{\partial^2}{\partial r^2} + \frac{2}{r}\frac{\partial}{\partial r} \right) + \frac{L^2}{2m_e r^2} + V \right] \phi = E\phi.$$

3.7.2 A brief note concerning the relation between classical rotations and the quantum angular momentum

First, we recall that the basis of the Lie algebra $so(3)$ of the rotation matrix group $SO(3)$ is given by $\{F_1, F_2, F_3\}$, where

$$F_1 = \begin{pmatrix} 0 & 0 & 0 \\ 0 & 0 & -1 \\ 0 & 1 & 0 \end{pmatrix},$$

$$F_2 = \begin{pmatrix} 0 & 0 & 1 \\ 0 & 0 & 0 \\ -1 & 0 & 0 \end{pmatrix}$$

and,

$$F_3 = \begin{pmatrix} 0 & -1 & 0 \\ 1 & 0 & 0 \\ 0 & 0 & 0 \end{pmatrix}.$$

At this point, consider a Gâteaux differentiable functional

$$J(R(\mathbf{r})) = \int_\Omega F(R(\mathbf{r}))\, d\mathbf{x},$$

where we have denoted,

$$d\mathbf{x} = dx\, dy\, dz.$$

About the variation of J relating $\mathbf{r}$, such a variation may be divided into pure rotations and radial translations.

Specifically, we define a variation of J in $\mathbf{r}$ on the direction φ, denoted by $\delta J(R(\mathbf{r}), \varphi))$, by

$$\delta J(R(\mathbf{r}), \varphi) = \lim_{\varepsilon \to 0} \frac{J(R(\mathbf{r} + \varepsilon\varphi)) - J(R(\mathbf{r}))}{\varepsilon}.$$

So, under relatively mild hypotheses, we may obtain,

$$\delta J(R(\mathbf{r}), \varphi) = \int_\Omega \frac{\partial F(R(\mathbf{r} + \varepsilon\varphi))}{\partial R} \frac{\partial R(\mathbf{r} + \varepsilon\varphi)}{\partial \varepsilon}\Big|_{\varepsilon=0}\, d\mathbf{x}.$$

We consider the specific case in which φ is a $\pi/2$ rotation related to $\mathbf{r} = (x_1, x_2, x_3) = (x, y, z)$, about the $x_3 = z$ axis, that is

$$\varphi = F_3 \mathbf{r}^T,$$

so that

$$\varphi = \begin{pmatrix} 0 & -1 & 0 \\ 1 & 0 & 0 \\ 0 & 0 & 0 \end{pmatrix} \begin{pmatrix} x \\ y \\ z \end{pmatrix} = (-y, x, 0)^T.$$

Hence, the last part of the above variation will stand for,

$$\frac{\partial R((x, y, z) + \varepsilon(-y, x, 0))}{\partial \varepsilon}\Big|_{\varepsilon=0} = -y\frac{\partial R(\mathbf{r})}{\partial x} + x\frac{\partial R(\mathbf{r})}{\partial y}.$$

Summarizing, a classical rotation concerning $\mathbf{r}$ about z will partially correspond in the quantum functional part, up to a constant, to an operator called the angular momentum in z denoted by L_z, which is specifically defined by

$$L_z(R(\mathbf{r})) = \frac{\hbar}{i_m}\left(-y\frac{\partial R(\mathbf{r})}{\partial x} + x\frac{\partial R(\mathbf{r})}{\partial y}\right).$$

With similar computations related to the matrices F_1 and F_2, we may also obtain the angular moments L_x and L_y, where,

$$L_x = \frac{h}{i_m}\left(y\frac{\partial}{\partial z} - z\frac{\partial}{\partial y}\right),$$

$$L_y = \frac{\hbar}{i_m}\left(z\frac{\partial}{\partial x} - x\frac{\partial}{\partial z}\right).$$

3.7.3 The eigenvalues of L_z

As indicated above, we have:

$$L_z\phi = \frac{\hbar}{i_m}\frac{\partial\phi}{\partial\varphi},$$

so that if $c \in \mathbb{R}$ is an eigenvalue of L_z we would have:

$$L_z\phi = c\phi.$$

Thus, the corresponding eigenvector would have the form:

$$\phi = e^{\frac{i_m c\varphi}{\hbar}}\hat{f}(r,\theta).$$

Let $l \in \mathbb{Z}^+$, and define $c = \hbar l$.
Also define

$$\phi_a = f(r)[\sin\theta]^l e^{i_m l\varphi}.$$

It may be verified that,

$$(L_x + i_m L_y)\phi_a = 0.$$

Now denote

$$L^+ = \frac{1}{\hbar}(L_x + i_m L_y),$$

and

$$L^- = \frac{1}{\hbar}(L_x - i_m L_y).$$

Denote also,

$$\phi_0 = e^{i_m l\varphi}[\sin\theta]^l.$$

From the above

$$L_z\phi_0 = \hbar l\phi_0.$$

One may prove that (see [37], for details),

$$[L_z, L^-] = -\hbar L^-,$$

so that

$$L^- L_z = L_z L^- + \hbar L^-.$$

Hence, from

$$L_z \phi_0 = \hbar l \phi_0,$$

we obtain

$$L^-(L_z \phi_0) = l\hbar(L^- \phi_0),$$

that is,

$$L_z(L^- \phi_0) + \hbar(L^- \phi_0) = \hbar l L^- \phi_0,$$

so that

$$L_z[L^- \phi_0] = \hbar(l-1)L^- \phi_0.$$

Inductively, we may obtain

$$L_z[(L^-)^j \phi_0] = \hbar(l-j)(L^-)^j \phi_0.$$

It may be proven by induction that

$$(L^-)^{2l+1}(\phi_0) = 0.$$

Hence,

$$v_j \equiv (L^-)^j \phi_0 \neq 0$$

are eigenvectors of $L_z \ \forall j \in \{0, ..., 2l\}$.

Moreover, since $(L_x + i_m L_y)\phi_0 = 0$, we may obtain

$$
\begin{aligned}
L^2 \phi_0 &= (L_x^2 + L_y^2 + L_z^2)\phi_0 \\
&= [(L_x - i_m L_y)(L_x + i_m L_y) - i_m(L_x L_y - L_y L_x) + L_z^2]\phi_0 \\
&= [\hbar L_z + L_z^2]\phi_0 = [\hbar^2 l + \hbar^2 l^2]\phi_0 \\
&= \hbar^2 l(l+1)\phi_0.
\end{aligned}
\tag{3.38}
$$

Hence, ϕ_0 is an eigenfunction of L^2 with eigenvalue $\hbar^2 l(l+1)$, that is

$$L^2 \phi_0 = \hbar^2[l(l+1)]\phi_0.$$

Observe that we have obtained

$$L_z v_j = \hbar(l-j)v_j,$$

and

$$L^- v_j = v_{j+1}, \text{ if } j < 2l,$$

and

$$L^- v_j = 0 = (L^-)^{2l+1}(\phi_0), \text{ if } j = 2l.$$

We may also obtain (we do not give the details here) that,

$$L^+ v_j = j(2l+1-j)v_{j-1}, \forall j \in \{0, ..., 2l\}.$$

From the above, we have obtained

$$
\begin{aligned}
L^2 &= L_x^2 + L_y^2 + L_z^2 \\
&= [(L_x - i_m L_y)(L_x + i_m L_y)] + \hbar L_z + L_z^2.
\end{aligned}
\tag{3.39}
$$

Therefore,

$$
\begin{aligned}
L^2 v_j &= \hbar^2 L^- L^+ v_j + \hbar L_z v_j + L_z^2 v_j \\
&= \hbar^2 j(2l+1-j)L^- v_{j-1} + \hbar^2 (l-j)v_j + \hbar^2 (l-j)^2 v_j \\
&= \hbar^2 j(2l+1-j)v_j + \hbar^2 (l-j)v_j + \hbar^2 (l-j)^2 v_j \\
&= \hbar^2 (2jl + j - j^2 + l - j + l^2 - 2jl + j^2)v_j \\
&= \hbar^2 l(l+1)v_j, \forall j \in \{0, 1, .., 2l\}.
\end{aligned}
\tag{3.40}
$$

Summarizing:

$$L^2 v_j = \hbar^2 l(l+1)v_j, \ \forall j \in \{0, 1, ..., 2l\},$$

that is, each v_j is an eigenfunction of L^2 with eigenvalue $\hbar l(l+1)$.

3.8 A numerical example

For $l = 1$ we have $j \in \{0, 1, 2\}$. Setting $m = l - j$ we have $m \in \{1, 0, -1\}$.
For $m = 1$ $g_1 = v_0 = \phi_0$ that is,

$$g_1 = e^{im\varphi} \sin \theta,$$

so that

$$g_0 = v_1 = L^-(v_0),$$

so that

$$g_0 = \frac{-2}{\hbar}(\cos \theta)$$

and

$$g_{-1} = v_2 = L^- v_1,$$

that is,

$$g_{-1} = \frac{-2}{\hbar^2}(e^{-im\varphi} \sin \theta)$$

We recall that for such a case, specifically with restriction

$$\int_\Omega |\phi|^2 \, dx = 1,$$

the Schrödinger's equations may be written by,

$$\left(\frac{-\hbar^2}{2m_e} \left(\frac{d^2}{dr^2} + \frac{2}{r}\frac{d}{dr} \right) - E + V(r) + \frac{L^2}{2m_e r^2} \right) f(r)g(\theta, \varphi) = 0.$$

In particular, for $g(\theta, \varphi) = g_k(\theta, \varphi), \forall k \in \{1, 0, -1\}$, from the above we have,

$$L^2 g_k = \hbar^2 l(l+1)g_k,$$

so that the last equation may be written as,

$$\left(\frac{-\hbar^2}{2m_e} \left(\frac{d^2}{dr^2} + \frac{2}{r}\frac{d}{dr} \right) - E + V(r) + \frac{\hbar^2 l(l+1)}{2m_e r^2} \right) f(r)g(\theta, \varphi) = 0.$$

and denoting $\tilde{f}(r) = rf(r)$, we obtain

$$\frac{d^2 \tilde{f}(r)}{dr^2} + \frac{2m_e}{\hbar^2} \left(E - V(r) - \frac{\hbar^2 l(l+1)}{2m_e r^2} \right) \tilde{f}(r) = 0.$$

In atomic units, for $V(r) = 1/r$, this last equation stands for,

$$\frac{d^2 \tilde{f}(r)}{dr^2} - \frac{l(l+1)}{r^2} \tilde{f}(r) + \frac{2\tilde{f}(r)}{r} - E\tilde{f}(r) = 0,$$

where the not relabeled constant E is obtained through the constraint

$$\int_0^\infty \tilde{f}(r)^2 \, dr = 1.$$

For the function $f = \tilde{f}/r$ obtained numerically, please see Figures 3.1 and 3.2 for the cases $l = 0$ and $l = 1$, respectively.

We have obtained the following energies,

1. $E = -0.4964$ for $l = 0$.

2. $E = 0.0076$ for $l = 1$.

The solutions for the wave function ϕ will be (up to appropriate normalizations): $m = 1$:

$$\phi_1 = f(r)g_1(\theta, \varphi) = f(r)e^{im\varphi} \sin\theta, \text{ spin up}$$

$m = 0$:

$$\phi_0 = f(r)g_0(\theta, \varphi) = f(r)\cos\theta, \text{ with } L_z\phi_0 = 0,$$

$m = -1$:

$$\phi_{-1} = f(r)g_1(\theta, \varphi) = f(r)e^{-im\varphi} \sin\theta, \text{ spin down}.$$

We have used finite differences to obtain the numerical results. Regarding the numerical algorithm used, please see [13] for similar developments.

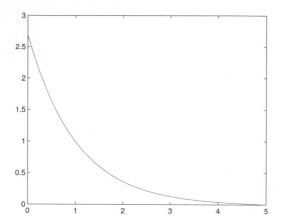

Figure 3.1: Radial solution $f(r)$ for l = 0.

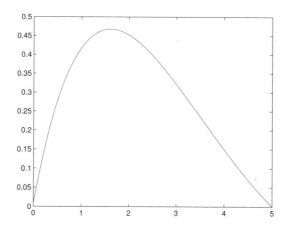

Figure 3.2: Radial solution $f(r)$ for l = 1.

3.9 A brief note on the many body quantum approach

Consider a system with N particles with corresponding position fields $\mathbf{r}_k : \Omega \times [0,T] \to \mathbb{R}^3$ and corresponding normal fields $\mathbf{n}_k$, where $k \in \{1,..,N\}$.

Denoting $\mathbf{r} = (\mathbf{r}_1, \cdots, \mathbf{r}_N)$, $\mathbf{n} = (\mathbf{n}_1, \cdots, \mathbf{n}_N)$ and $\rho = (\rho_1, \cdots, \rho_N)$, similarly as above, we would write the different parts of the system energy by:

1. Kinetics energy E_C (classical),

$$E_C = \frac{1}{2} \int_0^T \int_\Omega \sum_{j=1}^N \rho_j |R(\mathbf{r})|^2 \frac{\partial \mathbf{r}_j}{\partial t} \cdot \frac{\partial \mathbf{r}_j}{\partial t} \sqrt{g}\, dx\, dt,$$

where once more $x = (x_1, x_2, x_3) \in \mathbb{R}^3$ and $dx = dx_1 \, dx_2 \, dx_3$.

Also, $\rho_j : \Omega \times [0, T] \to [0, 1]$ corresponds to the proportion of density of the $j - th$ particle.

Finally $\mathbf{g}_k = \frac{\partial \mathbf{r}}{\partial x_k}$, $g_{kl} = \mathbf{g}_k \cdot \mathbf{g}_l$, and $g = \det\{g_{kl}\}$ where $k, l \in \{1, 2, 3\}$.

2. Energy relating the presence of electric potentials, E_p (classical),

$$E_p = \frac{1}{2} \int_0^T \int_\Omega \sum_{j=1}^N (K_1)_j V_j(\mathbf{r}) \rho_j |R(\mathbf{r})|^2 \sqrt{g} \, dx \, dt,$$

for appropriate real constants $(K_1)_j$.

3. Interacting Coulomb energy, E_{in} (classical),

$$E_{in}$$
$$= \sum_{j,k=1}^N \int_0^T \int_\Omega \int_\Omega (K_3)_{jk} \frac{\rho_j(x,t) |R(\mathbf{r}(x,t))|^2 \rho_k(\tilde{x},t) |R(\mathbf{r}(\tilde{x},t))|^2}{|\mathbf{r}_j(x,t) - \mathbf{r}_k(\tilde{x},t)|}$$
$$\times \sqrt{g(\mathbf{r}(x,t))} \, dx \, \sqrt{g(\mathbf{r}(\tilde{x},t))} \, d\tilde{x} \, dt,$$

for appropriate real constants $(K_3)_{jk}$.

4. Energy associated with the normal fields relating the motion of the N particles, E_q (quantum part),

$$E_q$$
$$= \sum_{j=1}^N \sum_{k=1}^3 \frac{\gamma_j}{2} \int_0^T \int_\Omega \left(\frac{\partial [\sqrt{\rho_j} R(\mathbf{r}) \mathbf{n}_j(\mathbf{r})]}{\partial x_k} - i_m \mu A_k \sqrt{\rho_j} R(\mathbf{r}) \mathbf{n}_j(\mathbf{r}) \right)$$
$$\cdot \left(\frac{\partial [\sqrt{\rho_j} R(\mathbf{r}) \mathbf{n}_j(\mathbf{r})]}{\partial x_k} - i_m \mu A_k \sqrt{\rho_j} R(\mathbf{r}) \mathbf{n}_j(\mathbf{r}) \right)^* \sqrt{g} \, dx \, dt \quad (3.41)$$

where $\{\gamma_j\}$ are appropriate real positive constants and $\mathbf{A}$ denotes the magnetic potential.

5. Functional energy part concerning the external magnetic field and respective potential, denoted by E_A where,

$$E_A = \frac{1}{8\pi} \|\operatorname{curl} \mathbf{A}(\mathbf{r}) - \mathbf{B}_0(\mathbf{r})\|_{\Omega \times [0,T],2}^2,$$

and where $\mathbf{B}_0$ is the external magnetic field.

6. Energy part relating the induced electric field $\mathbf{E}_{ind}$ work, denoted by E_{ind}, where,

$$E_{ind} = \frac{1}{2} \int_0^T \int_\Omega |R(\mathbf{r})|^2 \mathbf{E}_{ind} \cdot \left(\sum_{j=1}^N (K_1)_j \, \rho_j \frac{\partial \mathbf{r}_j}{\partial t} \right) \sqrt{g} \, dx \, dt,$$

where $\mathbf{E}_{ind}$ satisfies the following constraint (a Maxwell equation):

$$\operatorname{curl} \mathbf{E}_{ind}(\mathbf{r}) + \frac{1}{c} \operatorname{curl} \left(\sum_{j=1}^{N} \rho_j \frac{\partial \mathbf{r}_j}{\partial t} \times (\operatorname{curl}\mathbf{A}(\mathbf{r}) - \mathbf{B}_0(\mathbf{r})) \right)$$

$$- \frac{1}{c} \frac{\partial}{\partial t} (\operatorname{curl}\mathbf{A}(\mathbf{r}) - \mathbf{B}_0(\mathbf{r})) = 0, \text{ in } \Omega \times [0, T]. \tag{3.42}$$

7. Other Constraints: the system will be subject to the following other constraints:

   (a) $\int_{\Omega} \rho_j |R(\mathbf{r})|^2 \sqrt{g} \, dx = m_{p_j}$, on $[0, T]$, where m_{p_j} denotes the mass of particle j.

   (b) $\mathbf{n}_j(\mathbf{r}) \cdot \frac{\partial \mathbf{r}_j}{\partial t} = 0$, in $\Omega \times [0, T]$, $\forall j \in \{1, \cdots, N\}$.

   (c) $\mathbf{n}_j(\mathbf{r}) \cdot \mathbf{n}_j(\mathbf{r}) = 1$, in $\Omega \times [0, T]$, $\forall j \in \{1, \cdots, N\}$.

   (d) $0 \le \rho_j \le 1$ so that $\rho_j^2 - \rho_j \le 0$, in $\Omega \times [0, T]$, $\forall j \in \{1, \cdots, N\}$.

   (e) $\sum_{j=1}^{N} \rho_j = 1$, in $\Omega \times [0, T]$.

   (f) $\operatorname{div}\mathbf{B} = 0$, in $\Omega \times [0, T]$ (a Maxwell equation),

   (g) $\operatorname{div}\mathbf{E} = 4\pi \sum_{j=1}^{N} (K_1)_j \rho_j |R(\mathbf{r})|^2$, in $\Omega \times [0, T]$, (another Maxwell equation),

where $\mathbf{B}$ and $\mathbf{E}$ are specified in the subsequent lines.

Hence, we will be looking for critical points of the functional $J_\lambda : \tilde{U} \times U_A \to \mathbb{R}$, where $\tilde{U} = U \times U_1 \times U_2 \times V_1^N \times V_2^N \times V_2^N \times V_2^N \times V_2 \times V_2 \times V_2 \times V_2$,

$$\begin{aligned} U = \quad & \{\mathbf{r} = (\mathbf{r}_1, \cdots, \mathbf{r}_N) \in W^{1,2}(\Omega \times [0, T]; \mathbb{R}^{3N}) : \\ & \mathbf{r}(x, t) = \mathbf{r}_0(x, t), \text{ in } \partial\Omega \times [0, T], \\ & \mathbf{r}(x, 0) = \mathbf{r}_1(x), \ \mathbf{r}(x, T) = \mathbf{r}_2(x), \text{ in } \Omega\}, \end{aligned} \tag{3.43}$$

$$\begin{aligned} U_1 = \quad & \{R(\mathbf{r}) \in L^2([0, T]; W^{1,2}(\Omega; \mathbb{C})) : \mathbf{r} \in U, \\ & R(\mathbf{r}(x, t)) = R_0(x), \text{ on } \partial\Omega \times [0, T]\}, \end{aligned} \tag{3.44}$$

$$U_2 = \{\mathbf{n}(\mathbf{r}) \in L^2([0, T]; W^{1,2}(\Omega; \mathbb{R}^3)) : \mathbf{r} \in U\},$$

$$V_1 = L^2([0, T]),$$

$$V_2 = L^2(\Omega \times [0, T]),$$

$$U_A = \{\mathbf{A}(\mathbf{r}) \in W^{1,2}(\Omega \times [0, T]; \mathbb{R}^3) : \mathbf{r} \in U\},$$

and,

$$J_\lambda(\mathbf{r}, R, \mathbf{A}, \mathbf{n}, E, \lambda_1, \lambda_2, \lambda_3, \lambda_4, \lambda_5, \lambda_6, \lambda_{ind}) = J_1 + J_2 + J_3,$$

where,

$$J_1 = -\frac{1}{2} \int_0^T \int_\Omega \sum_{j=1}^N \rho_j |R(\mathbf{r})|^2 \frac{\partial \mathbf{r}_j}{\partial t} \cdot \frac{\partial \mathbf{r}_j}{\partial t} \sqrt{g}\, dx\, dt$$

$$+ \frac{1}{2} \int_0^T \int_\Omega \sum_{j=1}^N (K_1)_j V_j(\mathbf{r}) \rho_j |R(\mathbf{r})|^2 \sqrt{g}\, dx\, dt$$

$$+ \sum_{j,k=1}^N \int_0^T \int_\Omega \int_\Omega \frac{(K_3)_{jk} \rho_j(x,t) |R(\mathbf{r}(x,t))|^2 \rho_k(\tilde{x},t) |R(\mathbf{r}(\tilde{x},t))|^2}{|\mathbf{r}_j(x,t) - \mathbf{r}_k(\tilde{x},t)|}$$

$$\times \sqrt{g(\mathbf{r}(x,t))}\, dx\, \sqrt{g(\mathbf{r}(\tilde{x},t))}\, d\tilde{x}\, dt$$

$$+ \sum_{j=1}^N \sum_{k=1}^3 \frac{\gamma_j}{2} \int_0^T \int_\Omega \left(\frac{\partial[\sqrt{\rho_j} R(\mathbf{r}) \mathbf{n}_j(\mathbf{r})]}{\partial x_k} - i_m \mu A_k \sqrt{\rho_j} R(\mathbf{r}) \mathbf{n}_j(\mathbf{r}) \right)$$

$$\cdot \left(\frac{\partial[\sqrt{\rho_j} R(\mathbf{r}) \mathbf{n}_j(\mathbf{r})]}{\partial x_k} - i_m \mu A_k \sqrt{\rho_j} R(\mathbf{r}) \mathbf{n}_j(\mathbf{r}) \right)^* \sqrt{g}\, dx\, dt$$

$$+ \frac{1}{8\pi} \| \operatorname{curl} \mathbf{A}(\mathbf{r}) - \mathbf{B}_0(\mathbf{r}) \|^2_{\Omega \times [0,T], 2}$$

$$+ \int_0^T \int_\Omega |R(\mathbf{r})|^2 \mathbf{E}_{ind} \cdot \left(\sum_{j=1}^N (K_1)_j\, \rho_j \frac{\partial \mathbf{r}_j}{\partial t} \right) \sqrt{g}\, dx\, dt \qquad (3.45)$$

$$J_2 = \left\langle \lambda_{ind}, \operatorname{curl} \mathbf{E}_{ind}(\mathbf{r}) + \frac{1}{c} \operatorname{curl} \left(\sum_{j=1}^N \rho_j \frac{\partial \mathbf{r}_j}{\partial t} \times (\operatorname{curl} \mathbf{A}(\mathbf{r}) - \mathbf{B}_0(\mathbf{r})) \right) \right.$$

$$\left. - \frac{1}{c} \frac{\partial}{\partial t}(\operatorname{curl} \mathbf{A}(\mathbf{r}) - \mathbf{B}_0(\mathbf{r})) \right\rangle$$

$$- \sum_{j=1}^N \frac{1}{2} \int_0^T E_j(t) \left(\int_\Omega \rho_j |R(\mathbf{r})|^2 \sqrt{g}\, dx - m_{p_j} \right) dt$$

$$+ \sum_{j=1}^N \left\langle (\lambda_1)_j, \mathbf{n}_j(\mathbf{r}) \cdot \frac{\partial \mathbf{r}_j}{\partial t} \right\rangle$$

$$+ \sum_{j=1}^N \frac{1}{2} \langle (\lambda_2)_j, \mathbf{n}_j(\mathbf{r}) \cdot \mathbf{n}_j(\mathbf{r}) - 1 \rangle + \sum_{j=1}^N \langle (\lambda_3)_j^2, \rho_j^2 - \rho_j \rangle$$

$$\langle \lambda_4, \sum_{j=1}^N \rho_j - 1 \rangle \qquad (3.46)$$

and

$$J_3 = \langle \lambda_5, \operatorname{div}\mathbf{B} \rangle$$

$$+ \left\langle \lambda_6, \operatorname{div}\mathbf{E} - 4\pi \sum_{j=1}^N (K_1)_j \rho_j |R(\mathbf{r})|^2 \right\rangle$$

$$(3.47)$$

where

$$\mathbf{B} = \mathbf{B}_0 - \text{curl}\mathbf{A},$$

$$\mathbf{E} = \mathbf{E}_{ind} + \mathbf{E}_\rho + \frac{\sum_{j=1}^N \sum_{k=1}^N ((K_1)_j \rho_j(x) |R(\mathbf{r})|^2 \delta_{\mathbf{r}_k} V_j(\mathbf{r}))}{\sum_{j=1}^N (K_1)_j \rho_j(x) |R(\mathbf{r})|^2},$$

$$\mathbf{E}_\rho(\mathbf{r}) = \int_\Omega \sum_{j=1}^N \sum_{k=1}^N (K_3)_{jk} \frac{\rho_k(\tilde{x},t) |R(\mathbf{r}(\tilde{x},t))|^2 (\mathbf{r}_j(x,t) - \mathbf{r}_k(\tilde{x},t))}{|\mathbf{r}_j(x,t) - \mathbf{r}_k(\tilde{x},t)|^3} \sqrt{g(\mathbf{r}(\tilde{x},t))} \, d\tilde{x},$$

where $\delta_{\mathbf{r}_k}$ denotes the variation in $\mathbf{r}_k$ (indeed, the Gâteaux derivative) and $E_j, (\lambda_1)_j, (\lambda_2)_j, (\lambda_3)_j, \lambda_4, \lambda_5, \lambda_6, \lambda_{ind}$ are appropriate Lagrange multipliers for the concerning restrictions, $\forall j \in \{1, \cdots, N\}$.

3.10 A brief note on a formulation similar to those of density functional theory

Considering the functional $J_\lambda = J_1 + J_2 + J_3$, where J_1, J_2 and J_3 are indicated in (3.45), (3.46), (3.47), respectively, at this point we denote,

$$R(\mathbf{r}_1(x), \cdots, \mathbf{r}_N(x)) \equiv \tilde{R}(x).$$

Thus,

$$\sqrt{\rho_j(x)} R(\mathbf{r}_1(x), \cdots, \mathbf{r}_N(x)) = \sqrt{\rho_j(x)} \tilde{R}(x)$$
$$\equiv \phi_j(x), \tag{3.48}$$

$\forall j \in \{1, \cdots, N\}$.

Therefore, similarly as above, considering that

$$\frac{\partial \mathbf{n}_j}{\partial x_k} \cdot \mathbf{n}_j = 0, \text{ in } \Omega \times [0, T],$$

$\forall j \in \{1, \cdots, N\}$, and $k \in \{1, 2, 3\}$, we obtain

$$\int_0^T \int_\Omega \frac{\partial [\phi_j(x) \mathbf{n}_j(x,t)]}{\partial x_k} \cdot \frac{\partial [\phi_j(x) \mathbf{n}_j(x,t)]}{\partial x_k} \, dx \, dt$$
$$= \int_0^T \int_\Omega \frac{\partial \phi_j(x)}{\partial x_k} \cdot \frac{\partial \phi_j(x)}{\partial x_k} \, dx \, dt$$
$$+ \int_0^T \int_\Omega |\phi_j|^2 \frac{\partial \mathbf{n}_j(x,t)}{\partial x_k} \cdot \frac{\partial \mathbf{n}_j(x,t)}{\partial x_k} \, dx \, dt, \tag{3.49}$$

Neglecting the magnetic field and potential and considering also

$$\mathbf{r}_j(x) \approx x, \forall j \in \{1, \cdots, N\}$$

and

$$\{g_{ij}\} \approx \{\delta_{ij}\},$$

we could approximate J_λ by J_4, where

$$
\begin{aligned}
J_\lambda \\
= \quad & (J_1 + J_2 + J_3) \\
\approx \quad & J_4 \\
= \quad & \sum_{j=1}^{N} \sum_{k=1}^{3} \frac{\gamma_j T}{2} \int_\Omega \frac{\partial \phi_j(x)}{\partial x_k} \frac{\partial \phi_j(x)^*}{\partial x_k} \, dx \\
& + \sum_{j=1}^{N} \sum_{k=1}^{3} \frac{\gamma_j}{2} \int_0^T \int_\Omega |\phi_j|^2 \frac{\partial \mathbf{n}_j(x,t)}{\partial x_k} \cdot \frac{\partial \mathbf{n}_j(x,t)}{\partial x_k} \, dx \, dt \\
& + T \sum_{j=1}^{N} \sum_{k=1}^{N} \int_\Omega \int_\Omega (K_3)_{jk} \frac{|\phi_j(x)|^2 |\phi_k(\tilde{x})|^2}{|x - \tilde{x}|} \, dx \, d\tilde{x} \\
& + T \int_\Omega \sum_{j=1}^{N} (K_1)_j V_j(x) |\phi_j(x)|^2 \, dx \\
& - \sum_{j=1}^{N} E_j T \left(\int_\Omega |\phi_j(x)|^2 \, dx - m_{p_j} \right).
\end{aligned}
\tag{3.50}
$$

Remark 3.6   Here, we have relaxed the constraints in ρ so this is an approximation for the more general functional $J_\lambda = J_1 + J_2 + J_3$, similar to those of the Density Functional Theory.   ■

Chapter 4

Some Numerical Results and Examples

4.1   Some preliminary results

Firstly we highlight this chapter is also based on the book [22].

Consider the problem of finding an eigenvalue and corresponding eigenvector for a real positive definite symmetric matrix $n \times n$.

Theorem 4.1.1 *Let $M \in \mathbb{R}^{n \times n}$ be a real positive definite symmetric matrix $n \times n$. Denote $L = M^{-1}$.*

Under such hypotheses and notation, the sequence $\{L^k/|L|^k\}$ is convergent. Denote $M_0 = \lim_{k \to \infty} L^k/|L|^k$. Let $u_0 \in \mathbb{R}^{n \times 1}$ be such that

$$M_0 u_0 \neq \mathbf{0}.$$

Consider the sequence $\{\phi_k\}$ defined by

$$\phi_0 = u_0,$$

$$\phi_{k+1} = \frac{L\,\phi_k}{|\phi_k|}, \; \forall k \in \mathbb{N} \cup \{0\}.$$

Under such hypotheses, such a sequence is convergent so that

$$\tilde{\phi} = \lim_{k \to \infty} \phi_k,$$

is such that

$$M\,\tilde{\phi} = \lambda\,\tilde{\phi},$$

where

$$\lambda = \frac{1}{|\tilde{\phi}|}.$$

Finally, we also have

$$\tilde{\phi} = L\frac{M_0 u_0}{|M_0 u_0|}$$

and

$$\lambda = \frac{|M_0 u_0|}{|LM_0 u_0|}.$$

Proof 4.1   Observe that

$$\phi_1 = \frac{L\,\phi_0}{|\phi_0|},$$

$$\phi_2 = \frac{L^2\,\phi_0}{|L\,\phi_0|},$$

$$\phi_3 = \frac{L^3\,\phi_0}{|L^2\,\phi_0|},$$

and, inductively,

$$\phi_{k+1} = \frac{L^{k+1}\,\phi_0}{|L^k\,\phi_0|}, \; \forall k \in \mathbb{N}.$$

Hence,

$$\phi_{k+1} = \frac{L\,(L^k\phi_0)}{|L^k\,\phi_0|}, \; \forall k \in \mathbb{N},$$

so that

$$\phi_{k+1} = \frac{L\,(L^k\phi_0)/|L|^k}{|L^k\,\phi_0|/|L|^k}, \; \forall k \in \mathbb{N}.$$

Observe that

$$L = [P] \begin{pmatrix} \lambda_1 & 0 & \cdots & 0 \\ 0 & \lambda_2 & \cdots & 0 \\ \vdots & \vdots & \ddots & \vdots \\ 0 & 0 & \cdots & \lambda_n \end{pmatrix} [P^{-1}], \qquad (4.1)$$

where $\lambda_1, \cdots, \lambda_n$ are the eigenvalues of L and P is the matrix which the columns are the corresponding orthonormalized eigenvectors.

At this point we recall that, since L is positive definite, $\lambda_j > 0, \; \forall j \in \{1, \cdots, n\}$. Also,

$$|L| = \sup\{|Lv| : |v| = 1\} = \max\{\lambda_j : j \in \{1, \cdots, n\}\}.$$

Hence,

$$\frac{L^k}{|L|^k} = [P] \begin{pmatrix} \frac{\lambda_1^k}{|L|^k} & 0 & \cdots & 0 \\ 0 & \frac{\lambda_2^k}{|L|^k} & \cdots & 0 \\ \vdots & \vdots & \ddots & \vdots \\ 0 & 0 & \cdots & \frac{\lambda_n^k}{|L|^k} \end{pmatrix} [P^{-1}], \; \forall k \in \mathbb{N}. \qquad (4.2)$$

Moreover,

$$\alpha_j = \lim_{k\to\infty} \frac{\lambda_j^k}{|L|^k} = \begin{cases} 0, & \text{if } \lambda_j < |L| \\ 1, & \text{if } \lambda_j = |L|, \end{cases} \tag{4.3}$$

$\forall j \in \{1, \cdots, n\}$.

Therefore,

$$\lim_{k\to\infty} \frac{L^k}{|L|^k} = [P] \begin{pmatrix} \alpha_1 & 0 & \cdots & 0 \\ 0 & \alpha_2 & \cdots & 0 \\ \vdots & \vdots & \ddots & \vdots \\ 0 & 0 & \cdots & \alpha_n \end{pmatrix} [P^{-1}] \equiv M_0, \tag{4.4}$$

so that

$$
\begin{aligned}
\tilde{\phi} &= \lim_{k\to\infty} \frac{L\left(L^k \, \phi_0\right)}{|L^k \phi_0|} \\
&= \lim_{k\to\infty} \frac{[L\left(L^k \, \phi_0\right)]/|L|^k}{|L^k \phi_0|/|L|^k} \\
&= \frac{L\, M_0\, \phi_0}{|M_0 \phi_0|}.
\end{aligned} \tag{4.5}
$$

On the other hand, by construction,

$$
\begin{aligned}
\tilde{\phi} &= \lim_{k\to\infty} \phi_{k+1} \\
&= \lim_{k\to\infty} \frac{L\,\phi_k}{|\phi_k|} \\
&= \frac{L\,\tilde{\phi}}{|\tilde{\phi}|},
\end{aligned} \tag{4.6}
$$

so that,

$$L^{-1}\,\tilde{\phi} = M\,\tilde{\phi} = \lambda\,\tilde{\phi},$$

where

$$\lambda = \frac{1}{|\tilde{\phi}|} = \frac{|M_0 u_0|}{|LM_0 u_0|}.$$

The proof is complete.

We apply the last result to the two body system corresponding to a model analogous to the hydrogen atom in an appropriate sense.

So, in such a 4-dimensional model, at first both particles are are allowed to move but are confined in the box

$$\Omega = [0, 600] \times [0, 600] \times [0, 600] \times [0, 600].$$

The cartesian coordinates are $(x_1, y_1, x_2, y_2) \in \Omega$, where (x_1, y_1) corresponds to the electron coordinates and (x_2, y_2) corresponds to the proton ones.

The dimensionless equations for such a model are given by,

$$-\frac{1}{2}\frac{\partial^2\phi(x_1,y_1,x_2,y_2)}{\partial x_1^2} - \frac{1}{2}\frac{\partial^2\phi(x_1,y_1,x_2,y_2)}{\partial y_1^2}$$
$$-\frac{\alpha}{2}\frac{\partial^2\phi(x_1,y_1,x_2,y_2)}{\partial x_2^2} - \frac{\alpha}{2}\frac{\partial^2\phi(x_1,y_1,x_2,y_2)}{\partial y_2^2}$$
$$-\frac{\beta\phi(x_1,y_1,x_2,y_2)}{\sqrt{(x_1-y_1)^2+(x_2-y_2)^2}} = E\phi(x_1,y_1,x_2,y_2), \qquad (4.7)$$

where,

$$\alpha \approx \frac{0.067}{0.5},$$

and

$$\beta \approx \frac{0.067}{13.2}.$$

Denoting

$$M\,\phi(x_1,y_1,x_2,y_2) = -\frac{1}{2}\frac{\partial^2\phi(x_1,y_1,x_2,y_2)}{\partial x_1^2} - \frac{1}{2}\frac{\partial^2\phi(x_1,y_1,x_2,y_2)}{\partial y_1^2}$$
$$-\frac{\alpha}{2}\frac{\partial^2\phi(x_1,y_1,x_2,y_2)}{\partial x_2^2} - \frac{\alpha}{2}\frac{\partial^2\phi(x_1,y_1,x_2,y_2)}{\partial y_2^2}$$
$$-\frac{\beta\phi(x_1,y_1,x_2,y_2)}{\sqrt{(x_1-y_1)^2+(x_2-y_2)^2}}, \qquad (4.8)$$

we solve the eigenvalue problem,

$$M\,\phi(x_1,y_1,x_2,y_2) = E\,\phi(x_1,y_1,x_2,y_2),$$

by constructing the sequence $\{\phi_k\}$, where,

$$\phi_1(x_1,y_1,x_2,y_2) = 1, \text{ in } \Omega,$$

and given ϕ_k, we obtain ϕ_{k+1}, by solving the equation,

$$M\,\phi_{k+1} = \frac{\phi_k}{|\phi_k|},$$

where we solved such an equation using the Banach fixed point result.

Even though the matrix in question may not be positive definite, through such a sequence, using finite differences (see [49] for details on finite differences schemes), we have obtained

$$\phi_0 = \lim_{k\to\infty}\phi_{k+1}$$
$$= \lim_{k\to\infty}\frac{M^{-1}\phi_k}{|\phi_k|}$$
$$= \frac{M^{-1}\phi_0}{|\phi_0|}, \qquad (4.9)$$

so that

$$M\,\phi_0 = E\,\phi_0,$$

where

$$E = \frac{1}{|\phi_0|}.$$

Please see Figures 4.1 and 4.2 for the wave function ϕ_0^2 concerning the electron cross section corresponding to

$$\phi_0(x_1, y_1, 300, 300)^2,$$

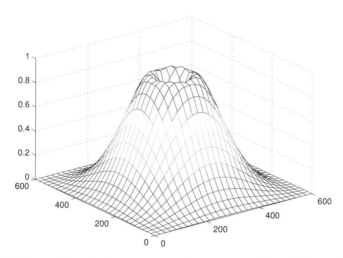

Figure 4.1: Solution relating the electron cross section corresponding to $\phi_0(x_1, y_1, 300, 300)^2$.

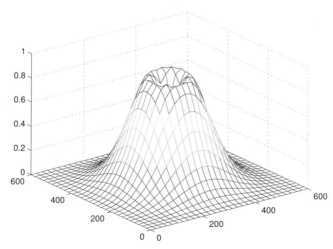

Figure 4.2: Solution relating the proton cross section corresponding to $\phi_0(300, 300, x_2, y_2)^2$.

and the proton one corresponding to

$$\phi_0(300,300,x_2,y_2)^2.$$

Also, we have obtained the value

$$E = 4.6772 \ 10^{-6} \times 0.067 \times 2,$$

for the eigenvalue of smallest absolute value.

From the figures, we see that the wave function ϕ_0^2 is close to zero at the set

$$\{(x_1,y_1,x_2,y_2) \in \Omega \ : \ x_1 = y_1 \text{ and } x_2 = y_2\},$$

since the potential goes to infinity on such a set.

4.2   A second numerical example

In order to illustrate the approach presented, we have developed a second numerical example.

Denoting $\Omega = [0,1]$ and a time interval by $[0,T]$, consider the energy given by the simplified functional $J : U \times U_1 \times U_2 \to \mathbb{R}$ where

$$
\begin{aligned}
J(Y,R,\mathbf{n}) \;=\; & -\frac{1}{2}\int_0^T \int_\Omega R(x)^2 \left(\frac{\partial Y(x,t)}{\partial t}\right)^2 dx\,dt \\
& +\frac{K}{2}\int_0^T \int_\Omega R(x)^2 Y(x,t)^2\,dx\,dt \\
& +\frac{\gamma}{2}\int_0^T \int_\Omega \frac{\partial[R(x)\mathbf{n}(x,t)]}{\partial x} \cdot \frac{\partial[R(x)\mathbf{n}(x,t)]}{\partial x}\,dx\,dt \quad (4.10)
\end{aligned}
$$

subject to

$$\int_\Omega R(x)^2\,dx = m,$$

$$\mathbf{n}(x,t) \cdot \frac{\partial \mathbf{r}(x,t)}{\partial t} = 0, \text{ in } \Omega \times [0,T],$$

where

$$\mathbf{r}(x,t) = x\mathbf{i} + Y(x,t)\mathbf{j},$$

so that

$$\frac{\partial \mathbf{r}(x,t)}{\partial t} = 0\mathbf{i} + \frac{\partial Y(x,t)}{\partial t}\mathbf{j} = \left(0, \frac{\partial Y(x,t)}{\partial t}\right), \text{ in } \Omega \times [0,T].$$

The final constraint is,

$$\mathbf{n}(x,t) \cdot \mathbf{n}(x,t) = 1, \text{ in } \Omega \times [0,T].$$

We also denote,

$$U = \{Y \in L^2(\Omega; W^{1,2}([0,T])) :$$
$$Y(x,0) = Y_0(x) \text{ and } Y(x,T) = Y_1(x), \text{ in } \Omega\}, \qquad (4.11)$$

$$U_1 = \{R \in W^{1,2}(\Omega) : R(0) = R(1) = 0\} = W_0^{1,2}(\Omega)$$

and,

$$U_2 = L^2([0,T]; W^{1,2}(\Omega; \mathbb{R}^2)).$$

Hence, we will be looking for critical points of the extended functional $J_\lambda : U \times U_1 \times U_2 \times V \times V \times \mathbb{R} \to \mathbb{R}$, where

$$
\begin{aligned}
J_\lambda(Y, R, \mathbf{n}, \lambda_1, \lambda_2, E) = & -\frac{1}{2} \int_0^T \int_\Omega R(x)^2 \left(\frac{\partial Y(x,t)}{\partial t} \right)^2 dx\, dt \\
& + \frac{K}{2} \int_0^T \int_\Omega R(x)^2 Y(x,t)^2 \, dx\, dt \\
& + \frac{\gamma}{2} \int_0^T \int_\Omega \frac{\partial [R(x)\mathbf{n}(x,t)]}{\partial x} \cdot \frac{\partial [R(x)\mathbf{n}(x,t)]}{\partial x} \, dx\, dt \\
& - E\, T \left(\int_\Omega R(x)^2 \, dx - m \right) \\
& + \int_0^T \int_\Omega \lambda_1(x,t) \mathbf{n}(x,t) \cdot \left(0, \frac{\partial Y(x,t)}{\partial t} \right) dx\, dt \\
& + \frac{1}{2} \int_0^T \int_\Omega \lambda_2(x,t)(\mathbf{n}(x,t) \cdot \mathbf{n}(x,t) - 1) \, dx\, dt. \quad (4.12)
\end{aligned}
$$

Here $V = L^2(\Omega \times [0,T])$, m denotes the total system mass and E, λ_1, λ_2 are appropriated Lagrange multipliers relating the concerning constraints.

The corresponding Euler-Lagrange equations are given by,

1. Variation in Y,

$$R(x)^2 \frac{\partial^2 Y(x,t)}{\partial t^2} + KR(x)^2 Y(x,t)$$
$$-\frac{\partial [\lambda_1(x,t)n_2(x,t)]}{\partial t} = 0, \text{ in } \Omega \times [0,T], \qquad (4.13)$$

and

$$\lambda_1(x,0)\mathbf{n}(x,0) = \lambda_1(x,T)\mathbf{n}(x,T) = 0, \text{ in } \Omega,$$

where we have denoted

$$\mathbf{n}(x,t) = (n_1(x,t), n_2(x,t)), \text{ in } \Omega \times [0,T].$$

2. Variation in R,

$$
-R(x) \int_0^T \left(\frac{\partial Y(x,t)}{\partial t} \right)^2 dt + KR(x) \int_0^T Y(x,t)^2 \, dt
$$
$$
-\gamma \int_0^T \frac{\partial^2 [R(x)\mathbf{n}(x,t)]}{\partial x^2} \cdot \mathbf{n}(x,t) \, dt - E \, T \, R(x) = 0,
$$
in Ω. \hfill (4.14)

3. Variation in $\mathbf{n}$,

$$
-\gamma R(x) \frac{\partial^2 [R(x)\mathbf{n}(x,t)]}{\partial x^2}
$$
$$
+\lambda_1(x,t) \left(0, \frac{\partial Y(x,t)}{\partial t} \right) + \lambda_2 \mathbf{n}(x,t) = 0, \text{ in } \Omega \times [0,T]. \quad (4.15)
$$

4. Variation in E,

$$
\int_\Omega R(x)^2 \, dx = m.
$$

5. Variation in λ_1,

$$
\mathbf{n}(x,t) \cdot \left(0, \frac{\partial Y(x,t)}{\partial t} \right) = 0, \text{ in } \Omega \times [0,T].
$$

6. Variation in λ_2,

$$
\mathbf{n}(x,t) \cdot \mathbf{n}(x,t) = 1, \text{ in } \Omega \times [0,T].
$$

From these last two equations we obtain,

$$
\mathbf{n}(x,t) = \mathbf{i} = (1,0) \in \mathbb{R}^2, \text{ in } \Omega \times [0,T],
$$

so that, the system to be solved is indeed summarized by the following equations,

$$
R(x)^2 \frac{\partial^2 Y(x,t)}{\partial t^2} + KR(x)^2 Y(x,t) = 0, \text{ in } \Omega \times [0,T], \quad (4.16)
$$

$$
-\frac{R(x)}{T} \int_0^T \left(\frac{\partial Y(x,t)}{\partial t} \right)^2 dt + \frac{KR(x)}{T} \int_0^T Y(x,t)^2 \, dt
$$
$$
-\gamma \frac{\partial^2 R(x)}{\partial x^2} - E \, R(x) = 0, \text{ in } \Omega \quad (4.17)
$$

and

$$
\int_\Omega R(x)^2 \, dx = m. \quad (4.18)
$$

We have solved this system for $T = \frac{\pi}{4}$, $K = 4$, $m = 1$,

$$Y_0(x) = \frac{\sqrt{5\pi}}{4}(x+c),$$

$$Y_1(x) = \frac{\sqrt{5\pi}}{4}(x+c),$$

where $c > 0$ is a constant to be specified.

The solution of equation (4.16) is given by,

$$Y(x,t) = \frac{\sqrt{5\pi}}{4}(x+c)(\cos(2t)+\sin(2t)), \text{ in } \Omega \times [0,T].$$

From such a result, equation (4.17) will stand for,

$$-\gamma\frac{\partial^2 R(x)}{\partial x^2} + 5(x+c)^2 R(x) - E\,R(x) = 0, \text{ in } \Omega. \tag{4.19}$$

For $c = 0.001$, we have solved this eigenvalue problem by using the result presented in Theorem 4.1.1.

The energies obtained are,

1. for $\gamma = 10^{-5}$, $E = 0.0212$,

2. for $\gamma = 10^{-4}$, $E = 0.0671$,

3. for $\gamma = 10^{-3}$, $E = 0.2121$,

4. for $\gamma = 10^{-2}$, $E = 0.6708$,

5. for $\gamma = 10^{-1}$, $E = 2.1605$,

6. for $\gamma = 1$, $E = 11.2554$.

We see the energies increase as γ increases, as expected.

The wave function converges to the standard one of quantum mechanics as γ increases, also as expected.

Finally, the square of wave function seems to converge point-wisely to the dirac function $\delta(0^+)$ as γ approaches zero, so that in such a case the system behavior becomes similar to a particle one.

4.3 A third example, the hydrogen atom

In this section, we present another numerical example.

Let $\Omega = B_{r_0}(\mathbf{0}) = \{\mathbf{r} \in \mathbb{R}^3 : |\mathbf{r}| \leq r_0\}$ and let $[0,T]$ be a time interval.

Consider the energy given by the functional $J : U \times U_1 \times U_2 \to \mathbb{R}$, related to a hydrogen atom with a fixed proton at $\mathbf{r} = \mathbf{0}$, where

$$J(R, \mathbf{r}, \mathbf{n}) \;=\; -\frac{1}{2} \int_0^T \int_\Omega m_e R(x)^2 \frac{\partial \mathbf{r}(x,t)}{\partial t} \cdot \frac{\partial \mathbf{r}(x,t)}{\partial t} \, dx \, dt$$

$$-K \int_0^T \int_\Omega \frac{R(x)^2}{|\mathbf{r}(x,t)|} \, dx \, dt$$

$$+\frac{\gamma}{2} \int_0^T \int_\Omega \frac{\partial [R(x)\mathbf{n}(x,t)]}{\partial x_k} \cdot \frac{\partial [R(x)\mathbf{n}(x,t)]}{\partial x_k} \, dx \, dt, \quad (4.20)$$

subject to

$$\int_\Omega R(x)^2 \, dx = 1,$$

where m_e denotes the electron mass,

$$\mathbf{n}(x,t) \cdot \frac{\partial \mathbf{r}(x,t)}{\partial t} = 0, \text{ in } \Omega \times [0, T],$$

$$\mathbf{n}(x,t) \cdot \mathbf{n}(x,t) = 1, \text{ in } \Omega \times [0, T].$$

Hence, we will be looking for critical points of the extended functional J_λ : $U \times U_1 \times U_2 \times [V]^2 \times \mathbb{R} \to \mathbb{R}$, given by

$$J(\mathbf{r}, R, \mathbf{n}) \;=\; -\frac{1}{2} \int_0^T \int_\Omega m_e R(x)^2 \frac{\partial \mathbf{r}(x,t)}{\partial t} \cdot \frac{\partial \mathbf{r}(x,t)}{\partial t} \, dx \, dt$$

$$-K \int_0^T \int_\Omega \frac{R(x)^2}{|\mathbf{r}(x,t)|} \, dx \, dt$$

$$+\frac{\gamma}{2} \int_0^T \int_\Omega \frac{\partial [R(x)\mathbf{n}(x,t)]}{\partial x_k} \cdot \frac{\partial [R(x)\mathbf{n}(x,t)]}{\partial x_k} \, dx \, dt$$

$$-ET \left(\int_\Omega R(x)^2 \, dx - 1 \right)$$

$$+\int_0^T \int_\Omega \lambda_1(x,t)\mathbf{n}(x,t) \cdot \frac{\partial \mathbf{r}(x,t)}{\partial t} \, dx \, dt$$

$$+\frac{1}{2} \int_0^T \int_\Omega \lambda_2(x,t)\mathbf{n}(x,t) \cdot \mathbf{n}(x,t) \, dx \, dt. \quad (4.21)$$

Here

$$U = W^{1,2}(\Omega \times [0, T]),$$

$$U_1 = \{R \in W^{1,2}(\Omega) \; : \; R(x) = 0 \text{ on } \partial\Omega\} = W_0^{1,2}(\Omega),$$

$$U_2 = L^2([0, T]; W^{1,2}(\Omega; \mathbb{R}^3)),$$

$$V = L^2(\Omega \times [0, T])$$

and E, λ_1, λ_2 are appropriated Lagrange multipliers relating the concerning constraints.

The corresponding Euler-Lagrange equations are given by,

1. Variation in **r**,

$$
m_e R(x)^2 \frac{\partial^2 \mathbf{r}(x,t)}{\partial t^2} + K \frac{R(x)^2}{|\mathbf{r}(x,t)|^3} \mathbf{r}
$$

$$
- \frac{\partial [\lambda_1(x,t)\mathbf{n}(x,t)]}{\partial t} = 0, \text{ in } \Omega \times [0,T], \tag{4.22}
$$

and

$$
\lambda_1(x,0)\mathbf{n}(x,0) = \lambda_1(x,T)\mathbf{n}(x,T) = 0, \text{ in } \Omega.
$$

2. Variation in R,

$$
-m_e R(x) \int_0^T \left(\frac{\partial \mathbf{r}(x,t)}{\partial t} \right)^2 dt - K R(x) \int_0^T \frac{1}{|\mathbf{r}(x,t)|} dt
$$

$$
- \gamma \sum_{k=1}^3 \int_0^T \frac{\partial^2 [R(x)\mathbf{n}(x,t)]}{\partial x_k^2} \cdot \mathbf{n}(x,t) \, dt - E \, T \, R(x) = 0,
$$

$$
\text{in } \Omega. \tag{4.23}
$$

3. Variation in **n**,

$$
-\gamma \sum_{k=1}^3 R(x) \frac{\partial^2 [R(x)\mathbf{n}(x,t)]}{\partial x_k^2}
$$

$$
+ \lambda_1(x,t) \frac{\partial \mathbf{r}(x,t)}{\partial t} + \lambda_2 \mathbf{n}(x,t) = 0, \text{ in } \Omega \times [0,T]. \tag{4.24}
$$

4. Variation in E,

$$
\int_\Omega R(x)^2 \, dx = 1.
$$

5. Variation in λ_1,

$$
\mathbf{n}(x,t) \cdot \frac{\partial \mathbf{r}(x,t)}{\partial t} = 0, \text{ in } \Omega \times [0,T].
$$

6. Variation in λ_2,

$$
\mathbf{n}(x,t) \cdot \mathbf{n}(x,t) = 1, \text{ in } \Omega \times [0,T].
$$

Using spherical coordinates, we denote $\mathbf{r}(x,t)$ by a not relabeled

$$
\mathbf{r}(r,\theta,\phi,t) = (\tilde{r}(r,\theta,\phi,t), \tilde{\theta}(r,\theta,\phi,t), \tilde{\phi}(r,\theta,\phi,t)),
$$

where for $\tilde{\phi}(r,\theta,\phi,t) = 0$, in $\Omega \times [0,T]$ we have

$$
\begin{aligned}
\mathbf{r}(r,\theta,\phi,t) &= X_3(r,\theta,\phi,t)\mathbf{k} + X_2(r,\theta,\phi,t)\mathbf{j} \\
&= \tilde{r}(r,\theta,\phi,t)\cos(\tilde{\theta}(r,\theta,\phi,t))\mathbf{k} \\
&\quad + \tilde{r}(r,\theta,\phi,t)\sin(\tilde{\theta}(r,\theta,\phi,t))\mathbf{j},
\end{aligned} \tag{4.25}
$$

where

$$X_3(r,\theta,\phi,t) = \tilde{r}(r,\theta,\phi,t)\cos(\tilde{\theta}(r,\theta,\phi,t)),$$

and

$$X_2(r,\theta,\phi,t) = \tilde{r}(r,\theta,\phi,t)\sin(\tilde{\theta}(r,\theta,\phi,t)), \text{ in } \Omega \times [0,T].$$

We also define,

$$\tilde{\mathbf{e}}_r = \cos(\tilde{\theta}(r,\theta,\phi,t))\mathbf{k} + \sin(\tilde{\theta}(r,\theta,\phi,t))\mathbf{j},$$

and

$$\tilde{\mathbf{e}}_\theta = -\sin(\tilde{\theta}(r,\theta,\phi,t))\mathbf{k} + \cos(\tilde{\theta}(r,\theta,\phi,t))\mathbf{j},$$

so that

$$\mathbf{r}(r,\theta,\phi,t) = \tilde{r}(r,\theta,\phi,t)\tilde{\mathbf{e}}_r,$$

$$\frac{\partial \tilde{\mathbf{e}}_r}{\partial t} = \frac{\partial \tilde{\theta}}{\partial t}\tilde{\mathbf{e}}_\theta,$$

and

$$\frac{\partial \tilde{\mathbf{e}}_\theta}{\partial t} = -\frac{\partial \tilde{\theta}}{\partial t}\tilde{\mathbf{e}}_r, \text{ in } \Omega \times [0,T].$$

With such a coordinate system, equation (4.22) would stand for,

$$m_e R(r,\theta,\phi)^2 \left[\left(\frac{\partial^2 \tilde{r}}{\partial t^2} - \tilde{r}\left(\frac{\partial \tilde{\theta}}{\partial t}\right)^2 \right) \tilde{\mathbf{e}}_r \right.$$

$$\left. + \left(\tilde{r}\frac{\partial^2 \tilde{\theta}}{\partial t^2} + 2\frac{\partial \tilde{r}}{\partial t}\frac{\partial \tilde{\theta}}{\partial t} \right) \tilde{\mathbf{e}}_\theta \right]$$

$$= -R(r,\theta,\phi)^2 \frac{K}{\tilde{r}^2}\tilde{\mathbf{e}}_r. \tag{4.26}$$

A solution for such an equation would be given by,

$$\tilde{r}(r,\theta,\phi,t) = r$$

so that

$$\frac{\partial \tilde{r}(r,\theta,\phi,t)}{\partial t} = 0,$$

and

$$\left(\frac{\partial \tilde{\theta}(r,\theta,\phi,t)}{\partial t} \right)^2 = \frac{K}{m_e \tilde{r}^3} = \frac{K}{m_e r^3},$$

so that

$$\frac{\partial \tilde{\theta}}{\partial t} = \frac{\sqrt{K/m_e}}{r^{3/2}},$$

and we could choose,

$$\tilde{\theta}(r,\theta,\phi,t) = \frac{\sqrt{K/m_e}\,t}{r^{3/2}}.$$

At this point, we denote

$$K_1 = K/m_e.$$

Observe that from

$$\mathbf{r} = \tilde{r}\cos(\tilde{\theta})\mathbf{k} + \tilde{r}\sin(\tilde{\theta})\mathbf{j}, \tag{4.27}$$

we obtain

$$\frac{\partial \mathbf{r}}{\partial t} = -\tilde{r}\sin(\tilde{\theta})\frac{\partial \tilde{\theta}}{\partial t}\mathbf{k} + \tilde{r}\cos(\tilde{\theta})\frac{\partial \tilde{\theta}}{\partial t}\mathbf{j}, \tag{4.28}$$

so that from

$$\mathbf{n} \cdot \frac{\partial \mathbf{r}}{\partial t} = 0, \text{ in } \Omega \times [0, T],$$

we obtain

$$\mathbf{n} = \cos(\tilde{\theta})\mathbf{k} + \sin(\tilde{\theta})\mathbf{j}, \text{ in } \Omega \times [0, T].$$

Therefore,

$$\mathbf{n} = \cos\left(\frac{\sqrt{K_1}t}{(x_1^2 + x_2^2 + x_3^2)^{3/4}}\right)\mathbf{k} + \sin\left(\frac{\sqrt{K_1}t}{(x_1^2 + x_2^2 + x_3^2)^{3/4}}\right)\mathbf{j},$$

so that,

$$\begin{aligned}
\frac{\partial \mathbf{n}}{\partial x_1} &= -\sin\left(\frac{\sqrt{K_1}t}{(x_1^2 + x_2^2 + x_3^2)^{3/4}}\right)\frac{-3\sqrt{K_1}tx_1}{2(x_1^2 + x_2^2 + x_3^2)^{7/4}}\mathbf{k} \\
&\quad + \cos\left(\frac{\sqrt{K_1}t}{(x_1^2 + x_2^2 + x_3^2)^{3/4}}\right)\frac{-3\sqrt{K_1}tx_1}{2(x_1^2 + x_2^2 + x_3^2)^{7/4}}\mathbf{j},
\end{aligned} \tag{4.29}$$

and therefore,

$$\begin{aligned}
&\frac{\partial \mathbf{n}}{\partial x_1} \cdot \frac{\partial \mathbf{n}}{\partial x_1} \\
&= \frac{9K_1t^2x_1^2}{4(x_1^2 + x_2^2 + x_3^2)^{7/2}}.
\end{aligned} \tag{4.30}$$

Similarly,

$$\begin{aligned}
&\frac{\partial \mathbf{n}}{\partial x_2} \cdot \frac{\partial \mathbf{n}}{\partial x_2} \\
&= \frac{9K_1t^2x_2^2}{4(x_1^2 + x_2^2 + x_3^2)^{7/2}},
\end{aligned} \tag{4.31}$$

and

$$\begin{aligned}
&\frac{\partial \mathbf{n}}{\partial x_3} \cdot \frac{\partial \mathbf{n}}{\partial x_3} \\
&= \frac{9K_1t^2x_3^2}{4(x_1^2 + x_2^2 + x_3^2)^{7/2}}.
\end{aligned} \tag{4.32}$$

Hence,

$$\sum_{k=1}^{3} \frac{\partial \mathbf{n}}{\partial x_k} \cdot \frac{\partial \mathbf{n}}{\partial x_k} = \frac{9K_1 t^2}{4(x_1^2 + x_2^2 + x_3^2)^{5/2}}$$

$$= \frac{9K_1 t^2}{4r^5}. \tag{4.33}$$

Also from

$$\mathbf{n} \cdot \mathbf{n} = 1,$$

we obtain

$$\frac{\partial \mathbf{n}}{\partial x_k} \cdot \mathbf{n} = 0,$$

and from this,

$$\frac{\partial^2 \mathbf{n}}{\partial x_k^2} \cdot \mathbf{n} = -\frac{\partial \mathbf{n}}{\partial x_k} \cdot \frac{\partial \mathbf{n}}{\partial x_k},$$

$\forall k \in \{1, 2, 3\}$, in $\Omega \times [0, T]$.

Hence,

$$\gamma \sum_{k=1}^{3} \left(\frac{\partial^2 [R(x)\mathbf{n}]}{\partial x_k^2} \right) \cdot \mathbf{n}$$

$$= \gamma \sum_{k=1}^{3} \frac{\partial^2 R(x)}{\partial x_k^2} \mathbf{n} \cdot \mathbf{n}$$

$$+ 2\gamma \sum_{k=1}^{3} \frac{\partial R(x)}{\partial x_k} \frac{\partial \mathbf{n}}{\partial x_k} \cdot \mathbf{n}$$

$$+ \gamma \sum_{k=1}^{3} R(x) \frac{\partial^2 \mathbf{n}}{\partial x_k^2} \cdot \mathbf{n}$$

$$= \gamma \sum_{k=1}^{3} \frac{\partial^2 R(x)}{\partial x_k^2} - \gamma R(x) \frac{9K_1 t^2}{4r^5}, \tag{4.34}$$

so that

$$\int_0^T \gamma \sum_{k=1}^{3} \left(\frac{\partial^2 [R(x)\mathbf{n}]}{\partial x_k^2} \right) \cdot \mathbf{n} \, dt$$

$$= \gamma \sum_{k=1}^{3} \frac{\partial^2 R(x)}{\partial x_k^2} T - \gamma R(x) \frac{3K_1 T^3}{4r^5}. \tag{4.35}$$

From this, equation (4.23) would stand for,

$$-\gamma \nabla^2 R(x) + \frac{3\gamma K_1 T^2}{4r^5} R(x) - \frac{1}{T} m_e \int_0^T \frac{\partial \mathbf{r}(x,t)}{\partial t} \cdot \frac{\partial \mathbf{r}(x,t)}{\partial t} \, dt$$

$$- \frac{K R(x)}{r} - E R(x) = 0, \text{ in } \Omega \times [0, T]. \tag{4.36}$$

We are going to neglect the classical energy part in this last equation to obtain,

$$-\gamma \nabla^2 R(x) + \frac{3\gamma K_1 T^2}{4r^5} R(x)$$

$$-\frac{K\,R(x)}{r} - E\,R(x) = 0, \text{ in } \Omega \times [0,T]. \tag{4.37}$$

Observe that, according to a Bohr postulate, for the classical hydrogen atom, the allowed angular moments are given by,

$$L = mrv = \hbar n, \forall n \in \mathbb{N}.$$

Hence, for the dimensionless version, we obtain

$$rv = r^2 \frac{\partial \tilde{\theta}}{\partial t} = \frac{r^2}{r^{3/2}} = r^{1/2} = n,$$

so that, in particular for $n = 1$, we obtain

$$r = 1.$$

In atomic units, we have developed numerical results for $r_0 = 5$ and some values for γ below specified. We have used the standard procedure of separation of variables to obtain a final eigenvalue problem only the radial variable r.

For the radial solutions R_0 obtained for $\gamma = 1.0$ with $T = 0$ and $\gamma = 1$, $\gamma = 10^{-3}$ and $\gamma = 10^{-5}$ with $T = 2\sqrt{1/(3\gamma)}$, please see Figures 4.3, 4.4, 4.5 and 4.6, respectively.

As $\gamma \to 0$, for the appropriate choice of T which as above indicated is given by

$$T = 2\sqrt{1/(3\gamma)},$$

we obtain a distribution of mass $R(x)$ concentrating about $r = 1$, that is, in an appropriate sense, $R(x)$ approaches the dirac delta function $\delta(1)$. In such a case, the classical part of the model approaches a particle (the electron) with circular orbit of radius $r = 1$. We emphasize that, for $\gamma = 1$ with $T = 0$, we obtain the standard ground state and ground energy for the hydrogen atom.

The energies obtained are,

1. for $\gamma = 1$, and $T = 0$ $E = -0.5674$.

2. for $\gamma = 1$ and $T = 2\sqrt{1/(3\gamma)}$, $E = -0.2167$,

3. for $\gamma = 10^{-3}$ and $T = 2\sqrt{1/(3\gamma)}$, $E = -4.9958\ 10^3$,

4. for $\gamma = 10^{-5}$ and $T = 2\sqrt{1/(3\gamma)}$, $E = -5.0000\ 10^5$.

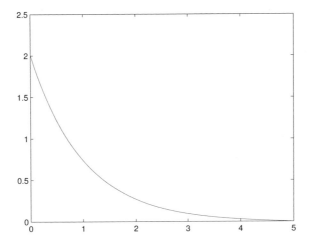

Figure 4.3: Radial solution R_0 for $\gamma = 1$ and $T = 0.0$.

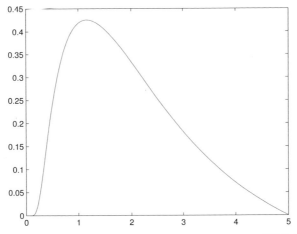

Figure 4.4: Radial solution R_0 for $\gamma = 1$ and $T = 2\sqrt{1/(3\gamma)}$.

4.4 A related control problem

Consider again a system with $Z = \sum_{l=0}^{n-1} 2l + 1$ moving electrons and Z protons resting at $r = 0$, in which the electrons are located at the n layers l, where $l \in \{0, \ldots, n-1\}$. More specifically, the position field of the electron j at the layer l is denoted by

$$\mathbf{r}_j^l : \Omega \times [0, T] \to \mathbb{R}^3,$$

$\forall j \in \{1, \ldots, 2l+1\}$, where

$$\Omega = B_{R_0}(\mathbf{0}) \subset \mathbb{R}^3.$$

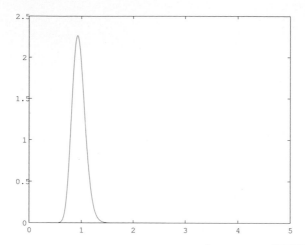

Figure 4.5: Radial solution R_0 for $\gamma = 10^{-3}$ and $T = 2\sqrt{1/(3\gamma)}$.

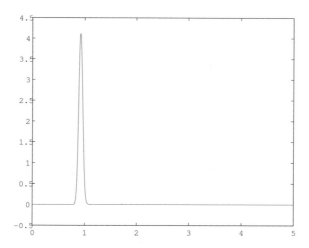

Figure 4.6: Radial solution R_0 for $\gamma = 10^{-5}$ and $T = 2\sqrt{1/(3\gamma)}$.

for some $R_0 > 0$ denotes the system domain and $[0, T]$ is a time interval. To each particle is associated a distribution of mass/charge denoted by

$$m^l_{p_j} |\varphi^l_j|^2 : \Omega \to \mathbb{R},$$

where $m^l_{p_j}$ denotes the mass of particle j at the layer l and ϕ^l_j is such that

$$\int_\Omega |\varphi^l_j(x)|^2 \, dx = 1, \forall j \in \{1, \ldots, 2l+1\}, \, l \in \{0, 1, \ldots, n-1\}.$$

The energy for such a system, denoted by

$$J_\lambda : U \times U \times U_1 \times \mathbb{R}^Z \to \mathbb{R},$$

is given

$$J(\mathbf{r}, \varphi, \mathbf{n}, E)$$

$$= \sum_{l=0}^{n-1} \sum_{j=1}^{2l+1} \int_0^T \int_\Omega m_{p_j}^l \frac{\partial \mathbf{r}_j^l}{\partial t} \cdot \frac{\partial \mathbf{r}_j^l}{\partial t}\, dx dt$$

$$+ \sum_{l=0}^{n-1} \sum_{j=1}^{2l+1} \sum_{k=1}^{3} A_j^l \frac{\partial \mathbf{r}_j^l}{\partial x_k} \cdot \frac{\partial \mathbf{r}_j^l}{\partial x_k}\, dx dt$$

$$+ \sum_{l,l_1=0}^{n-1} \sum_{j,k=1}^{2l+1} Ke^2 \int_0^T \int_\Omega \int_\Omega \frac{|\varphi_j^l(x)|\,|\varphi_k^{l_1}(\tilde{x})|^2}{|\mathbf{r}_j^l(x,t) - \mathbf{r}_k^{l_1}(\tilde{x},t)|}\, dx d\tilde{x} dt$$

$$- \sum_{l=0}^{n-1} \sum_{j=1}^{2l+1} Ke^2 Z \frac{\varphi_j^l(x)}{\mathbf{r}_j^l(x,t)}\, dx dt$$

$$+ \sum_{l=0}^{n-1} \sum_{j=1}^{2l+1} \sum_{k=1}^{3} \frac{\gamma}{2} \int_0^T \int_\Omega \frac{\partial[\varphi_j^l \mathbf{n}_j^l]}{\partial x_k} \cdot \frac{\partial[\varphi_j^l \mathbf{n}_j^l]}{\partial x_k}\, dx\, dt$$

$$- \sum_{l=0}^{n} \sum_{j=1}^{2l+1} E_j^l T \left(\int_\Omega |\varphi_j^l(x)|^2\, dx - 1 \right)$$

$$\sum_{l=0}^{n} \sum_{j=1}^{2l+1} \left\langle (\lambda_1)_j^l, \mathbf{n}_j^l \cdot \frac{\partial \mathbf{r}_j^l}{\partial t} \right\rangle_{L^2}$$

$$\sum_{l=0}^{n-1} \sum_{j=1}^{2l+1} \left\langle (\lambda_2)_j^l, \mathbf{n}_j^l \cdot \mathbf{n}_j^l - 1 \right\rangle_{L^2}, \tag{4.38}$$

where $\{E_j^l, (\lambda_1)_j^l, (\lambda_2)_j^l\}$ are appropriate Lagrange multipliers concerning the respective constraints, for $j \in \{1, \cdots, 2l+1\}$ and $\forall l \in \{0, \ldots, n-1\}$. Moreover, generically,

$$\langle f, g \rangle_{L^2} = \int_0^T \int_\Omega fg\, dx dt, \quad \forall f, g \in L^2(\Omega \times [0, T]),$$

and

$$\langle \mathbf{f}, \mathbf{g} \rangle_{L^2} = \int_0^T \int_\Omega \mathbf{f} \cdot \mathbf{g}\, dx dt,$$

$\forall \mathbf{f}, \mathbf{g} \in L^2(\Omega \times [0, T]; \mathbb{R}^3)$.

$$U = \{\mathbf{r} = (\{\mathbf{r}_j^l\} \in W^{1,2}(\Omega \times [0, T]; \mathbb{R}^{3Z})$$

$$\text{such that } \mathbf{r}_k^l = \tilde{\mathbf{r}}_0 \text{ on } \partial\Omega, \ \forall l \in \{0, \cdots, n-1\}, \ k \in \{1, \ldots, 2l+1\}\},$$

$$U_1 = \{\varphi = \{\varphi_j^l\} \in W^{1,2}(\Omega; \mathbb{R}^Z) : \varphi = \mathbf{0}, \text{ on } \partial\Omega\} = W_0^{1,2}(\Omega; \mathbb{R}^Z),$$

$$\mathbf{n} = \{\mathbf{n}_j^l\}, \in U_2 = L^2([0, T]; W^{1,2}(\Omega; \mathbb{R}^{3Z}))$$

and

$$E = \{E_j^l\} \in \mathbb{R}^{3Z}.$$

Considering the Euler-Lagrange equation related to the variation in $\mathbf{r}_j^l$ for such a functional, we get

$$m_{p_j}^l |\varphi_j^l(x)|^2 \frac{\partial^2 \mathbf{r}_j^l(x,t)}{\partial t^2}$$

$$+ A_j^l \nabla^2 \mathbf{r}_j^l$$

$$- Ke^2 |\varphi_j^l(x)|^2 \sum_{l_1=0}^{n-1} \sum_{k=1}^{2l_1+1} \int_\Omega \frac{|\varphi_k^{l_1}(\tilde{x})|^2 (\mathbf{r}_j^l(x,t) - \mathbf{r}_k^{l_1}(\tilde{x},t))}{|\mathbf{r}_j^l(x,t) - \mathbf{r}_k^{l_1}(\tilde{x},t)|^3} d\tilde{x}$$

$$+ Ke^2 Z \frac{|\varphi_j^l(x)|^2 \mathbf{r}_j^l(x,t)}{|\mathbf{r}_j^l(x,t)|^3}$$

$$- \frac{\partial ((\lambda_1)_j^l \mathbf{n}_j^l)}{\partial t} = 0, \text{ in } \Omega \times [0,T]. \tag{4.39}$$

We shall also look for a solution such that, in spherical coordinates

$$\mathbf{r}_j(r,\theta,\phi,t)$$
$$= (r_j(r,\theta,\phi,t), \theta_j(r,\theta,\phi,t), \phi_j(r,\theta,\phi,t)). \tag{4.40}$$

If A_j^l is big enough, we may obtain the approximate solution

$$r_j^l \approx r,$$

so that an approximate solution may be given by

$$\mathbf{r}_j^l = r\mathbf{e}_{\mathbf{r}_j^l},$$

where

$$\mathbf{e}_{\mathbf{r}_j^l} = \cos(\omega_l t + \alpha_j^l)\mathbf{k} + \sin(\omega_l t + \alpha_j^l)\sin\phi_j^l \mathbf{j} + \sin(\omega_l t + \alpha_j^l)\cos\phi_j^l \mathbf{i},$$

and thus

$$\mathbf{e}_{\mathbf{r}_j^l} = \mathbf{n}_j^l.$$

Hence,

$$\mathbf{n}_j^l \cdot \frac{\partial \mathbf{r}_j^l}{\partial t} = 0, \text{ in } \Omega \times [0,T].$$

With such results in mind, the variation in φ_j^l gives us

$$-\gamma \sum_{k=1}^{3} \frac{\partial^2 \varphi_j^l(x)}{\partial x_k^2}$$

$$+m_{p_j} \varphi_j^l(x) \frac{1}{T} \int_0^T \frac{\partial \mathbf{r}_j^l}{\partial t} \cdot \frac{\partial \mathbf{r}_j^l}{\partial t} \, dt$$

$$+\frac{K}{T} e^2 \varphi_j^l(x) \sum_{l_1=0}^{n-1} \sum_{k=1}^{2l_1+1} \int_0^T \int_\Omega \frac{|\varphi_j^l(\tilde{x})|^2}{|\mathbf{r}_j^l(x,t) - \mathbf{r}_k^{l_1}(\tilde{x},t)|} \, d\tilde{x} \, dt$$

$$-2\frac{KeZ}{T} \int_0^T \frac{\varphi_j^l(x)}{|\mathbf{r}_j^l(x,t)|} \, dt + \frac{1}{T} \int_0^T V_j^l(\mathbf{r}_j^l(x,t)) \varphi_j(x) \, dt$$

$$-E_j^l \varphi_j^l(x) = 0, \text{ in } \Omega, \ \forall j \in \{1, \cdots, N\}. \tag{4.41}$$

At this point, we would recall that,

$$|\mathbf{r}_j^l(x,t) - \mathbf{r}_k^l(\tilde{x},t)| = r^2 + \tilde{r}^2 - 2r\tilde{r}\cos(\alpha_{jk}^l), \ \forall j,k \in \{1, \cdots, 2l+1\}, \tag{4.42}$$

Considering such statements, we shall look for solutions φ_j in the form,

$$\varphi_j^l(r, \theta, \phi) = h_j \hat{\varphi}_j^l(r) [L^-]^{j-1} \{[\sin(\theta)]^l e^{i\,l\phi}\},$$

where $l \in \{0, \ldots, n-1\}$, $\forall j \in \{1, \cdots, 2l+1\}$.

Here, h_j arc such that

$$h_j \sqrt{\int_{\phi=0}^{\phi=2\pi} \int_{\theta=0}^{\theta=\pi} |[L^-]^{j-1}\{[\sin(\theta)]^l e^{il\phi}\}|^2 \sin\theta \, d\theta \, d\phi} = 1.$$

From this and equation (4.41), neglecting the classical part, also from (4.42), we get

$$-\frac{\gamma}{r^2} \frac{\partial}{\partial r} \left[r^2 \frac{\partial \hat{\varphi}_j(r)}{\partial r} \right] + \frac{\gamma l(l+1)}{r^2} \hat{\varphi}_j^l(r)$$

$$+Ke^2 \hat{\varphi}_j^l(r) \sum_{k=1}^{2l+1} \int_\Omega \frac{|\hat{\varphi}_k^l(\tilde{r})|^2 ||h_k^l L^- [\sin(\tilde{\theta})]^l e^{il\tilde{\phi}}]|^2}{\sqrt{r^2 + \tilde{r}^2 - 2r\tilde{r}\cos(\alpha_{jk}^l)}}$$

$$\times \tilde{r}^2 \sin(\tilde{\theta}) \, d\tilde{r} \, d\tilde{\theta} \, d\tilde{\phi} T$$

$$+Ke^2 \hat{\varphi}_j^l(r) \sum_{l_1=0, l_1 \neq l}^{n-1} \sum_{k=1}^{2l_1+1} \frac{1}{T} \int_0^T \int_\Omega \frac{|\hat{\varphi}_k^{l_1}(\tilde{r})|^2 ||h_k^{l_1} L^- [\sin(\tilde{\theta})]^l e^{il\tilde{\phi}}]|^2}{\sqrt{r^2 + \tilde{r}^2 - 2r\tilde{r}\cos(\alpha_{j^l k^{l_1}})}}$$

$$\times \tilde{r}^2 \sin(\tilde{\theta}) \, d\tilde{r} \, d\tilde{\theta} \, d\tilde{\phi} \, dt$$

$$-2Ke^2 Z \frac{\hat{\varphi}_j^l(r)}{r} + \frac{1}{T} \int_0^T V_j^l(r) \hat{\varphi}_j^l(r) \, dt$$

$$-E_j^l \hat{\varphi}_j^l(r) = 0, \text{ in } \Omega, \ \forall l \in \{0, \ldots, n-1\}, \ j \in \{1, \ldots, 2l+1\}. \tag{4.43}$$

With such assumptions and statements in mind, denoting $W \sum_{l=0}^{n-1}(2l+1)(l)$ for an appropriate set $\hat{\Omega}$, we define the following control problem.

$$\min_{\{\alpha_{jk}^l\} \in \mathbb{R}^W} J_1(\varphi, \alpha)$$

where

$$J_1(\varphi, \alpha)$$
$$= K_3 \sum_{l=0}^{n-1} \sum_{j,k=1}^{2l+1} \int_0^{R_0} |\frac{|\hat{\varphi}(r)|^2 \hat{\varphi}(\tilde{r})|^2}{\sqrt{r^2 + \tilde{r}^2 - 2r\tilde{r}\cos(\alpha_{jk}^l)}} r^2 \tilde{r}^2 \, dr d\tilde{r}, \qquad (4.44)$$

subject to

$$-\frac{\gamma}{r^2}\frac{\partial}{\partial r}\left[r^2\frac{\partial \hat{\varphi}_j^l(r)}{\partial r}\right] + \frac{\gamma l(l+1)}{r^2}\hat{\varphi}_j^l(r)$$

$$+Ke^2\hat{\varphi}_j^l(r)\sum_{k=1}^{2l+1}\int_{\Omega}\frac{|\hat{\varphi}_k^l(\tilde{r})|^2 ||h_k^l L^-[\sin(\tilde{\theta})^l e^{il\tilde{\phi}}]|^2}{\sqrt{r^2 + \tilde{r}^2 - 2r\tilde{r}\cos(\alpha_{jk}^l)}}$$

$$\times \tilde{r}^2 \sin(\tilde{\theta}) \, d\tilde{r} \, d\tilde{\theta} \, d\tilde{\phi} T$$

$$+Ke^2\hat{\varphi}_j^l(r)\sum_{l_1=0, l_1 \neq l}^{n-1}\sum_{k=1}^{2l_1+1}\frac{1}{T}\int_0^T\int_{\Omega}\frac{|\hat{\varphi}_k^{l_1}(\tilde{r})|^2 ||h_k^{l_1}L^-[\sin(\tilde{\theta})^l e^{il\tilde{\phi}}]|^2}{\sqrt{r^2 + \tilde{r}^2 - 2r\tilde{r}\cos(\alpha_{j^l k^{l_1}})}}$$

$$\times \tilde{r}^2 \sin(\tilde{\theta}) \, d\tilde{r} \, d\tilde{\theta} \, d\tilde{\phi} \, dt$$

$$-2Ke^2 Z\frac{\hat{\varphi}_j^l(r)}{r} + \frac{1}{T}\int_0^T V_j^l(r)\hat{\varphi}_j^l(r) \, dt$$

$$-E_j^l\hat{\varphi}_j^l(r) = 0, \text{ in } \Omega, \forall l \in \{0, \ldots, n-1\}, j \in \{1, \ldots, 2l+1\}. \quad (4.45)$$

and

$$\int_0^{R_0} |\hat{\varphi}_j^l(r)|^2 r^2 \, dr = 1, \forall l \in \{0, \ldots, n-1\}, j \in \{1, \ldots, 2l+1\}. \qquad (4.46)$$

4.4.1 A numerical example

We have computed the wave functions for a hydrogen atom with both proton and electron allowed to move.

For the wave functions $\hat{\varphi}_1(r)$ and $\hat{\varphi}_2(r)$ obtained, please see Figures 4.7 and 4.8.

For the energies, we have obtained,

1. $E_1 = 4.7500$, for the electron,

2. $E_2 = 4.3220$, for the proton.

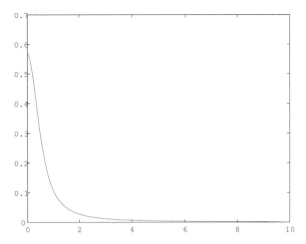

Figure 4.7: Radial solution $\hat{\varphi}_1(r)^2$ for the electron.

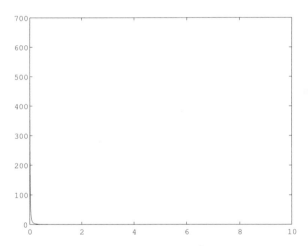

Figure 4.8: Radial solution $\hat{\varphi}_2(r)^2$ for the proton.

Remark 4.1 Observe that the mass of the heavier proton concentrates about $\mathbf{r} = 0$, as expected. ∎

4.4.2 Another numerical example

We have also computed the wave functions for an atom with three protons localized at $\mathbf{r} = 0$, and three electrons with wave functions $\hat{\varphi}_1(r)$, $\hat{\varphi}_2(r)$ and $\hat{\varphi}_3(r)$ and energies E_1, E_2 and E_3, respectively.

For the solutions for $\hat{\varphi}_1(r)^2$, $\hat{\varphi}_2(r)^2$ and $\hat{\varphi}_3(r)^2$ with $l_1 = l_2 = l_3 = 1$, please see Figures 4.9, 4.10 and 4.11, respectively.

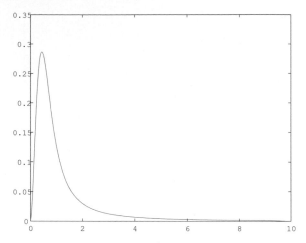

Figure 4.9: Radial solution $\hat{\varphi}_1(r)^2$ for the first electron with $l_1 = 1$.

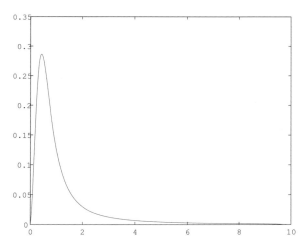

Figure 4.10: Radial solution $\hat{\varphi}_2(r)^2$ for the second electron with $l_2 = 1$.

For the energies, we have obtained,

1. $E_1 = 4.4692$,

2. $E_2 = 4.4692$,

3. $E_3 = 4.4693$.

For the angles α_1, α_2 and α_3 we have obtained,

1. $\alpha_1 = 242.35°$,

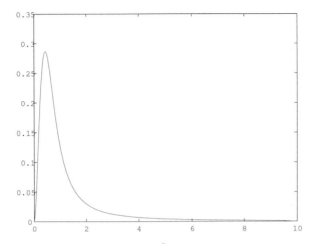

Figure 4.11: Radial solution $\hat{\phi}_3(r)^2$ for the third electron with $l_3 = 1$.

2. $\alpha_2 = 2.47°$,

3. $\alpha_3 = 122.39°$.

Remark 4.2 Observe that the angle between each pair of electrons is about $120°$, as expected. ∎

4.5 An approximation for the standard many body problem

Consider a many body system of N particles (protons/electrons). Denoting by $\Omega = [B_{r_0}(\mathbf{0})]^N$ the system domain, the standard many body system energy is given by $J : U \times \mathbb{R} \to \mathbb{R}$, where

$$
\begin{aligned}
J(R,E) \;=\; & \frac{1}{2} \sum_{j=1}^{N} \gamma_j \int_{\Omega} \nabla_j R(\mathbf{r}_1, \cdots, \mathbf{r}_N) \cdot \nabla_j R(\mathbf{r}_1, \cdots, \mathbf{r}_N)^* \, d\mathbf{r}_1 \cdots d\mathbf{r}_N \\
& + \frac{1}{2} \sum_{j,k=1, \, j \neq k}^{N} \int_{\Omega} K \frac{q_j q_k |R(\mathbf{r}_1, \cdots, \mathbf{r}_N)|^2}{|\mathbf{r}_j - \mathbf{r}_k|} \, d\mathbf{r}_1 \cdots d\mathbf{r}_N \\
& - E \left(\int_{\Omega} |R(\mathbf{r}_1, \cdots, \mathbf{r}_N)|^2 \, d\mathbf{r}_1 \cdots d\mathbf{r}_N - 1 \right),
\end{aligned}
\tag{4.47}
$$

where

$$
U = \{ R \in W^{1,2}(\Omega) \; : \; R = 0, \text{ on } \partial\Omega \}.
$$

Here, in spherical coordinates $\mathbf{r}_k = (x_k, y_k, z_k) \in \mathbb{R}^3$ would correspond to $(r_k, \theta_k, \phi_k), \forall k \in \{1, \cdots, N\}$. Concerning the functional indicated in (4.47), the corresponding Euler-Lagrange equations are given by,

$$- \sum_{j=1}^{N} \gamma_j \nabla_j^2 R(\mathbf{r}_1, \cdots, \mathbf{r}_N)$$

$$-K \sum_{j,k=1,\, j \neq k}^{N} \frac{q_j\, q_k R(\mathbf{r}_1, \cdots, \mathbf{r}_N)}{|\mathbf{r}_j - \mathbf{r}_k|} - ER(\mathbf{r}_1, \cdots, \mathbf{r}_N) = 0, \quad \text{in } \Omega. \quad (4.48)$$

Observe that denoting by α_{jk} the angle between $\mathbf{r}_j$ and $\mathbf{r}_k$, we have

$$\begin{aligned}
\mathbf{r}_j \cdot \mathbf{r}_k &= r_j r_k \cos \alpha_{jk} \\
&= r_j r_k [\cos \theta_j \mathbf{k} + \sin \theta_j \sin \phi_j \mathbf{j} \\
&\quad + \sin \theta_j \cos \phi_j \mathbf{i}] \cdot [\cos \theta_k \mathbf{k} + \sin \theta_k \sin \phi_k \mathbf{j} \\
&\quad + \sin \theta_k \cos \phi_k \mathbf{i}],
\end{aligned} \quad (4.49)$$

so that

$$\begin{aligned}
\cos \alpha_{jk} &= \cos \theta_j \cos \theta_k \\
&\quad + \sin \theta_j \sin \phi_j \sin \theta_k \sin \phi_k \\
&\quad + \sin \theta_j \cos \phi_j \sin \theta_k \cos \phi_k.
\end{aligned} \quad (4.50)$$

From a classical result in multi-variable calculus, there exist functions $\beta_{jk}(r_1, \cdots, r_N)$ such that

$$\int_\Omega \frac{|R(\mathbf{r}_1, \cdots, \mathbf{r}_N)|^2}{|\mathbf{r}_j - \mathbf{r}_k|}\, d\mathbf{r}_1 \cdots d\mathbf{r}_N$$

$$= \int_{[0,r_0]^N} \int_{\theta_1=0}^{\theta_1=\pi} \cdots \int_{\theta_N=0}^{\theta_N=\pi} \int_{\phi_1=0}^{\phi_1=2\pi} \cdots \int_{\phi_N=0}^{\phi_N=2\pi}$$

$$\left(\frac{|R(\mathbf{r}_1, \cdots, \mathbf{r}_N)|^2}{\sqrt{r_j^2 + r_k^2 - 2r_j r_k \cos(\alpha_{jk})}} \right)$$

$$\times \prod_{l=1}^{N} r_l^2 \, \sin \theta_l \, d\theta_l \, d\phi_l \, dr_l$$

$$= \int_{[0,r_0]^N} \int_{\theta_1=0}^{\theta_1=\pi} \cdots \int_{\theta_N=0}^{\theta_N=\pi} \int_{\phi_1=0}^{\phi_1=2\pi} \cdots \int_{\phi_N=0}^{\phi_N=2\pi}$$

$$\left(\frac{|R(\mathbf{r}_1, \cdots, \mathbf{r}_N)|^2}{\sqrt{r_j^2 + r_k^2 - 2r_j r_k \cos(\beta_{jk}(r_1, \cdots, r_N))}} \right)$$

$$\times \prod_{l=1}^{N} r_l^2 \, \sin \theta_l \, d\theta_l \, d\phi_l \, dr_l. \quad (4.51)$$

Hence,

$$
\begin{aligned}
J(R,E) \;=\; & \frac{1}{2}\sum_{j=1}^{N}\gamma_j\int_{\Omega}\nabla_j R(\mathbf{r}_1,\cdots,\mathbf{r}_N)\cdot\nabla_j R(\mathbf{r}_1,\cdots,\mathbf{r}_N)^*\,d\mathbf{r}_1\cdots d\mathbf{r}_N \\
& +\frac{1}{2}\sum_{j,k=1,\,j\neq k}^{N}\int_{\Omega}K\frac{q_j q_k|R(\mathbf{r}_1,\cdots,\mathbf{r}_N)|^2}{\sqrt{r_j^2+r_k^2-2r_j r_k\cos(\beta_{jk}(\mathbf{r}_1,\cdots,\mathbf{r}_N))}}\,d\mathbf{r}_1\cdots d\mathbf{r}_N \\
& -E\left(\int_{\Omega}|R(\mathbf{r}_1,\cdots,\mathbf{r}_N)|^2\,d\mathbf{r}_1\cdots d\mathbf{r}_N-1\right),
\end{aligned}
\tag{4.52}
$$

With such a functional in mind, we shall look for a solution such that,

$$
R(\mathbf{r}_1,\cdots,\mathbf{r}_N)=\hat{R}(r_1,\cdots,r_N)\prod_{j=1}^{N}h_j[\sin(\theta_j)]^{l_j}e^{i\,l_j\phi_j},
\tag{4.53}
$$

where $l_j\in\mathbb{N}$, $\forall j\in\{1,\cdots,N\}$, and h_j is given by

$$
h_j=\left(\sqrt{\int_0^{\pi}\int_0^{2\pi}|[\sin(\theta_j)]^{l_j}e^{i\,l_j\phi_j}|^2\sin(\theta_j)\,d\theta_j\,d\phi_j}\right)^{-1}.
$$

Observe that, in spherical coordinates,

$$
\begin{aligned}
& \nabla_j^2 R(\mathbf{r}_1,\cdots,\mathbf{r}_N) \\
=\;& \frac{1}{r_j^2}\frac{\partial}{\partial r_j}\left[\frac{r_j^2\partial R(\mathbf{r}_1,\cdots,\mathbf{r}_N)}{\partial r_j}\right] \\
& +\frac{1}{r_j^2}\frac{1}{\sin(\theta_j)}\frac{\partial}{\partial\theta_j}\left[\sin(\theta_j)\frac{\partial R(\mathbf{r}_1,\cdots,\mathbf{r}_N)}{\partial\theta_j}\right] \\
& +\frac{1}{r_j^2[\sin(\theta_j)]^2}\frac{\partial^2 R(\mathbf{r}_1,\cdots,\mathbf{r}_N)}{\partial\phi_j^2},
\end{aligned}
\tag{4.54}
$$

so that for $R(\mathbf{r}_1,\cdots,\mathbf{r}_N)$ indicated in (4.53), the corresponding Euler Lagrange equation would stand for,

$$
\begin{aligned}
& -\sum_{j=1}^{N}\frac{\gamma_j}{r_j^2}\frac{\partial}{\partial r_j}\left[\frac{r_j^2\partial\hat{R}(r_1,\cdots,r_N)}{\partial r_j}\right]+\sum_{j=1}^{N}\frac{\gamma_j}{r_j^2}l_j(l_j+1)\hat{R}(r_1,\cdots,r_N) \\
& +K\sum_{j,k=1,\,j\neq k}^{N}\frac{q_j\,q_k\,\hat{R}(r_1,\cdots,r_N)}{\sqrt{r_j^2+r_k^2-2r_j r_k\cos(\beta_{jk}(r_1,\cdots,r_N))}} \\
& -E\hat{R}(r_1,\cdots,r_N)=0,\ \text{in }\Omega.
\end{aligned}
\tag{4.55}
$$

With such results and statements in mind, we shall define the following control problem.

$$\min_{(\hat{R},\{\beta_{jk}\})\in U\times[W^{1,2}([0,r_0]^N)]^{N\times N}} J_2(\hat{R},\{\beta_{jk}\}),$$

where,

$$J_2(\hat{R},\{\beta_{jk}\})$$

$$= \left| \int_\Omega \sum_{j,k=1,\,j\neq k} q_j\, q_k \right.$$

$$\times \left(\frac{|\hat{R}(r_1,\cdots,r_N)|^2 \prod_{s=1}^N |(\sin\theta_s)^{l_s} e^{il x_s \phi_s}|^2}{\sqrt{r_j^2 + r_k^2 - 2r_j r_k \cos(\beta_{jk}(r_1,\cdots,r_N))}} \right.$$

$$\left. \left. - \frac{|\hat{R}(r_1,\cdots,r_N)|^2 \prod_{s=1}^N |(\sin\theta_s)^{l_s} e^{il_s\phi_s}|^2}{|\mathbf{r}_j - \mathbf{r}_k|}\, d\mathbf{r}_1\cdots d\mathbf{r}_N \right) \right|^2 \quad (4.56)$$

subject to

$$-\sum_{j=1}^N \frac{\gamma_j}{r_j^2}\frac{\partial}{\partial r_j}\left[\frac{r_j^2 \partial \hat{R}(r_1,\cdots,r_N)}{\partial r_j}\right] + \sum_{j=1}^N \frac{\gamma_j}{r_j^2} l_j(l_j+1)\hat{R}(r_1,\cdots,r_N)$$

$$+K \sum_{j,k=1,\,j\neq k}^N \frac{q_j\, q_k\, \hat{R}(r_1,\cdots,r_N)}{\sqrt{r_j^2 + r_k^2 - 2r_j r_k \cos(\beta_{jk}(r_1,\cdots,r_N))}}$$

$$-E\hat{R}(r_1,\cdots,r_N) = 0,\ \text{in}\ \Omega, \quad (4.57)$$

and

$$\int_0^{r_0}\cdots\int_0^{r_0} |\hat{R}(r_1,\cdots,r_N)|^2 \prod_{j=1}^N r_j^2\, dr_j = 1.$$

4.6 Conclusion

In this chapter, we have developed a variational formulation for the relativistic Klein-Gordon equation by extending the standard classical mechanics energy to a more general functional, obtained thorough the introduction of the normal field concept.

We believe the results presented here may be applied and extended to other models in mechanics, including the quantum and relativistic approaches for the study of atoms and molecules.

Finally, we have the objective and interest in applying such results also for the case in which electromagnetic fields are included.

Chapter 5

A Variational Formulation for Relativistic Mechanics Based on Riemannian Geometry and its Application to the Quantum Mechanics Context

5.1 Introduction

In this article, we develop a variational formulation suitable for the relativistic quantum mechanics approach in a free particle context. The results are based on fundamental concepts of Riemannian geometry and suitable extensions for the relativistic case. Definitions such as vector fields, connection, Lie Bracket and Riemann tensor are addressed in the subsequent sections for the main energy construction.

Indeed, the action developed in this article, in some sense, generalizes and extends the one presented in the Weinberg book [53], in Chapter 12 on page 358. In such a book, the concerned action is denoted by $I = I_M + I_G$, where I_M, the matter action, for N particles with mass m_n and charge e_n, $\forall n \in \{1,\dots,N\}$, is given by

$$
\begin{aligned}
I_M = & -\sum_{n=1}^{N} m_n \int_{-\infty}^{\infty} \sqrt{-g_{\mu\nu}(x(p))\frac{dx_n^{\mu}(p)}{dp}\frac{dx_n^{\nu}(p)}{dp}}\, dp \\
& -\frac{1}{4}\int_{\Omega}\sqrt{g}F_{\mu\nu}F^{\mu\nu}\, d^4x \\
& +\sum_{n=1}^{N} e_n \int_{-\infty}^{\infty} \frac{dx_n^{\mu}(p)}{dp} A_{\mu}(x(p))\, dp,
\end{aligned} \tag{5.1}
$$

where $\{x_n(p)\}$ is the position field with concerning metrics $\{g_{\mu\nu}(x(p))\}$ and

$$
F_{\mu\nu} = \frac{\partial A_\nu}{\partial x_\mu} - \frac{\partial A_\mu}{\partial x_\nu}
$$

represents the electromagnetic tensor field through a vectorial potential $\{A_\mu\}$.

Moreover, the gravitational action I_G is defined by

$$
I_G = -\frac{1}{16\pi G}\int_{\Omega} R(x)\sqrt{g}\, d^4x,
$$

where

$$
R = g^{\mu\nu}R_{\mu\nu}.
$$

Here,

$$
R_{\mu\nu} = R^{\sigma}_{\mu\sigma\nu},
$$

where

$$
R^{\eta}_{\mu\sigma\nu}
$$

are the components of the well known Riemann curvature tensor.

According to [53], the Euler-Lagrange equations for I correspond to the Einstein field equations,

$$
R^{\mu\nu} - \frac{1}{2}g^{\mu\nu}R + 8\pi G T^{\mu\nu} = 0,
$$

where the energy-momentum tensor $T^{\mu\nu}$ is expressed by

$$
\begin{aligned}
T^{\lambda\kappa} = & \sqrt{g}\sum_{n=1}^{N}\int_{-\infty}^{\infty}\frac{dx_n^{\lambda}(p)}{d\tau_n}\frac{dx_n^{\kappa}(p)}{d\tau_n}\delta^4(x-x_n)\, d\tau_n \\
& +F^{\lambda}_{\mu}(x)F^{\mu\kappa}(x) - \frac{1}{4}g^{\lambda\kappa}F_{\mu\nu}F^{\mu\nu}.
\end{aligned} \tag{5.2}
$$

One of the main differences of our model from this previous one, is that we consider a possible variation in the density along the mechanical system.

Also, in our model, the motion of the system in question is specified by a four-dimensional manifold given by the function

$$\mathbf{r}(\hat{\mathbf{u}}(\mathbf{x},t)) = (ct, X_1(\mathbf{u}(\mathbf{x},t)), X_2(\mathbf{u}(\mathbf{x},t)), X_3(\mathbf{u}(\mathbf{x},t))),$$

with corresponding mass density

$$(\rho \circ \hat{\mathbf{u}}) : \Omega \times [0,T] \to \mathbb{R}^+,$$

where $\Omega \subset \mathbb{R}^3$ and $[0,T]$ is a time interval. At this point, we define $\phi(\hat{\mathbf{u}}(\mathbf{x},t))$ as a complex function such that

$$|\phi(\hat{\mathbf{u}}(\mathbf{x},t))|^2 = \frac{\rho(\hat{\mathbf{u}}(\mathbf{x},t))}{m},$$

where m denotes the total system mass at rest. We emphasize that it seems to be clear in the previous book that the parametrization of the position field, through the parameter p, is one-dimensional.

In this work, we do not consider the presence of electromagnetic fields.

Anyway, the final expression of the related new action developed here is given by

$$
\begin{aligned}
&J(\phi, \mathbf{r}, \hat{\mathbf{u}}, E) \\
&= \int_0^T \int_\Omega mc \sqrt{-g_{ij} \frac{\partial u_i}{\partial t} \frac{\partial u_j}{\partial t}} \, |\phi(\hat{\mathbf{u}}(\mathbf{x},t))|^2 \, \sqrt{-g} \, |\det(\hat{\mathbf{u}}'(\mathbf{x},t))| \, d\mathbf{x} \, dt \\
&+ \frac{\gamma}{2} \int_0^T \int_\Omega g^{jk} \frac{\partial \phi}{\partial u_j} \frac{\partial \phi^*}{\partial u_k} \, \sqrt{-g} \, |\det(\hat{\mathbf{u}}'(\mathbf{x},t))| \, d\mathbf{x} dt \\
&+ \frac{\gamma}{4} \int_0^T \int_\Omega g^{jk} \left(\phi^* \frac{\partial \phi}{\partial u_l} + \phi \frac{\partial \phi^*}{\partial u_l} \right) \Gamma^l_{jk} \, \sqrt{-g} \, |\det(\hat{\mathbf{u}}'(\mathbf{x},t))| \, d\mathbf{x} dt \\
&+ \frac{\gamma}{2} \int_0^T \int_\Omega |\phi|^2 \, g^{jk} \left(\frac{\partial \Gamma^l_{lk}}{\partial u_j} - \frac{\partial \Gamma^l_{jk}}{\partial u_l} + \Gamma^p_{lk} \Gamma^l_{jp} - \Gamma^p_{jk} \Gamma^l_{lp} \right) \sqrt{-g} \, |\det(\hat{\mathbf{u}}'(\mathbf{x},t))| \, d\mathbf{x} dt \\
&- \int_0^T E(t) \left(\int_\Omega |\phi(\hat{\mathbf{u}}(\mathbf{x},t))|^2 \, \sqrt{-g} \, |\det(\hat{\mathbf{u}}'(\mathbf{x},t))| \, d\mathbf{x}/c - 1 \right) c \, dt.
\end{aligned}
\tag{5.3}
$$

Observe that the action part

$$
\begin{aligned}
&\frac{\gamma}{2} \int_0^T \int_\Omega |\phi|^2 \, g^{jk} \left(\frac{\partial \Gamma^l_{lk}}{\partial u_j} - \frac{\partial \Gamma^l_{jk}}{\partial u_l} + \Gamma^p_{lk} \Gamma^l_{jp} - \Gamma^p_{jk} \Gamma^l_{lp} \right) \sqrt{-g} \, |\det(\hat{\mathbf{u}}'(\mathbf{x},t))| \, d\mathbf{x} dt \\
&= \frac{\gamma}{2} \int_0^T \int_\Omega |\phi|^2 \hat{R} \sqrt{-g} \, |\det(\hat{\mathbf{u}}'(\mathbf{x},t))| \, d\mathbf{x} dt,
\end{aligned}
\tag{5.4}
$$

where

$$\hat{R} = g^{jk} \hat{R}_{jk},$$
$$\hat{R}_{jk} = \hat{R}^l_{jlk}$$

and

$$\hat{R}^l_{ijk} = \frac{\partial \Gamma^l_{jk}}{\partial u_i} - \frac{\partial \Gamma^l_{ik}}{\partial u_j} + \Gamma^p_{jk}\,\Gamma^l_{ip} - \Gamma^p_{ik}\,\Gamma^l_{jp}$$

represents the Riemann curvature tensor, corresponds to the Hilbert-Einstein one, as specified in the subsequent sections.

In the last section, we show how such a formulation may result, as an approximation, the well known relativistic Klein-Gordon one and the respective Euler-Lagrange equations. We believe the main results obtained may be extended to more complex mechanical systems, including in some extent, the quantum mechanics approach.

Finally, about the references, details on the Sobolev Spaces involved may be found in [1, 13]. For standard references in quantum mechanics, we refer to [40, 37, 12] and the non-standard [11].

5.2 Some introductory topics on vector analysis and Riemannian geometry

In this section, we present some introductory remarks on Riemannian geometry. The results here developed have been presented in detail in [23]. For the sake of completeness, we repeat some of the proofs.

We start with the definition of surface in $\mathbb{R}^n$.

Definition 5.2.1 (Surface in $\mathbb{R}^n$, the respective tangent space and the dual one)
Let $D \subset \mathbb{R}^m$ be an open, bounded, connected set with a regular (Lipschitzian) boundary denoted by ∂D. We define an m-dimensional C^1 class surface $M \subset \mathbb{R}^n$, where $1 \leq m < n$, as the range of a function $\mathbf{r} : D \subset \mathbb{R}^m \to \mathbb{R}^n$, where

$$M = \{\mathbf{r}(\mathbf{u}) \ : \ \mathbf{u} = (u_1, \dots, u_m) \in D\}$$

and

$$\mathbf{r}(\mathbf{u}) = \hat{X}_1(\mathbf{u})\mathbf{e}_1 + \hat{X}_2(\mathbf{u})\mathbf{e}_2 + \cdots + \hat{X}_n(\mathbf{u})\mathbf{e}_n,$$

where $\hat{X}_k : D \to \mathbb{R}$ is a C^1 class function, $\forall k \in \{1, \dots, n\}$, and $\{\mathbf{e}_1, \dots, \mathbf{e}_n\}$ is the canonical basis of $\mathbb{R}^n$.

Let $\mathbf{u} \in D$ and $p = \mathbf{r}(\mathbf{u}) \in M$. We also define the tangent space of M at p, denoted by $T_p(M)$, as

$$T_p(M) = \left\{\alpha_1 \frac{\partial \mathbf{r}(\mathbf{u})}{\partial u_1} + \cdots + \alpha_m \frac{\partial \mathbf{r}(\mathbf{u})}{\partial u_m} \ : \ \alpha_1, \dots, \alpha_m \in \mathbb{R}\right\}.$$

We assume

$$\left\{\frac{\partial \mathbf{r}(\mathbf{u})}{\partial u_1}, \cdots, \frac{\partial \mathbf{r}(\mathbf{u})}{\partial u_m}\right\}$$

to be a linearly independent set $\forall \mathbf{u} \in D$.

Finally, we define the dual space to $T_p(M)$, denoted by $T_p(M)^*$, as the set of all continuous and linear functionals (in fact real functions) defined on $T_p(M)$, that is,

$$T_p(M)^* = \left\{ f : T_p(M) \to \mathbb{R} \ : \ f(\mathbf{v}) = \alpha \cdot \mathbf{v}, \ \text{for some } \alpha \in \mathbb{R}^n, \ \forall \mathbf{v} = v_i \frac{\partial \mathbf{r}(\mathbf{u})}{\partial u_i} \in T_p(M) \right\}.$$

Theorem 5.2.2 *Let $M \subset \mathbb{R}^n$ be a m-dimensional C^1 class surface, where*

$$M = \{ \mathbf{r}(\mathbf{u}) \in \mathbb{R}^n \ : \ \mathbf{u} \in D \subset \mathbb{R}^m \}.$$

Let $\mathbf{u} \in D$, $p = \mathbf{r}(\mathbf{u}) \in D$ and $f \in C^1(M)$.
Define $df : T_p(M) \to \mathbb{R}$ by

$$df(\mathbf{v}) = \lim_{\varepsilon \to 0} \frac{(f \circ \mathbf{r})(\{u_i\} + \varepsilon\{v_i\}) - (f \circ \mathbf{r})(\{u_i\})}{\varepsilon},$$

$\forall \mathbf{v} = v_i \frac{\partial \mathbf{r}(\mathbf{u})}{\partial u_i} \in T_p(M)$.
Under such hypotheses,

$$df \in T_p(M)^*.$$

Reciprocally, let $F \in T_p(M)^*$.
Under such assumption, there exists $f \in C^1(M)$ such that

$$F(\mathbf{v}) - df(\mathbf{v}), \ \forall \mathbf{v} \in T_p(M).$$

Proof 5.1   Let $\mathbf{v} = v_i \frac{\partial \mathbf{r}(\mathbf{u})}{\partial u_i} \in T_p(M)$.
Thus,

$$
\begin{aligned}
df(\mathbf{v}) &= \lim_{\varepsilon \to 0} \frac{(f \circ \mathbf{r})(\{u_i\} + \varepsilon\{v_i\}) - (f \circ \mathbf{r})(\{u_i\})}{\varepsilon} \\
&= \frac{\partial (f \circ \mathbf{r})(\mathbf{u})}{\partial \hat{X}_j} \frac{\partial \hat{X}_j(\mathbf{u})}{\partial u_i} v_i \\
&= \nabla f(\mathbf{r}(\mathbf{u})) \cdot \mathbf{v} \\
&= \alpha \cdot \mathbf{v}, \quad\quad\quad\quad\quad\quad\quad\quad (5.5)
\end{aligned}
$$

where

$$\alpha = \nabla f(\mathbf{r}(\mathbf{u})),$$

so that $df \in T_p(M)^*$.
   Reciprocally, assume $F \in T_p(M)^*$, that is, suppose there exists $\alpha \in \mathbb{R}^n$ such that

$$F(\mathbf{v}) = \alpha \cdot \mathbf{v},$$

$\forall \mathbf{v} = v_i \frac{\partial \mathbf{r}(\mathbf{u})}{\partial u_i} \in T_p(M)$.
   Define $f : M \to \mathbb{R}$ by

$$f(\mathbf{w}) = \alpha \cdot \mathbf{w}, \ \forall \mathbf{w} \in M.$$

In particular,
$$f(\mathbf{r}(\mathbf{u})) = \alpha \cdot \mathbf{r}(\mathbf{u}) = \alpha_j \hat{X}_j(\mathbf{u}), \ \forall \mathbf{u} \in D.$$

For $p = \mathbf{r}(\mathbf{u}) \in M$ and $\mathbf{v} = v_i \frac{\partial \mathbf{r}(\mathbf{u})}{\partial u_i} \in T_p(M)$, we have

$$
\begin{aligned}
df(\mathbf{v}) &= \lim_{\varepsilon \to 0} \frac{(f \circ \mathbf{r})(\{u_i\} + \varepsilon\{v_i\}) - (f \circ \mathbf{r})(\{u_i\})}{\varepsilon} \\
&= \frac{\partial (f \circ \mathbf{r})(\mathbf{u})}{\partial \hat{X}_j} \frac{\partial \hat{X}_j(\mathbf{u})}{\partial u_i} v_i \\
&= \alpha_j \frac{\partial \hat{X}_j(\mathbf{u})}{\partial u_i} v_i \\
&= \alpha \cdot \mathbf{v},
\end{aligned}
\tag{5.6}
$$

Therefore,
$$F(\mathbf{v}) = df(\mathbf{v}), \ \forall \mathbf{v} \in T_p(M).$$

The proof is complete.

At this point, we present the tangential vector field definition, to be addressed in the subsequent results and sections.

Definition 5.2.3 (Vector field) *Let $M \subset \mathbb{R}^n$ be an m-dimensional C^1 class surface, where $1 \leq m < n$. We define the set of C^1 class tangential vector fields in M, denoted by $\mathscr{X}(M)$, as*

$$\mathscr{X}(M) = \left\{ X = X_i(\mathbf{u}) \frac{\partial \mathbf{r}(\mathbf{u})}{\partial u_i} \in T(M) = \{T_p(M) \ : \ p = \mathbf{r}(\mathbf{u}) \in M\} \right\},$$

where $X_i : D \to \mathbb{R}$ is a C^1 class function, $\forall i \in \{1, \ldots, m\}$.

Let $f \in C^1(M)$ and $X \in \mathscr{X}(M)$. We define the derivative of f on the direction X at $\mathbf{u}$, denoted by $(X \cdot f)(p)$, where $p = \mathbf{r}(\mathbf{u})$, as

$$
\begin{aligned}
(X \cdot f)(p) &= df(X(\mathbf{u})) \\
&= \lim_{\varepsilon \to 0} \frac{(f \circ \mathbf{r})(\{u_i\} + \varepsilon\{X_i(\mathbf{u})\}) - (f \circ \mathbf{r})(\{u_i\})}{\varepsilon} \\
&= \frac{\partial (f \circ \mathbf{r})(\mathbf{u})}{\partial u_i} X_i(\mathbf{u}).
\end{aligned}
\tag{5.7}
$$

The next definition is also very important for this work, namely, the connection one.

Definition 5.2.4 (Connection) *Let $M \subset \mathbb{R}^n$ be a m-dimensional C^1 class surface, where*

$$M = \{\mathbf{r}(\mathbf{u}) \in \mathbb{R}^n \ : \ \mathbf{u} \in D \subset \mathbb{R}^m\}$$

and

$$\mathbf{r}(\mathbf{u}) = \hat{X}_1(\mathbf{u})\mathbf{e}_1 + \cdots + \hat{X}_n(\mathbf{u})\mathbf{e}_n.$$

We define an affine connection on M, as a map $\nabla : \mathscr{X}(M) \times \mathscr{X}(M) \to \mathscr{X}(M)$ *such that*

1.
$$\nabla_{fX+gY} Z = f\nabla_X Z + g\nabla_Y Z,$$

2.
$$\nabla_X (Y+Z) = \nabla_X Y + \nabla_X Z$$

   and

3.
$$\nabla_X (fY) = (X \cdot f)Y + f\nabla_X Y,$$

$\forall X, Y, Z \in \mathscr{X}(M)$, $f, g \in C^{\infty}(M)$.

About the connection representation, we have the following result.

Theorem 5.2.5 *Let* $M \subset \mathbb{R}^n$ *be a m-dimensional* C^1 *class surface, where*

$$M = \{\mathbf{r}(\mathbf{u}) \in \mathbb{R}^n : \mathbf{u} \in D \subset \mathbb{R}^m\}$$

and

$$\mathbf{r}(\mathbf{u}) = \hat{X}_1(\mathbf{u})\mathbf{e}_1 + \cdots + \hat{X}_n(\mathbf{u})\mathbf{e}_n.$$

Let $\nabla : \mathscr{X}(M) \times \mathscr{X}(M) \to \mathscr{X}(M)$ *be an affine connection on M. Let* $\mathbf{u} \in D$, $p = \mathbf{r}(\mathbf{u}) \in M$ *and* $X, Y \in \mathscr{X}(M)$ *be such that*

$$X = X_i(\mathbf{u})\frac{\partial \mathbf{r}(\mathbf{u})}{\partial u_i},$$

and

$$Y = Y_i(\mathbf{u})\frac{\partial \mathbf{r}(\mathbf{u})}{\partial u_i}.$$

Under such hypotheses, we have

$$\nabla_X Y = \sum_{i=1}^m \left(X \cdot Y_i + \sum_{j,k=1}^m \Gamma^i_{jk} X_j Y_k \right) \frac{\partial \mathbf{r}(\mathbf{u})}{\partial u_i} \in T_p(M), \qquad (5.8)$$

where Γ^i_{jk} *are defined through the relations,*

$$\nabla_{\frac{\partial \mathbf{r}(\mathbf{u})}{\partial u_j}} \frac{\partial \mathbf{r}(\mathbf{u})}{\partial u_k} = \Gamma^i_{jk}(\mathbf{u})\frac{\partial \mathbf{r}(\mathbf{u})}{\partial u_i}.$$

Proof 5.2   Observe that

$$
\begin{aligned}
\nabla_X Y &= \nabla_{X_i \frac{\partial \mathbf{r(u)}}{\partial u_i}} \left(Y_j \frac{\partial \mathbf{r(u)}}{\partial u_j} \right) \\
&= X_i \nabla_{\frac{\partial \mathbf{r(u)}}{\partial u_i}} \left(Y_j \frac{\partial \mathbf{r(u)}}{\partial u_j} \right) \\
&= X_i \left(\frac{\partial \mathbf{r(u)}}{\partial u_i} \cdot Y_j \right) \frac{\partial \mathbf{r(u)}}{\partial u_j} + X_i Y_j \nabla_{\frac{\partial \mathbf{r(u)}}{\partial u_i}} \frac{\partial \mathbf{r(u)}}{\partial u_j} \\
&= \left(X_i \frac{\partial \mathbf{r(u)}}{\partial u_i} \cdot Y_j \right) \frac{\partial \mathbf{r(u)}}{\partial u_j} + X_i Y_j \Gamma^k_{ij} \frac{\partial \mathbf{r(u)}}{\partial u_k} \\
&= \sum_{i=1}^m \left(X \cdot Y_i + \sum_{j,k=1}^m \Gamma^i_{jk} X_j Y_k \right) \frac{\partial \mathbf{r(u)}}{\partial u_i}.
\end{aligned}
\tag{5.9}
$$

The proof is complete.

Remark 5.1   If the connection in question is such that

$$
\Gamma^i_{jk} = \frac{1}{2} g^{il} \left(\frac{\partial g_{kl}}{\partial u_j} + \frac{\partial g_{jl}}{\partial u_k} - \frac{\partial g_{jk}}{\partial u_l} \right)
$$

such a connection is said to be the Levi-Civita one. In the next lines, we assume the concerning connection is indeed the Levi-Civita one.   ■

We finish this section with the Lie Bracket definition.

Definition 5.2.6 (Lie bracket) *Let $M \subset \mathbb{R}^n$ be a C^1 class m-dimensional surface where $1 \leq m < n$.*

Let $X, Y \in \mathscr{X}(M)$, where $\mathscr{X}(M)$ denotes the set of the $C^\infty(M) = \cap_{k \in \mathbb{N}} C^k(M)$ class vector fields. We define the Lie bracket of X and Y, denoted by $[X, Y] \in \mathscr{X}(M)$, by

$$
[X, Y] = (X \cdot Y_i - Y \cdot X_i) \frac{\partial \mathbf{r(u)}}{\partial u_i},
$$

where

$$
X = X_i \frac{\partial \mathbf{r(u)}}{\partial u_i}
$$

and

$$
Y = Y_i \frac{\partial \mathbf{r(u)}}{\partial u_i}.
$$

5.3   A relativistic quantum mechanics action

In this section, we present a proposal for a relativistic quantum mechanics action.

Let $\Omega \subset \mathbb{R}^3$ be an open, bounded, connected set with a C^1 class boundary denoted by $\partial \Omega$. Denoting by c the speed of light, in a free volume context, for a

C^1 class function $\mathbf{r}$ and $\hat{\mathbf{u}} \in W^{1,2}(\Omega \times [0,T]; \mathbb{R}^4)$, let $(\mathbf{r} \circ \hat{\mathbf{u}}) : \Omega \times [0,T] \to \mathbb{R}^4$ be a particle position field where

$$\mathbf{r}(\hat{\mathbf{u}}(\mathbf{x},t)) = (ct, X_1(\mathbf{u}(\mathbf{x},t)), X_2(\mathbf{u}(\mathbf{x},t)), X_3(\mathbf{u}(\mathbf{x},t))),$$

with corresponding mass density

$$(\rho \circ \hat{\mathbf{u}}) : \Omega \times [0,T] \to \mathbb{R}^+,$$

where $[0,T]$ is a time interval.

We denote $\hat{\mathbf{u}} : \Omega \times [0,T] \to \mathbb{R}^4$ point-wise as

$$\hat{\mathbf{u}}(\mathbf{x},t) = (u_0(\mathbf{x},t), \mathbf{u}(\mathbf{x},t)),$$

where

$$u_0(\mathbf{x},t) = ct,$$

and

$$\mathbf{u}(\mathbf{x},t) = (u_1(\mathbf{x},t), u_2(\mathbf{x},t), u_3(\mathbf{x},t)),$$

$\forall (\mathbf{x},t) = ((x_1, x_2, x_3), t) \in \Omega \times [0,T]$.

At this point, we recall to have defined $\phi(\hat{\mathbf{u}}(\mathbf{x},t))$ as a complex function such that

$$|\phi(\hat{\mathbf{u}}(\mathbf{x},t))|^2 = \frac{\rho(\hat{\mathbf{u}}(\mathbf{x},t))}{m},$$

where m denotes the total system mass at rest. Also, we assume ϕ to be a C^2 class function and define

$$d\tau^2 = c^2 dt^2 - dX_1(\mathbf{u}(\mathbf{x},t))^2 - dX_2(\mathbf{u}(\mathbf{x},t))^2 - dX_3(\mathbf{u}(\mathbf{x},t))^2,$$

so that the mass differential will be denoted by

$$
\begin{aligned}
dm &= \frac{\rho(\hat{\mathbf{u}}(\mathbf{x},t))}{\sqrt{1 - \frac{v^2}{c^2}}} \sqrt{-g} \, |\det(\hat{\mathbf{u}}'(\mathbf{x},t))| \, d\mathbf{x} \\
&= \frac{m|\phi(\hat{\mathbf{u}}(\mathbf{x},t))|^2}{\sqrt{1 - \frac{v^2}{c^2}}} \sqrt{-g} \, |\det(\hat{\mathbf{u}}'(\mathbf{x},t))| \, d\mathbf{x}, \quad (5.10)
\end{aligned}
$$

where $d\mathbf{x} = dx_1 dx_2 dx_3$ and $\hat{\mathbf{u}}'(\mathbf{x},t)$ denotes the Jacobian matrix of the vectorial function $\hat{\mathbf{u}}(\mathbf{x},t)$.

Also,

$$\mathbf{g}_i = \frac{\partial \mathbf{r}(\hat{\mathbf{u}})}{\partial u_i}, \ \forall i \in \{0,1,2,3\},$$

$$g_{ij} = \mathbf{g}_i \cdot \mathbf{g}_j, \ \forall i,j \in \{0,1,2,3\},$$

and

$$g = \det\{g_{ij}\}.$$

Moreover,

$$\{g^{ij}\} = \{g_{ij}\}^{-1}.$$

5.3.1 The kinetics energy

Observe that

$$
\begin{aligned}
c^2 - v^2 &= -\frac{d\mathbf{r}(\hat{\mathbf{u}})}{dt} \cdot \frac{d\mathbf{r}(\hat{\mathbf{u}})}{dt} \\
&= -\left(\frac{\partial \mathbf{r}(\hat{\mathbf{u}})}{\partial u_i} \frac{\partial u_i}{\partial t}\right) \cdot \left(\frac{\partial \mathbf{r}(\hat{\mathbf{u}})}{\partial u_j} \frac{\partial u_j}{\partial t}\right) \\
&= -\frac{\partial \mathbf{r}(\hat{\mathbf{u}})}{\partial u_i} \cdot \frac{\partial \mathbf{r}(\hat{\mathbf{u}})}{\partial u_j} \frac{\partial u_i}{\partial t} \frac{\partial u_j}{\partial t} \\
&= -g_{ij} \frac{\partial u_i}{\partial t} \frac{\partial u_j}{\partial t},
\end{aligned} \tag{5.11}
$$

where the product in question is generically given by

$$
\mathbf{y} \cdot \mathbf{z} = -y_0 z_0 + \sum_{i=1}^{3} y_i z_i, \ \forall \mathbf{y} = (y_0, y_1, y_2, y_3), \ \mathbf{z} = (z_0, z_1, z_2, z_3) \in \mathbb{R}^4
$$

and

$$
v = \sqrt{\left(\frac{dX_1(\mathbf{u}(\mathbf{x},t))}{dt}\right)^2 + \left(\frac{dX_2(\mathbf{u}(\mathbf{x},t))}{dt}\right)^2 + \left(\frac{dX_3(\mathbf{u}(\mathbf{x},t))}{dt}\right)^2}.
$$

The semi-classical kinetics energy differential is given by

$$
\begin{aligned}
dE_c &= \frac{d\mathbf{r}(\hat{\mathbf{u}})}{dt} \cdot \frac{d\mathbf{r}(\hat{\mathbf{u}})}{dt} \, dm \\
&= -\left(\frac{d\tau}{dt}\right)^2 dm \\
&= -(c^2 - v^2) \, dm,
\end{aligned} \tag{5.12}
$$

so that

$$
\begin{aligned}
dE_c &= -m \frac{(c^2 - v^2)}{\sqrt{1 - \frac{v^2}{c^2}}} |\phi(\hat{\mathbf{u}})|^2 \sqrt{-g} \, |\det(\hat{\mathbf{u}}'(\mathbf{x},t))| \, d\mathbf{x} \\
&= -mc^2 \sqrt{1 - \frac{v^2}{c^2}} |\phi(\hat{\mathbf{u}})|^2 \sqrt{-g} \, |\det(\hat{\mathbf{u}}'(\mathbf{x},t))| \, d\mathbf{x} \\
&= -mc \sqrt{c^2 - v^2} |\phi(\hat{\mathbf{u}})|^2 \sqrt{-g} \, |\det(\hat{\mathbf{u}}'(\mathbf{x},t))| \, d\mathbf{x} \\
&= -mc \sqrt{-g_{ij} \frac{\partial u_i}{\partial t} \frac{\partial u_j}{\partial t}} |\phi(\hat{\mathbf{u}})|^2 \sqrt{-g} \, |\det(\hat{\mathbf{u}}'(\mathbf{x},t))| \, d\mathbf{x}, \tag{5.13}
\end{aligned}
$$

and thus, the semi-classical kinetics energy E_c is given by

$$
E_c = \int_0^T \int_\Omega dE_c \, dt,
$$

that is,

$$
E_c = -\int_0^T \int_\Omega mc \sqrt{-g_{ij} \frac{\partial u_i}{\partial t} \frac{\partial u_j}{\partial t}} |\phi(\hat{\mathbf{u}})|^2 \sqrt{-g} \, |\det(\hat{\mathbf{u}}'(\mathbf{x},t))| \, d\mathbf{x} \, dt.
$$

5.3.2 The energy part relating the curvature and wave function

At this point we define an energy part, related to the Riemann curvature tensor, denoted by E_q, where

$$E_q = \frac{\gamma}{2} \int_0^T \int_\Omega g^{jk} R_{jk} \sqrt{-g} \, |\det(\hat{\mathbf{u}}'(\mathbf{x},t))| \, d\mathbf{x} \, dt.$$

and

$$R_{jk} = Re[R^i_{jik}(\phi)].$$

Also, generically $Re[z]$ denotes the real part of $z \in \mathbb{C}$ and $R^l_{ijk}(\phi)$ is such that

$$\nabla_{\left(\phi \frac{\partial \mathbf{r}(\hat{\mathbf{u}})}{\partial u_i}\right)} \nabla_{\frac{\partial \mathbf{r}(\hat{\mathbf{u}})}{\partial u_j}} \left(\phi^* \frac{\partial \mathbf{r}(\hat{\mathbf{u}})}{\partial u_k}\right) - \nabla_{\left(\phi \frac{\partial \mathbf{r}(\hat{\mathbf{u}})}{\partial u_j}\right)} \nabla_{\frac{\partial \mathbf{r}(\hat{\mathbf{u}})}{\partial u_i}} \left(\phi^* \frac{\partial \mathbf{r}(\hat{\mathbf{u}})}{\partial u_k}\right)$$

$$-\nabla_{\left[\phi \frac{\partial \mathbf{r}(\hat{\mathbf{u}})}{\partial u_i}, \frac{\partial \mathbf{r}(\hat{\mathbf{u}})}{\partial u_j}\right]} \left(\phi^* \frac{\partial \mathbf{r}(\hat{\mathbf{u}})}{\partial u_k}\right) = R^l_{ijk}(\phi) \frac{\partial \mathbf{r}(\hat{\mathbf{u}})}{\partial u_l}. \tag{5.14}$$

More specifically, we have

$$\nabla_{\left(\phi \frac{\partial \mathbf{r}(\hat{\mathbf{u}})}{\partial u_i}\right)} \nabla_{\frac{\partial \mathbf{r}(\hat{\mathbf{u}})}{\partial u_j}} \left(\phi^* \frac{\partial \mathbf{r}(\hat{\mathbf{u}})}{\partial u_k}\right)$$

$$= \nabla_{\left(\phi \frac{\partial \mathbf{r}(\hat{\mathbf{u}})}{\partial u_i}\right)} \left(\frac{\partial \phi^*}{\partial u_j} \frac{\partial \mathbf{r}(\hat{\mathbf{u}})}{\partial u_k} + \phi^* \Gamma^l_{jk} \frac{\partial \mathbf{r}(\hat{\mathbf{u}})}{\partial u_l}\right)$$

$$= \phi \frac{\partial^2 \phi^*}{\partial u_i \partial u_j} \frac{\partial \mathbf{r}(\hat{\mathbf{u}})}{\partial u_k} + \phi \frac{\partial \phi^*}{\partial u_j} \Gamma^p_{ik} \frac{\partial \mathbf{r}(\mathbf{u})}{\partial u_p}$$

$$+ \phi \frac{\partial \left(\phi^* \Gamma^l_{jk}\right)}{\partial u_i} \frac{\partial \mathbf{r}(\hat{\mathbf{u}})}{\partial u_l}$$

$$+ |\phi|^2 \Gamma^l_{ij} \Gamma^p_{il} \frac{\partial \mathbf{r}(\hat{\mathbf{u}})}{\partial u_p}$$

$$= \phi \frac{\partial^2 \phi^*}{\partial u_i \partial u_j} \delta_{kl} \frac{\partial \mathbf{r}(\hat{\mathbf{u}})}{\partial u_l} + \phi \frac{\partial \phi^*}{\partial u_j} \Gamma^l_{ik} \frac{\partial \mathbf{r}(\hat{\mathbf{u}})}{\partial u_l}$$

$$+ \phi \frac{\partial (\phi^* \Gamma^l_{jk})}{\partial u_i} \frac{\partial \mathbf{r}(\hat{\mathbf{u}})}{\partial u_l}$$

$$+ |\phi|^2 \Gamma^p_{jk} \Gamma^l_{ip} \frac{\partial \mathbf{r}(\hat{\mathbf{u}})}{\partial u_l} \tag{5.15}$$

and similarly

$$
\nabla_{\left(\phi \frac{\partial \mathbf{r}(\hat{\mathbf{u}})}{\partial u_j}\right)} \nabla_{\frac{\partial \mathbf{r}(\hat{\mathbf{u}})}{\partial u_i}} \left(\phi^* \frac{\partial \mathbf{r}(\hat{\mathbf{u}})}{\partial u_k}\right)
$$

$$
= \quad \phi \, \frac{\partial^2 \phi^*}{\partial u_i \partial u_j} \, \delta_{kl} \, \frac{\partial \mathbf{r}(\hat{\mathbf{u}})}{\partial u_l} + \phi \, \frac{\partial \phi^*}{\partial u_i} \Gamma_{jk}^l \, \frac{\partial \mathbf{r}(\hat{\mathbf{u}})}{\partial u_l}
$$

$$
+ \phi \, \frac{\partial (\phi^* \Gamma_{ik}^l)}{\partial u_j} \, \frac{\partial \mathbf{r}(\hat{\mathbf{u}})}{\partial u_l}
$$

$$
+ |\phi|^2 \, \Gamma_{ik}^p \, \Gamma_{jp}^l \, \frac{\partial \mathbf{r}(\hat{\mathbf{u}})}{\partial u_l}. \tag{5.16}
$$

Moreover,

$$
\nabla_{\left[\phi \frac{\partial \mathbf{r}(\hat{\mathbf{u}})}{\partial u_i}, \frac{\partial \mathbf{r}(\hat{\mathbf{u}})}{\partial u_j}\right]} \left(\phi^* \frac{\partial \mathbf{r}(\hat{\mathbf{u}})}{\partial u_k}\right)
$$

$$
= \quad \nabla_{\left(-\frac{\partial \mathbf{r}(\hat{\mathbf{u}})}{\partial u_j} \cdot \phi\right) \frac{\partial \mathbf{r}(\hat{\mathbf{u}})}{\partial u_i}} \left(\phi^* \frac{\partial \mathbf{r}(\hat{\mathbf{u}})}{\partial u_k}\right)
$$

$$
= \quad -\nabla_{\left(\frac{\partial \phi}{\partial u_j} \frac{\partial \mathbf{r}(\hat{\mathbf{u}})}{\partial u_i}\right)} \left(\phi^* \frac{\partial \mathbf{r}(\hat{\mathbf{u}})}{\partial u_k}\right)
$$

$$
= \quad -\frac{\partial \phi}{\partial u_j} \nabla_{\frac{\partial \mathbf{r}(\hat{\mathbf{u}})}{\partial u_i}} \left(\phi^* \frac{\partial \mathbf{r}(\hat{\mathbf{u}})}{\partial u_k}\right)
$$

$$
= \quad -\frac{\partial \phi}{\partial u_j} \frac{\partial \phi^*}{\partial u_i} \frac{\partial \mathbf{r}(\hat{\mathbf{u}})}{\partial u_k} - \frac{\partial \phi}{\partial u_j} \, \phi^* \nabla_{\frac{\partial \mathbf{r}(\hat{\mathbf{u}})}{\partial u_i}} \frac{\partial \mathbf{r}(\hat{\mathbf{u}})}{\partial u_k}
$$

$$
= \quad -\frac{\partial \phi}{\partial u_j} \frac{\partial \phi^*}{\partial u_i} \frac{\partial \mathbf{r}(\hat{\mathbf{u}})}{\partial u_k} - \frac{\partial \phi}{\partial u_j} \, \phi^* \Gamma_{ik}^l \frac{\partial \mathbf{r}(\hat{\mathbf{u}})}{\partial u_l}
$$

$$
= \quad -\frac{\partial \phi}{\partial u_j} \frac{\partial \phi^*}{\partial u_i} \delta_{kl} \frac{\partial \mathbf{r}(\hat{\mathbf{u}})}{\partial u_l} - \frac{\partial \phi}{\partial u_j} \, \phi^* \Gamma_{ik}^l \frac{\partial \mathbf{r}(\hat{\mathbf{u}})}{\partial u_l}. \tag{5.17}
$$

Thus,

$$
R_{ijk}^l(\phi) = \phi \, \frac{\partial^2 \phi^*}{\partial u_i \partial u_j} \, \delta_{kl} + \phi \, \frac{\partial \phi^*}{\partial u_j} \Gamma_{ik}^l + \phi \, \frac{\partial (\phi^* \Gamma_{jk}^l)}{\partial u_i} + |\phi|^2 \, \Gamma_{jk}^p \, \Gamma_{ip}^l
$$

$$
- \phi \, \frac{\partial^2 \phi^*}{\partial u_i \partial u_j} \, \delta_{kl} - \phi \, \frac{\partial \phi^*}{\partial u_i} \Gamma_{jk}^l - \phi \, \frac{\partial (\phi^* \Gamma_{ik}^l)}{\partial u_j} - |\phi|^2 \, \Gamma_{ik}^p \, \Gamma_{jp}^l
$$

$$
+ \frac{\partial \phi}{\partial u_j} \frac{\partial \phi^*}{\partial u_i} \delta_{kl} + \frac{\partial \phi}{\partial u_j} \, \phi^* \Gamma_{ik}^l. \tag{5.18}
$$

Simplifying this last result, we obtain

$$
\begin{aligned}
R^l_{ijk}(\phi) &= \phi \, \frac{\partial \phi^*}{\partial u_j} \Gamma^l_{ik} + \phi \, \frac{\partial (\phi^* \, \Gamma^l_{jk})}{\partial u_i} - \phi \, \frac{\partial \phi^*}{\partial u_i} \Gamma^l_{jk} - \phi \, \frac{\partial (\phi^* \Gamma^l_{ik})}{\partial u_j} \\
&\quad + |\phi|^2 \left(\Gamma^p_{jk} \, \Gamma^l_{ip} - \Gamma^p_{ik} \, \Gamma^l_{jp} \right) + \frac{\partial \phi}{\partial u_j} \frac{\partial \phi^*}{\partial u_i} \delta_{kl} + \frac{\partial \phi}{\partial u_j} \, \phi^* \Gamma^l_{ik} \\
&= |\phi|^2 \left(\frac{\partial \Gamma^l_{jk}}{\partial u_i} - \frac{\partial \Gamma^l_{ik}}{\partial u_j} + \Gamma^p_{jk} \, \Gamma^l_{ip} - \Gamma^p_{ik} \, \Gamma^l_{jp} \right) \\
&\quad + \frac{\partial \phi}{\partial u_j} \frac{\partial \phi^*}{\partial u_i} \delta_{kl} + \frac{\partial \phi}{\partial u_j} \, \phi^* \Gamma^l_{ik} \\
&= |\phi|^2 \hat{R}^l_{ijk} \\
&\quad + \frac{\partial \phi}{\partial u_j} \frac{\partial \phi^*}{\partial u_i} \delta_{kl} + \frac{\partial \phi}{\partial u_j} \, \phi^* \Gamma^l_{ik},
\end{aligned}
\tag{5.19}
$$

where

$$
\hat{R}^l_{ijk} = \frac{\partial \Gamma^l_{jk}}{\partial u_i} - \frac{\partial \Gamma^l_{ik}}{\partial u_j} + \Gamma^p_{jk} \, \Gamma^l_{ip} - \Gamma^p_{ik} \, \Gamma^l_{jp}
$$

represents the Riemann curvature tensor.

At this point, we recall to have defined this energy part by

$$
E_q = \frac{\gamma}{2} \int_0^T \int_\Omega R \, \sqrt{-g} \, |\det(\hat{u}'(x,t))| \, dx dt,
$$

where

$$
R = g^{jk} R_{jk},
$$

and, as indicated above, $R_{jk} = Re[R^i_{jik}(\phi)]$.

Hence, the final expression for the energy (action) is given by

$$
\begin{aligned}
J(\phi, \mathbf{r}, \hat{\mathbf{u}}, E) &= -E_c + E_q \\
&\quad - \int_0^T E(t) \left(\int_\Omega |\phi(\hat{\mathbf{u}}(\mathbf{x},t))|^2 \, \sqrt{-g} \, |\det(\mathbf{u}'(\mathbf{x},t))| \, d\mathbf{x}/c - 1 \right) cdt,
\end{aligned}
$$

where $E(t)$ is a Lagrange multiplier related to the total mass constraint.

More explicitly, the final action (the generalized Einstein-Hilbert one), would be given by

$$J(\phi, \mathbf{r}, \hat{\mathbf{u}}, E)$$

$$= \int_0^T \int_\Omega mc \sqrt{-g_{ij} \frac{\partial u_i}{\partial t} \frac{\partial u_j}{\partial t}} \, |\phi(\hat{\mathbf{u}}(\mathbf{x},t))|^2 \, \sqrt{-g} \, |\det(\hat{\mathbf{u}}'(\mathbf{x},t))| \, d\mathbf{x} \, dt$$

$$+ \frac{\gamma}{2} \int_0^T \int_\Omega g^{jk} \frac{\partial \phi}{\partial u_j} \frac{\partial \phi^*}{\partial u_k} \sqrt{-g} \, |\det(\hat{\mathbf{u}}'(\mathbf{x},t))| \, d\mathbf{x} dt$$

$$+ \frac{\gamma}{4} \int_0^T \int_\Omega g^{jk} \left(\phi^* \frac{\partial \phi}{\partial u_l} + \phi \frac{\partial \phi^*}{\partial u_l} \right) \Gamma_{jk}^l \sqrt{-g} \, |\det(\hat{\mathbf{u}}'(\mathbf{x},t))| \, d\mathbf{x} dt$$

$$+ \frac{\gamma}{2} \int_0^T \int_\Omega |\phi|^2 \, g^{jk} \left(\frac{\partial \Gamma_{lk}^l}{\partial u_j} - \frac{\partial \Gamma_{jk}^l}{\partial u_l} + \Gamma_{lk}^p \Gamma_{jp}^l - \Gamma_{jk}^p \Gamma_{lp}^l \right) \sqrt{-g} \, |\det(\hat{\mathbf{u}}'(\mathbf{x},t))| \, d\mathbf{x} dt$$

$$- \int_0^T E(t) \left(\int_\Omega |\phi(\hat{\mathbf{u}}(\mathbf{x},t))|^2 \, \sqrt{-g} \, |\det(\hat{\mathbf{u}}'(\mathbf{x},t))| \, d\mathbf{x}/c - 1 \right) c \, dt,$$

where γ is an appropriate positive constant to be specified.

5.4 Obtaining the relativistic Klein-Gordon equation as an approximation of the previous action

In particular for the special case in which

$$\mathbf{r}(\hat{\mathbf{u}}(\mathbf{x},t)) = \hat{\mathbf{u}}(\mathbf{x},t) \approx (ct, \mathbf{x}),$$

so that

$$\frac{d\mathbf{r}(\hat{\mathbf{u}}(\mathbf{x},t))}{dt} \approx (c,0,0,0),$$

we would obtain

$$\mathbf{g}_0 \approx (1,0,0,0), \ \mathbf{g}_1 \approx (0,1,0,0), \ \mathbf{g}_2 \approx (0,0,1,0) \text{ and } \mathbf{g}_3 \approx (0,0,0,1) \in \mathbb{R}^4,$$

and $\Gamma_{ij}^k \approx 0, \ \forall i,j,k \in \{0,1,2,3\}$.

Therefore, denoting $\phi(\hat{\mathbf{u}}(\mathbf{x},t)) \approx \phi(ct, \mathbf{x})$ simply by a not relabeled $\phi(\mathbf{x},t)$, we may obtain

$$E_q/c \approx \frac{\gamma}{2} \int_0^T \int_\Omega \left(-\frac{1}{c^2} \frac{\partial \phi(\mathbf{x},t)}{\partial t} \frac{\partial \phi^*(\mathbf{x},t)}{\partial t} \right.$$

$$\left. + \sum_{k=1}^3 \frac{\partial \phi(\mathbf{x},t)}{\partial x_k} \frac{\partial \phi^*(\mathbf{x},t)}{\partial x_k} \right) d\mathbf{x} dt, \tag{5.20}$$

and

$$E_c/c = m c^2 \int_0^T \int_\Omega |\phi|^2 \sqrt{1 - v^2/c^2} \, \sqrt{-g} |\det(\hat{\mathbf{u}}'(\mathbf{x},t))| \, d\mathbf{x} \, dt/c$$

$$\approx mc^2 \int_0^T \int_\Omega |\phi(\mathbf{x},t)|^2 \, d\mathbf{x} dt. \tag{5.21}$$

Hence, we would also obtain

$$
J(\phi,\mathbf{r},\hat{\mathbf{u}},E)/c \;\approx\; \frac{\gamma}{2}\left(\int_0^T \int_\Omega -\frac{1}{c^2}\frac{\partial \phi(\mathbf{x},t)}{\partial t}\frac{\partial \phi^*(\mathbf{x},t)}{\partial t}\,d\mathbf{x}dt \right.
$$

$$
\left. + \sum_{k=1}^{3}\int_\Omega \int_0^T \frac{\partial \phi(\mathbf{x},t)}{\partial x_k}\frac{\partial \phi^*(\mathbf{x},t)}{\partial x_k}\,d\mathbf{x}dt \right)
$$

$$
+ mc^2 \int_0^T \int_\Omega |\phi(\mathbf{x},t)|^2\,d\mathbf{x}dt
$$

$$
- \int_0^T E(t)\left(\int_\Omega |\phi(\mathbf{x},t)|^2 d\mathbf{x} - 1 \right) dt. \tag{5.22}
$$

The Euler Lagrange equations for such an energy are given by

$$
\frac{\gamma}{2}\left(\frac{1}{c^2}\frac{\partial^2 \phi(\mathbf{x},t)}{\partial t^2} - \sum_{k=1}^{3}\frac{\partial^2 \phi(\mathbf{x},t)}{\partial x_k^2} \right)
$$

$$
+ mc^2 \phi(\mathbf{x},t) - E(t)\phi(\mathbf{x},t) = 0, \text{ in } \Omega, \tag{5.23}
$$

where we assume the space of admissible functions is given by $C^1(\Omega \times [0,T];\mathbb{C})$ with the following time and spatial boundary conditions,

$$
\phi(\mathbf{x},0) = \phi_0(\mathbf{x}), \text{ in } \Omega,
$$

$$
\phi(\mathbf{x},T) = \phi_1(\mathbf{x}), \text{ in } \Omega,
$$

$$
\phi(\mathbf{x},t) = 0, \text{ on } \partial\Omega \times [0,T].
$$

Equation (7.14) is the relativistic Klein-Gordon one.

For $E(t) = E \in \mathbb{R}$ (not time dependent), at this point, we suggest a solution (and implicitly related time boundary conditions) $\phi(\mathbf{x},t) = e^{-\frac{iEt}{\hbar}}\phi_2(\mathbf{x})$, where

$$
\phi_2(\mathbf{x}) = 0, \text{ on } \partial\Omega.
$$

Therefore, replacing this solution into equation (7.14), we would obtain

$$
\left(\frac{\gamma}{2}\left(-\frac{E^2}{c^2\hbar^2}\phi_2(\mathbf{x}) - \sum_{k=1}^{3}\frac{\partial^2 \phi_2(\mathbf{x})}{\partial x_k^2} \right) + mc^2\phi_2(\mathbf{x}) - E\phi_2(\mathbf{x}) \right) e^{-\frac{iEt}{\hbar}} = 0,
$$

in Ω.

Denoting

$$
E_1 = -\frac{\gamma E^2}{2c^2\hbar^2} + mc^2 - E,
$$

the final eigenvalue problem would stand for

$$
-\frac{\gamma}{2}\sum_{k=1}^{3}\frac{\partial^2 \phi_2(\mathbf{x})}{\partial x_k^2} + E_1\phi_2(\mathbf{x}) = 0, \text{ in } \Omega
$$

where E_1 is such that

$$\int_\Omega |\phi_2(\mathbf{x})|^2 \, d\mathbf{x} = 1.$$

Moreover, from (7.14), such a solution $\phi(\mathbf{x},t) = e^{-\frac{iEt}{\hbar}} \phi_2(\mathbf{x})$ is also such that

$$\frac{\gamma}{2} \left(\frac{1}{c^2} \frac{\partial^2 \phi(\mathbf{x},t)}{\partial t^2} - \sum_{k=1}^{3} \frac{\partial^2 \phi(\mathbf{x},t)}{\partial x_k^2} \right)$$
$$+ mc^2 \phi(\mathbf{x},t) = i\hbar \frac{\partial \phi(\mathbf{x},t)}{\partial t}, \text{ in } \Omega. \qquad (5.24)$$

At this point, we recall that in quantum mechanics,

$$\gamma = \hbar^2/m.$$

Finally, we remark that this last equation (7.15) is a kind of relativistic Schrödinger-Klein-Gordon equation.

5.5   A note on the Einstein field equations in a vacuum

In this section, we obtain the Einstein field equations for a field of position in a vacuum.

Let $\Omega \subset \mathbb{R}^3$ be an open, bounded and connected set with a regular boundary denoted by $\partial\Omega$. Let $[0,T]$ be a time interval and consider the Hilbert-Einstein action given by $J : U \to \mathbb{R}$, where for an appropriate constant $\gamma > 0$

$$J(\mathbf{r}) = \frac{\gamma}{2} \int_0^T \int_\Omega \hat{R} \sqrt{-g} \, d\mathbf{x}dt,$$

where again

$$\mathbf{u} = (u_0, u_1, u_2, u_3) = (t, x_1, x_2, x_3) = (x_0, x_1, x_2, x_3).$$

Also,

$$g_{jk} = \frac{\partial \mathbf{r}(\mathbf{u})}{\partial u_j} \cdot \frac{\partial \mathbf{r}(\mathbf{u})}{\partial u_k},$$
$$g = det\{g_{jk}\},$$
$$\hat{R} = g^{jk} R_{jk},$$
$$R_{ik} = R^j_{ijk},$$

and

$$R^l_{ijk} = \frac{\partial \Gamma^l_{jk}}{\partial u_i} - \frac{\partial \Gamma^l_{ik}}{\partial u_j} + \Gamma^p_{jk} \Gamma^l_{ip} - \Gamma^p_{ik} \Gamma^l_{jp}$$

represents the Riemann curvature tensor.

Finally, as indicated above,

$$\mathbf{r} : \Omega \times [0,T] \to \mathbb{R}^4$$

stands for

$$\mathbf{r}(\mathbf{u}) = (ct, X_1(\mathbf{u}), X_2(\mathbf{u}), X_3(\mathbf{u}))$$

and

$$U = \{\mathbf{r} \in W^{2,2}(\Omega; \mathbb{R}^4) : \mathbf{r}(0, u_1, u_2, u_3) = \mathbf{r}_1(u_1, u_2, u_3),$$
$$\mathbf{r}(cT, u_1, u_2, u_3) = \mathbf{r}_2(u_1, u_2, u_3) \text{ in } \Omega, \ \mathbf{r}|_{\partial\Omega} = \mathbf{r}_0 \text{ on } [0,T]\} . (5.25)$$

Hence, already including the Lagrange multipliers, considering $\mathbf{r}$ and $\{g_{jk}\}$ as independent variables, such a functional again denoted by $J(\mathbf{r}, \{g_{jk}\}, \lambda)$ is expressed as

$$J(\mathbf{r}, \{g_{jk}\}, \lambda) = \frac{\gamma}{2} \int_0^T \int_\Omega \hat{R}\sqrt{-g} \, d\mathbf{x} dt + \int_0^T \int_\Omega \lambda_{jk} \left(\frac{\partial \mathbf{r}}{\partial u_j} \cdot \frac{\partial \mathbf{r}}{\partial u_k} - g_{jk} \right) d\mathbf{x} dt.$$

The variation of such a functional in g gives us

$$\gamma \left(R_{jk} - \frac{1}{2} g_{jk} \hat{R} \right) \sqrt{-g} - \lambda_{jk} = 0, \text{ in } \Omega.$$

The variation in $\mathbf{r}$ gives us

$$\frac{\partial^2 X_l(\mathbf{u})}{\partial u_j \partial u_k} \lambda_{jk} + \frac{\partial X_l(\mathbf{u})}{\partial u_j} \frac{\partial \lambda_{jk}}{\partial u_k} = 0, \text{ in } \Omega, \tag{5.26}$$

so that

$$\frac{\partial^2 X_l(\mathbf{u})}{\partial u_j \partial u_k} \left([R_{jk} - \frac{1}{2} g_{jk}\hat{R}]\sqrt{-g} \right) + \frac{\partial X_l(\mathbf{u})}{\partial u_j} \frac{\partial [(R_{jk} - \frac{1}{2} g_{jk}\hat{R})\sqrt{-g}]}{\partial u_k} = 0, \text{ in } \Omega, \tag{5.27}$$

$\forall l \in \{1,2,3\}$.

Observe that the condition $R_{jk} = 0$ in $\Omega \times [0,T]$, $\forall j,k \in \{0,1,2,3\}$, it is sufficient to solve the system indicated in (5.27) but it is not necessary.

The system indicated in (5.27) is the Einstein field one. It is my understanding that the actual variable for this system is $\mathbf{r}$ not $\{g_{jk}\}$.

However, in some situations, it is possible to solve (5.27) through a specific metric $\{(g_0)_{jk}\}$, but one question remains, how to obtain a corresponding $\mathbf{r}$.

With such an issue in mind, given a specific metric $\{(g_0)_{jk}\}$, we suggest the following control problem,

$$\text{Find } \mathbf{r} \in U \text{ which minimizes } J_1(\mathbf{r}) = \sum_{j,k=0}^3 \left\| \frac{\partial \mathbf{r}(\mathbf{u})}{\partial u_j} \cdot \frac{\partial \mathbf{r}(\mathbf{u})}{\partial u_k} - (g_0)_{jk} \right\|_2^2,$$

subject to

$$\frac{\partial^2 X_l(\mathbf{u})}{\partial u_j \partial u_k} \left([R_{jk} - \frac{1}{2} g_{jk} \hat{R}] \sqrt{-g} \right) + \frac{\partial X_l(\mathbf{u})}{\partial u_j} \frac{\partial [(R_{jk} - \frac{1}{2} g_{jk} \hat{R}) \sqrt{-g}]}{\partial u_k} = 0, \text{ in } \Omega,$$

(5.28)

$\forall l \in \{1, 2, 3\}$.

5.6 Conclusion

This work proposes an action (energy) suitable for the relativistic quantum mechanics context. The Riemann tensor represents an important part of the action in question, but now including the density distribution of mass in its expression. In one of the last sections, we obtain the relativistic Klein-Gordon equation as an approximation of the main action, under specific properly described conditions.

We believe the results obtained may be applied to more general models, such as those involving atoms and molecules subject to the presence of electromagnetic fields.

Anyway, we postpone the development of such studies for a future research.

Chapter 6

A General Variational Formulation for Relativistic Mechanics Based on Fundamentals of Differential Geometry

6.1 Introduction

Let $\Omega \subset \mathbb{R}^3$ be an open, bounded, connected set with a smooth boundary (at least C^1 class) denoted by $\partial \Omega$ and let $[0, T]$ be a time interval. Consider a relativistic motion given by a position field

$$(\mathbf{r} \circ \hat{\mathbf{u}}) : \Omega \times [0, T] \to \mathbb{R}^{n+1}.$$

Here, for an open, bounded and connected set D with a smooth boundary, we consider a world sheet smooth (C^3 class) manifold $\mathbf{r} : D \subset \mathbb{R}^{m+1} \to \mathbb{R}^{n+1}$, where point-wise

$$\mathbf{r}(\hat{\mathbf{u}}) = (ct, X_1(\mathbf{u}), \dots, X_n(\mathbf{u}))$$

and where

$$\mathbf{r}(\hat{\mathbf{u}}(\mathbf{x}, t)) = (u_0(\mathbf{x}, t), X_1(\mathbf{u}(\mathbf{x}, t)), \dots, X_n(\mathbf{u}(\mathbf{x}, t))),$$

$$\hat{\mathbf{u}}(\mathbf{x}, t) = (u_0(\mathbf{x}, t), u_1(\mathbf{x}, t), \dots, u_m(\mathbf{x}, t)),$$

$1 \leq m < n$ and

$$u_0(\mathbf{x}, t) = ct.$$

Consider also a density scalar field given by

$$m|\phi(\mathbf{u})|^2 : \Omega \times [0, T] \to \mathbb{R}^+,$$

where m is the total system mass and

$$\phi : D \subset \mathbb{R}^{m+1} \to \mathbb{C}$$

is a wave function.

At this point, we highlight that

$$\frac{d\mathbf{r}(\mathbf{u}(\mathbf{x}, t))}{dt} \cdot \frac{d\mathbf{r}(\mathbf{u}(\mathbf{x}, t))}{dt} = -c^2 + \sum_{j=1}^{n} \left(\frac{dX_j(\mathbf{u}(\mathbf{x}, t))}{dt} \right)^2 = -c^2 + v^2,$$

where c denotes the speed of light in a vacuum and

$$v = \sqrt{\sum_{j=1}^{n} \left(\frac{dX_j(\mathbf{u}(\mathbf{x}, t))}{dt} \right)^2}.$$

We also emphasize that generically, for $a = (\hat{x}_0, x_1, x_2, x_3) \in \mathbb{R}^4$ and $b = (\hat{y}_0, y_1, y_2, y_3) \in \mathbb{R}^4$ we have

$$a \cdot b = -\hat{x}_0 \hat{y}_0 + \sum_{j=1}^{3} x_i y_i.$$

Moreover, $x_0 = t$, $\mathbf{x} = (x_1, x_2, x_3)$ and

$$d\mathbf{x} = dx_1 dx_2 dx_3.$$

Finally, we generically refer to

$$(\mathbf{r} \circ \hat{\mathbf{u}}) : \Omega \times [0, T] \to \mathbb{R}^{n+1}$$

as a space-time manifold. Furthermore, with such a notation in mind we denote

$$\begin{aligned} ds^2 &= d\mathbf{r}(\mathbf{u}(\mathbf{x}, t)) \cdot d\mathbf{r}(\mathbf{u}(\mathbf{x}, t)) \\ &= -c^2 dt^2 + ([dX_1(\mathbf{u}(\mathbf{x}, t))]^2 + [dX_2(\mathbf{u}(\mathbf{x}, t))]^2 + \ldots + [dX_n(\mathbf{u}(\mathbf{x}, t))]^2). \end{aligned}$$

Remark 6.1 Regarding the references, the mathematical background necessary may be found in [23, 13]. For the part on relativistic physics, we follow to some extent the references [52, 53]. ■

6.2 The system energy

Consider first the mass differential, given by,

$$dm = \frac{m|\phi(\mathbf{u}(\mathbf{x},t))|^2}{\sqrt{1 - \frac{v^2}{c^2}}}\sqrt{-g}\sqrt{U}\,d\mathbf{x},$$

so that the kinetics energy differential is defined by

$$
\begin{aligned}
dE_c &= \frac{d\mathbf{r}}{dt} \cdot \frac{d\mathbf{r}}{dt}\,dm \\
&= -\frac{c^2 - v^2}{\sqrt{1 - \frac{v^2}{c^2}}}m|\phi|^2\sqrt{g}\sqrt{U}\,d\mathbf{x} \\
&= -mc\sqrt{c^2 - v^2}|\phi|^2\sqrt{g}\sqrt{U}\,d\mathbf{x} \\
&= -mc\sqrt{-\frac{d\mathbf{r}}{dt} \cdot \frac{d\mathbf{r}}{dt}}|\phi|^2\sqrt{-g}\sqrt{U}\,d\mathbf{x} \\
&= -mc\sqrt{-\frac{\partial \mathbf{r}}{\partial u_j}\frac{\partial u_j}{\partial t} \cdot \frac{\partial \mathbf{r}}{\partial u_k}\frac{\partial u_k}{\partial t}}|\phi|^2\sqrt{-g}\sqrt{U}\,d\mathbf{x} \\
&= -mc|\phi|^2\sqrt{-g_{jk}\frac{\partial u_j}{\partial t}\frac{\partial u_k}{\partial t}}\sqrt{-g}\sqrt{U}\,d\mathbf{x}.
\end{aligned}
\tag{6.1}
$$

Where

$$\mathbf{g}_j = \frac{\partial \mathbf{r}(\mathbf{u})}{\partial u_j}, \ \forall j \in \{0,\dots,m\},$$

$$g_{jk} = \mathbf{g}_j \cdot \mathbf{g}_k, \ \forall j,k \in \{0,\dots,m\},$$

$$\{g^{jk}\} = \{g_{jk}\}^{-1},$$

$$g = \det\{g_{ij}\}$$

and

$$U_{ij} = \frac{\partial \mathbf{u}(\mathbf{x},t)}{\partial x_i} \cdot \frac{\partial \mathbf{u}(\mathbf{x},t)}{\partial x_j}, \ \forall i,j \in \{0,1,2,3\}.$$

Moreover, we define

$$U = |\det\{U_{ij}\}|.$$

At this point, we assume that there exists a smooth normal field $\mathbf{n}$ such that

$$\text{Span}\left\{\left\{\frac{\partial \mathbf{r}(\mathbf{u})}{\partial u_j}, \ \forall j \in \{0,\dots,m\}\right\}, \mathbf{n}(\mathbf{u})\right\} \subset \mathbb{R}^{n+1}, \ \forall \mathbf{u} \in D$$

and

$$\frac{\partial^2 \mathbf{r}(\mathbf{u})}{\partial u_j \partial u_k} = \Gamma^l_{jk}(\mathbf{u})\frac{\partial \mathbf{r}(\mathbf{u})}{\partial u_l} + b_{jk}(\mathbf{u})\mathbf{n}(\mathbf{u}), \ \forall \mathbf{u} \in D,$$

where $\{\Gamma^l_{jk}\}$ are the Christoffel symbols and the concerning normal field $\mathbf{n}(\mathbf{u})$ is also such that

$$\mathbf{n}(\mathbf{u}) \cdot \mathbf{n}(\mathbf{u}) = 1, \ \forall \mathbf{u} \in D,$$

$$\frac{\partial \mathbf{r}(\mathbf{u})}{\partial u_l} \cdot \mathbf{n}(\mathbf{u}) = 0, \ \text{in } D, \forall l \in \{0, \dots, m\}$$

and

$$b_{jk}(\mathbf{u}) = \frac{\partial^2 \mathbf{r}(\mathbf{u})}{\partial u_j \partial u_k} \cdot \mathbf{n}(\mathbf{u}), \ \forall \mathbf{u} \in D, \ \forall j, k \in \{0, \dots, m\}.$$

Suppose also the concerning world sheet position field is such that there exist smooth normal fields

$$\hat{\mathbf{n}}_1, \dots, \hat{\mathbf{n}}_s$$

where $m + 1 + s \geq n + 1$ such that

$$\text{Span}\left\{\left\{\frac{\partial \mathbf{r}(\mathbf{u})}{\partial u_j}, \ \forall j \in \{0, \dots, m\}\right\}, \hat{\mathbf{n}}_1(\mathbf{u}), \dots, \hat{\mathbf{n}}_s(\mathbf{u})\right\} = \mathbb{R}^{n+1}, \ \forall \mathbf{u} \in D$$

so that

$$\mathbf{n}(\mathbf{u}) = f_q(\mathbf{u}) \hat{\mathbf{n}}_q(\mathbf{u}), \ \forall \mathbf{u} \in D$$

for an appropriate field $\{f_q\}_{q=1}^s$.

Moreover, we assume

$$\hat{\mathbf{n}}_j(\mathbf{u}) \cdot \hat{\mathbf{n}}_k(\mathbf{u}) = \delta_{jk}, \ \forall \mathbf{u} \in D, \ j, k \in \{1, \dots, s\}$$

and

$$\frac{\partial \mathbf{r}(\mathbf{u})}{\partial u_j} \cdot \hat{\mathbf{n}}_k(\mathbf{u}) = 0,$$

$\forall \mathbf{u} \in D, \ \forall j \in \{0, \dots, m\}, \ k \in \{1, \dots, s\}$.

Here, we recall that

$$\mathbf{n}(\mathbf{u}) \cdot \frac{\partial \mathbf{r}(\mathbf{u})}{\partial u_k} = 0, \ \text{in } D.$$

Hence,

$$\frac{\partial \mathbf{n}(\mathbf{u})}{\partial u_j} \cdot \frac{\partial \mathbf{r}(\mathbf{u})}{\partial u_k} + \mathbf{n}(\mathbf{u}) \cdot \frac{\partial^2 \mathbf{r}(\mathbf{u})}{\partial u_j \partial u_k} = 0,$$

that is,

$$\frac{\partial \mathbf{n}(\mathbf{u})}{\partial u_j} \cdot \frac{\partial \mathbf{r}(\mathbf{u})}{\partial u_k} = -b_{jk}. \tag{6.2}$$

We may also denote

$$\frac{\partial \mathbf{n}(\mathbf{u})}{\partial u_j} = c_j^s \frac{\partial \mathbf{r}(\mathbf{u})}{\partial u_s} + e_j^q \hat{\mathbf{n}}_q,$$

for an appropriate $\{c_j^s\}$ and where

$$e_j^q = \frac{\partial \mathbf{n}(\mathbf{u})}{\partial u_j} \cdot \hat{\mathbf{n}}_q.$$

From this and (6.2), we obtain

$$c_j^s \frac{\partial \mathbf{r}(\mathbf{u})}{\partial u_s} \cdot \mathbf{g}_k = c_j^s g_{sk} = -b_{jk},$$

so that

$$c_j^s g_{sk} g^{kl} = -b_{jk} g^{kl} = -b_j^l,$$

that is,

$$c_j^l = c_j^s \delta_s^l = -b_j^l,$$

where

$$b_j^l = b_{jk} g^{kl}.$$

Summarizing, we have

$$\frac{\partial \mathbf{n}(\mathbf{u})}{\partial u_j} = -b_j^l \frac{\partial \mathbf{r}(\mathbf{u})}{\partial u_l} + e_j^q \hat{\mathbf{n}}_q.$$

Observe now that

$$
\begin{aligned}
\frac{\partial^3 \mathbf{r}(\mathbf{u})}{\partial u_i \partial u_j \partial u_k} &= \frac{\partial}{\partial u_i}\left(\Gamma_{jk}^l \frac{\partial \mathbf{r}(\mathbf{u})}{\partial u_l} + b_{jk} \mathbf{n} \right) \\
&= \left(\frac{\partial \Gamma_{jk}^l}{\partial u_i} + \Gamma_{jk}^p \Gamma_{pi}^l \right) \frac{\partial \mathbf{r}(\mathbf{u})}{\partial u_l} \\
&\quad + \Gamma_{jk}^p b_{pi} \mathbf{n} + \frac{\partial b_{jk}}{\partial u_i} \mathbf{n} - b_{jk} b_i^l \frac{\partial \mathbf{r}(\mathbf{u})}{\partial u_l} \\
&\quad + b_{jk} e_i^i \hat{\mathbf{n}}_l.
\end{aligned}
\tag{6.3}
$$

Similarly,

$$
\begin{aligned}
\frac{\partial^3 \mathbf{r}(\mathbf{u})}{\partial u_j \partial u_i \partial u_k} &= \frac{\partial}{\partial u_j}\left(\Gamma_{ik}^l \frac{\partial \mathbf{r}(\mathbf{u})}{\partial u_l} + b_{ik} \mathbf{n} \right) \\
&= \left(\frac{\partial \Gamma_{ik}^l}{\partial u_j} + \Gamma_{ik}^p \Gamma_{pj}^l \right) \frac{\partial \mathbf{r}(\mathbf{u})}{\partial u_l} \\
&\quad + \Gamma_{jk}^p b_{pi} \mathbf{n} + \frac{\partial b_{ik}}{\partial u_j} \mathbf{n} - b_{ik} b_j^l \frac{\partial \mathbf{r}(\mathbf{u})}{\partial u_l} \\
&\quad + b_{ik} e_l^j \hat{\mathbf{n}}_l.
\end{aligned}
\tag{6.4}
$$

Thus, for such a smooth (C^3 class) manifold, from

$$\frac{\partial^3 \mathbf{r}(\mathbf{u})}{\partial u_i \partial u_j \partial u_k} = \frac{\partial^3 \mathbf{r}(\mathbf{u})}{\partial u_j \partial u_i \partial u_k},$$

assuming a concerning linear independence and equating the terms in

$$\frac{\partial \mathbf{r}(\mathbf{u})}{\partial u_l},$$

we get

$$
\begin{aligned}
W^l_{ijk} &= b_{jk} b^l_i \\
&= \frac{\partial \Gamma^l_{jk}}{\partial u_i} - \frac{\partial \Gamma^l_{ik}}{\partial u_j} \\
&\quad + \Gamma^p_{jk} \Gamma^l_{pi} - \Gamma^p_{ik} \Gamma^l_{pj} + b_{ik} b^l_j.
\end{aligned}
\tag{6.5}
$$

Defining the Riemann curvature tensor by

$$R^l_{ijk} = \frac{\partial \Gamma^l_{jk}}{\partial u_i} - \frac{\partial \Gamma^l_{ik}}{\partial u_j} + \Gamma^p_{jk} \Gamma^l_{pi} - \Gamma^p_{ik} \Gamma^l_{pj}, \tag{6.6}$$

we also define the energy part $J_1(\phi, \mathbf{r}, \mathbf{u}, \mathbf{n})$ as

$$
\begin{aligned}
J_1(\phi, \mathbf{r}, \mathbf{u}, \mathbf{n}) &= \frac{1}{2} \int_0^T \int_\Omega |\phi|^2 g^{jk} b_{jl} b^l_k \sqrt{-g} \sqrt{U} \, d\mathbf{x} \, dt \\
&= \frac{1}{2} \int_0^T \int_\Omega |\phi|^2 g^{jk} R^l_{jlk} \sqrt{-g} \sqrt{U} \, d\mathbf{x} \, dt \\
&\quad + \frac{1}{2} \int_0^T \int_\Omega |\phi|^2 g^{jk} b_{jk} b^l_l \sqrt{-g} \sqrt{U} \, d\mathbf{x} \, dt.
\end{aligned}
\tag{6.7}
$$

The next energy part is defined through the tensor S^l_{ijk} which, considering the Levi-Civita connection ∇ and the standard Lie Bracket $[\cdot, \cdot]$ (see [23, 52] for more details), is such that

$$\nabla_{\left[\phi \frac{\partial \mathbf{r}(\mathbf{u})}{\partial u_i}, \frac{\partial \mathbf{r}(\mathbf{u})}{\partial u_j}\right]} \left(\phi^* \frac{\partial \mathbf{r}(\mathbf{u})}{\partial u_k}\right) = S^l_{ijk} \frac{\partial \mathbf{r}(\mathbf{u})}{\partial u_l} + \hat{b}_{ijk} \mathbf{n}.$$

Observe that

$$\nabla_{\left(\frac{\partial \phi}{\partial u_j} \frac{\partial \mathbf{r(u)}}{\partial u_i}\right)} \left(\phi^* \frac{\partial \mathbf{r(u)}}{\partial u_k}\right)$$

$$= \frac{\partial \phi}{\partial u_j} \frac{\partial \phi^*}{\partial u_i} \frac{\partial \mathbf{r(u)}}{\partial u_k}$$

$$+ \frac{\partial \phi}{\partial u_j} \phi^* \frac{\partial^2 \mathbf{r(u)}}{\partial u_i \partial u_k}$$

$$= \frac{\partial \phi}{\partial u_j} \frac{\partial \phi^*}{\partial u_i} \frac{\partial \mathbf{r(u)}}{\partial u_l} \delta_{lk}$$

$$+ \frac{\partial \phi}{\partial u_j} \phi^* \left(\Gamma^l_{ik} \frac{\partial \mathbf{r(u)}}{\partial u_l} + b_{ik}\mathbf{n}\right) \qquad (6.8)$$

Thus,

$$S^l_{ijk} = \frac{\partial \phi}{\partial u_j} \frac{\partial \phi^*}{\partial u_i} \delta_{kl} + \frac{\partial \phi}{\partial u_i} \phi^* \Gamma^l_{jk}.$$

With such results in mind, we define this energy part as

$$J_2(\phi, \mathbf{r}, \mathbf{u}, \mathbf{n}) = \frac{1}{2} \int_0^T \int_\Omega g^{jk} Re[S^l_{jlk}] \sqrt{-g} \sqrt{U} \, d\mathbf{x} \, dt,$$

where generically $Re[z]$ and z^* denote the real part and complex conjugation, respectively, of $z \in \mathbb{C}$.

6.3 The final energy expression

The expression for the energy, already including the Lagrange multiplier concerning the mass restriction, is given by

$$J(\phi, \mathbf{r}, \mathbf{u}, \mathbf{n}, E) = -\int_0^T \int_\Omega dE_c \, dt + J_1(\phi, \mathbf{r}, \mathbf{u}, \mathbf{n}) + J_2(\phi, \mathbf{r}, \mathbf{u}, \mathbf{n})$$

$$-\int_0^T E(t) \left(\int_\Omega |\phi|^2 \sqrt{-g}\sqrt{U} \, d\mathbf{x} - 1\right) dt, \qquad (6.9)$$

so that

$$J(\phi, \mathbf{r}, \mathbf{n}, \mathbf{u}, E) = \int_0^T \int_\Omega mc|\phi|^2 \sqrt{-g_{jk} \frac{\partial u_j}{\partial t} \frac{\partial u_k}{\partial t}} \sqrt{-g}\sqrt{U} \, d\mathbf{x} \, dt$$

$$+ \frac{1}{2} \int_0^T \int_\Omega |\phi|^2 g^{jk} b_{jl} b^l_k \sqrt{-g}\sqrt{U} \, d\mathbf{x} \, dt$$

$$+ \frac{1}{2} \int_0^T \int_\Omega g^{jk} \frac{\partial \phi}{\partial u_j} \frac{\partial \phi^*}{\partial u_k} \sqrt{-g}\sqrt{U} \, d\mathbf{x} \, dt$$

$$+ \frac{1}{4} \int_0^T \int_\Omega \left(\frac{\partial \phi}{\partial u_l} \phi^* + \frac{\partial \phi^*}{\partial u_l} \phi\right) \Gamma^l_{jk} g^{jk} \sqrt{-g}\sqrt{U} \, d\mathbf{x} \, dt$$

$$- \int_0^T E(t) \left(\int_\Omega |\phi|^2 \sqrt{-g}\sqrt{U} \, d\mathbf{x} - 1\right) dt$$

We shall look for critical points subject to

$$\mathbf{n}(\mathbf{u}(\mathbf{x},t)) \cdot \mathbf{n}(\mathbf{u}(\mathbf{x},t)) = 1, \text{ in } \Omega \times [0,T]$$

and

$$\frac{\partial \mathbf{r}(\mathbf{u}(\mathbf{x},t))}{\partial u_j} \cdot \mathbf{n}(\mathbf{u}(\mathbf{x},t)) = 0, \text{ in } \Omega \times [0,T], \ \forall j \in \{0,\dots,m\}.$$

Already including the concerning Lagrange multipliers, the final functional expression would be

$$
\begin{aligned}
J(\phi,\mathbf{r},\mathbf{u},\mathbf{n},E,\lambda) \;=\;& \int_0^T \int_\Omega mc|\phi|^2 \sqrt{-g_{jk}\frac{\partial u_j}{\partial t}\frac{\partial u_k}{\partial t}}\sqrt{-g}\sqrt{U}\,d\mathbf{x}\,dt \\
&+ \frac{1}{2}\int_0^T \int_\Omega |\phi|^2 g^{jk} b_{jl} b_k^l \sqrt{-g}\sqrt{U}\,d\mathbf{x}\,dt \\
&+ \frac{1}{2}\int_0^T \int_\Omega g^{jk}\frac{\partial \phi}{\partial u_j}\frac{\partial \phi^*}{\partial u_k}\sqrt{-g}\sqrt{U}\,d\mathbf{x}\,dt \\
&+ \frac{1}{4}\int_0^T \int_\Omega \left(\frac{\partial \phi}{\partial u_l}\phi^* + \frac{\partial \phi^*}{\partial u_l}\phi\right)\Gamma^l_{jk}g^{jk}\sqrt{-g}\sqrt{U}\,d\mathbf{x}\,dt \\
&- \int_0^T E(t)\left(\int_\Omega |\phi|^2 \sqrt{-g}\sqrt{U}\,d\mathbf{x}-1\right)dt \\
&+ \sum_{j=0}^m \int_0^T \int_\Omega \lambda_j(\mathbf{x},t)\frac{\partial \mathbf{r}(\mathbf{u}(\mathbf{x},t))}{\partial u_j}\cdot \mathbf{n}(\mathbf{u}(\mathbf{x},t))\sqrt{-g}\sqrt{U}\,d\mathbf{x}\,dt \\
&+ \int_0^T \int_\Omega \lambda_{m+1}(\mathbf{x},t)(\mathbf{n}(\mathbf{u}(\mathbf{x},t))\cdot \mathbf{n}(\mathbf{u}(\mathbf{x},t))-1)\sqrt{-g}\sqrt{U}\,d\mathbf{x}\,dt
\end{aligned}
$$

Remark 6.2 We must consider such a functional defined on a space of sufficiently smooth functions with appropriate boundary and initial conditions prescribed.

Finally, the main difference concerning standard differential geometry in $\mathbb{R}^3$ is that, since

$$1 \leq m < n,$$

we have to obtain through the variation of J, the optimal normal field $\mathbf{n}$. Summarizing, at first we do not have an explicit expression for such a field. ∎

6.4 Causal structure

In this section, we develop some formalism concerning the causal structure in a space-time manifold defined by a function

$$(\mathbf{r} \circ \hat{\mathbf{u}}) : \Omega \times (-\infty,+\infty) \to \mathbb{R}^{n+1},$$

where $t \in (-\infty,+\infty)$ denotes time.

We follow, to some extent, the content in Wald's book [52], where more details may be found.

Definition 6.4.1 *Let M be a space-time manifold time oriented, in the sense that the light cone related to the tangent spaces varies smoothly along M. A C^1 class curve $\lambda : [a,b] \to M$ is said to be time-like future directed if for each $p \in \lambda$ the respective tangent vector is time-like future directed, that is,*

$$\frac{d\lambda(s)}{ds} \cdot \frac{d\lambda(s)}{ds} < 0, \ \forall s \in [a,b], \ (\text{time-like condition})$$

and

$$\frac{\partial t(s)}{\partial s} > 0, \forall s \in [a,b], (\text{future directed condition}).$$

Here,

$$\lambda(s) = \mathbf{r}(\hat{\mathbf{u}}(\mathbf{x}(s), t(s)))$$

for appropriate smooth functions

$$\mathbf{x}(s), t(s).$$

Similarly, we say that such a curve is causal future directed, if the tangent vector is a time-like future directed or is a null vector, $\forall s \in [a,b]$.

Finally, in an analogous fashion, we may define a continuous and piece-wise C^1 class time-like future directed curve.

Remark 6.3   At this point, we highlight that in the next lines the norm $\| \cdot \|$ refers to the standard Euclidean one in $\mathbb{R}^{n+1}$. ■

Definition 6.4.2 *The chronological future of $p \in M$, denoted by $I^+(p)$, is defined as*

$$
\begin{aligned}
I^+(p) \ = \ &\{q \in M : \\
&\text{there exists a continuous and piece-wise } C^1 \text{ class time-like} \\
&\text{future directed curve } \lambda : [a,b] \to M \\
&\text{such that } \lambda(a) = p \text{ and } \lambda(b) = q\}.
\end{aligned}
\tag{6.10}
$$

Observe that, if M is smooth (as previously indicated, the world sheet manifold in question is at least C^3 class) by continuity, if $q \in I^+(p)$ there exists a neighborhood $\mathcal{O}(q)$ such that

$$\mathcal{O}(q) \cap M \subset I^+(p).$$

From now and on we always assume any space-time mentioned is always smooth and time-oriented.

Also, for $S \subset M$, we define

$$I^+(S) = \cup_{p \in S} I^+(p),$$

so that since $I^+(p)$ is open for each $p \in M$, we may infer that $I^+(S)$ is open.

Remark 6.4 Similarly, we define the chronological pasts $I^-(p)$ and $I^-(S)$. Moreover the causal future of $p \in M$, denoted by $J^+(p)$ is defined as

$$
\begin{aligned}
J^+(p) \quad = \quad \{q \in M : \\
& \text{there exists a continuous and piece-wise } C^1 \text{ class} \\
& \text{causal future directed curve } \lambda : [a,b] \to M \\
& \text{such that } \lambda(a) = p \text{ and } \lambda(b) = q\}. \quad\quad (6.11)
\end{aligned}
$$

Also, we define

$$J^+(S) = \cup_{p \in S} J^+(p),$$

and similarly define the causal pasts $J^-(p)$ and $J^-(S)$. ∎

Definition 6.4.3 *Let M be a space time manifold. We say the M is normal if for each connected set $S \subset M$, there exists $r > 0$ such that if $p, q \in I^+(S)$ and $0 < \|p - q\| < r$, then, interchanging the roles of p and q if necessary, there exists a smooth time-like future directed curve $\lambda : [a,b] \to I^+(S)$ such that*

$$\lambda(a) = p$$

and

$$\lambda(b) = q.$$

Moreover, for each $U \subset M$ open in M, $I^+(p)|_U$ consists of all point reach by time like future directed geodesics starting in p and contained in U, so that $I^+(p)|_U$ denotes the chronological future of the space-time $U \subset M$.

Definition 6.4.4 *A set $S \subset M$ is said to be achronal if $p, q \in S$ does not exist such that $q \in I^+(p)$, that is if*

$$I^+(S) \cap S = \emptyset.$$

Theorem 6.4.5 *Let M be a space-time manifold. Let $S \subset M$. Under such assumptions, $\partial I^+(S)$ is achronal.*

Proof 6.1 Let $q \in \partial I^+(S)$. Assume $p \in I^+(q)$. Thus, $q \in I^-(p)$ and, since $I^-(p)$ is open in M, there exists U open in M, such that $U \subset I^-(p)$ and also such that $q \in U$.

Note that, since $q \in \partial I^+(S)$, we have that

$$I^+(q) \cap I^+(S) \neq \emptyset.$$

Let $q_1 \in U \cap I^+(S)$.

From this, there exists $p_1 \in S$ and a continuous and piece-wise C^1 class time-like future directed curve $\lambda : [a,b] \to M$ such that $\lambda(a) = p_1$ and $\lambda(b) = q_1 \in U \subset I^-(p)$.

From such a result we may obtain a continuous and piece-wise C^1 class time like future directed curve $\lambda_1 : [b,c] \to M$ such that $\lambda_1(b) = q_1$ and $\lambda_1(c) = p$ so that $\lambda_2 : [a,c] \to M$ such that

$$\lambda_2(s) = \begin{cases} \lambda(s), & \text{if } s \in [a,b] \\ \lambda_1(s), & \text{if } s \in [b,c] \end{cases} \tag{6.12}$$

is a continuous and piece-wise C^1 time-like future directed curve such that $\lambda_2(a) = p_1 \in S$ and $\lambda_2(c) = p$.

Therefore, we may infer that $p \in I^+(S), \ \forall p \in I^+(q)$, so that

$$I^+(q) \subset I^+(S).$$

Suppose, to obtain contradiction, that $\partial I^+(S)$ is not achronal. Thus, there exist $q, r \in \partial I^+(S)$ such that

$$r \in I^+(q) \subset I^+(S).$$

From this we may infer that

$$r \in \partial I^+(S) \cap I^+(S),$$

which contradicts $I^+(S)$ to be open.

Therefore, $\partial I^+(S)$ is achronal.

Definition 6.4.6 *Let M be a space-time manifold and let $\lambda \subset M$ be a causal future directed curve. We say that a point $p \in M$ is a final point of λ if for each open set U such that $p \in U$, there exists $s_0 \in \mathbb{R}$ such that if $s > s_0$, then*

$$\lambda(s) \in U.$$

Moreover, we say that a curve is inextensible if does not have any final point. Past inextensibility is defined similarly.

Theorem 6.4.7 *Let M be a closed space-time manifold. Let*

$$\lambda_n : (-\infty, b] \to M$$

be a sequence of differentiable past inextensible curves such that for each $m \in \mathbb{N}$ there exists $K_m, \hat{K}_m \in \mathbb{R}^+$ such that

$$\|\lambda_n(s)\| \leq K_m, \ \forall s \in [-m, b],$$

and

$$\|\lambda_n'(s)\| \leq \hat{K}_m, \ \forall s \in [-m, b].$$

Assume there exists $p \in M$ that for each open U such that $p \in U$, there exists $n_0 \in \mathbb{N}$ such that if $n > n_0$ then there exists $s_n \in (-\infty, b)$, such that

$$\lambda_n(s) \subset U, \ \forall s \in (s_n, b].$$

Under such hypotheses, there exists a subsequence $\{\lambda_{n_k}\}$ of $\{\lambda_n\}$ and a continuous curve $\lambda : (-\infty, b] \to M$ such that

$$\lambda_{n_k} \to \lambda, \text{ uniformly in } [-m, b], \ \forall m \in \mathbb{N}$$

and

$$\lambda(b) = p.$$

Proof 6.2   Let

$$\{\alpha_n\} = \{q \in \mathbb{Q} \ : \ q \le b\}.$$

Observe that, from the hypotheses $\{\lambda_n(\alpha_1)\} \subset M$ is a bounded sequence, so that there exists a subsequence

$$\{\lambda_{n_k}(\alpha_1)\}$$

and a vector which we shall denote by $\lambda(\alpha_1)$ such that

$$\lambda_{n_k}(\alpha_1) \to \lambda(\alpha_1), \text{ as } k \to \infty.$$

We shall also denote

$$\lambda_{n_k}(\alpha_1) = L_k^1(\alpha_1).$$

Similarly, $\{L_1^k(\alpha_2)\}$ is bounded so that there exists a subsequence $\{L_{n_k}^1\}$ of $\{L_k^1\}$ and a vector in M, which we will denote by $\lambda(\alpha_2)$ such that

$$L_{n_k}^1(\alpha_2) \to \lambda(\alpha_2), \text{ as } k \to \infty.$$

Denoting $L_{n_k}^1 = L_k^2$, we have obtained

$$L_k^2(\alpha_1) \to \lambda(\alpha_1),$$

and

$$L_k^2(\alpha_2) \to \lambda(\alpha_2), \text{ as } k \to \infty.$$

Proceeding in this fashion, we may inductively obtain a subsequences $\{L_k^j\}_{k \in \mathbb{N}}$ of λ_n such that

$$L_k^j(\alpha_l) \to \lambda(\alpha_l) \text{ as } k \to \infty, \ \forall l \in \{1, \dots, j\}.$$

Let $\varepsilon > 0$, $l \in \mathbb{N}$ and $j \ge l$. Hence, there exists $K_j \in \mathbb{N}$ such that if $k \ge K_j$ then

$$\|L_k^j(\alpha_l) - \lambda(\alpha_l)\| < \varepsilon.$$

In particular

$$\|L_{K_j}^j(\alpha_l) - \lambda(\alpha_l)\| < \varepsilon, \forall j > l.$$

Hence, denoting

$$\Lambda_j = L_{K_j}^j,$$

we have obtained that $\{\Lambda_j\}$ is a subsequence of $\{\lambda_n\}$ such that

$$\Lambda_k(\alpha_j) \to \lambda(\alpha_j), \ \forall j \in \mathbb{N}.$$

Fix $m \in \mathbb{N}$ such that $-m < b$ and let $s \in [-m, b]$. We are going to prove that

$$\{\Lambda_k(s)\}$$

is a Cauchy sequence.

Let $\{\alpha_{n_l}\}$ be a subsequence of $\{\alpha_n\}$ such that

$$\alpha_{n_l} \to s, \text{ as } l \to \infty.$$

Hence, there exists $l_0 \in \mathbb{N}$ such that if $l > l_0$, then

$$|\alpha_{n_l} - s| < \frac{\varepsilon}{3\hat{K}_m}.$$

Choose $l > l_0$. Since $\{\Lambda_k(\alpha_{n_l})\}$ is a Cauchy sequence, there exists $k_0 \in \mathbb{N}$ such that if $k, p > k_0$, then

$$\|\Lambda_k(\alpha_{n_l}) - \Lambda_p(\alpha_{n_l})\| < \frac{\varepsilon}{3}.$$

Thus, if $k, p > k_0$, we obtain

$$
\begin{aligned}
&\|\Lambda_k(s) - \Lambda_p(s)\| \\
=\ & \|\Lambda_k(s) - \Lambda_k(\alpha_{n_l}) + \Lambda_k(\alpha_{n_l}) - \Lambda_p(\alpha_{n_l}) + \Lambda_p(\alpha_{n_l}) - \Lambda_p(s)\| \\
\leq\ & \|\Lambda_k(s) - \Lambda_k(\alpha_{n_l})\| + \|\Lambda_k(\alpha_{n_l}) - \Lambda_p(\alpha_{n_l})\| + \|\Lambda_p(\alpha_{n_l}) - \Lambda_p(s)\| \\
\leq\ & \hat{K}_m|s - \alpha_{n_l}| + \frac{\varepsilon}{3} + \hat{K}_m|s - \alpha_{n_l}| \\
<\ & \frac{\varepsilon}{3} + \frac{\varepsilon}{3} + \frac{\varepsilon}{3} \\
=\ & \varepsilon.
\end{aligned}
\tag{6.13}
$$

From this, we may infer that $\{\Lambda_k(s)\}$ is a Cauchy sequence so that we may define

$$\lambda(s) = \lim_{k \to \infty} \Lambda_k(s), \ \forall s \in [-m, b].$$

We claim that this last convergence, up to a subsequence, is uniform on $[-m, b]$. Indeed, let

$$c_k = \sup_{s \in [-m,b]} \{\|\Lambda_k(s) - \lambda(s)\|\}.$$

Let $s_k \in [-m, b]$ be such that

$$c_k - 1/k < \|\Lambda_k(s_k) - \lambda(s_k)\| \leq c_k.$$

Since $[-m, b]$ is compact, there exists a subsequence $\{s_{k_l}\}$ of $\{s_k\}$ and $s \in [-m, b]$ such that

$$s_{k_l} \to s, \text{ as } l \to \infty.$$

At this point, we shall prove that

$$\|\lambda(s_{k_l}) - \lambda(s)\| \to 0.$$

Indeed, there exists $l_0 \in \mathbb{N}$ such that if $l > l_0$, then

$$|s_{k_l} - s| < \frac{\varepsilon}{\hat{K}}.$$

$$\|\Lambda_p(s_{k_l}) - \Lambda_p(s)\| \le \hat{K}|s_{k_l} - s| < \varepsilon, \forall l > l_0, \forall p \in \mathbb{N}.$$

From this, we get

$$\|\lambda(s_{k_l}) - \lambda(s)\| = \lim_{p \to \infty} \|\Lambda_p(s_{k_l}) - \Lambda_p(s)\| \le \varepsilon, \forall l > l_0.$$

Observe that from such a result we may, in a similar fashion, infer that λ is continuous.

From these last results, observing that there exists $l_1 \in \mathbb{N}$ such that if $l > l_1$, then

$$\|\Lambda_{k_l}(s) - \lambda(s)\| < \varepsilon,$$

we have that

$$
\begin{aligned}
&\|\Lambda_{k_l}(s_{k_l}) - \lambda(s_{k_l})\| \\
=\ & \|\Lambda_{k_l}(s_{k_l}) - \Lambda_{k_l}(s) + \Lambda_{k_l}(s) - \lambda(s) + \lambda(s) - \lambda(s_{k_l})\| \\
\le\ & \|\Lambda_{k_l}(s_{k_l}) - \Lambda_{k_l}(s)\| + \|\Lambda_{k_l}(s) - \lambda(s)\| + \|\lambda(s) - \lambda(s_{k_l})\| \\
\le\ & \varepsilon + \varepsilon + \varepsilon \\
=\ & 3\varepsilon, \forall l > \max\{l_0, l_1\}.
\end{aligned}
\tag{6.14}
$$

From this, we may infer that $c_{k_l} \to 0$ as $l \to \infty$, so that the convergence in question of the subsequence $\{\Lambda_{k_l}\}$ of $\{\lambda_n\}$ is uniform. We claim now that $c_k \to 0$ as $k \to \infty$.

Suppose, to obtain contradiction, that the claim is false. Thus, $\{c_k\}$ does not converge to 0.

Hence, there exists $\varepsilon_0 > 0$ such that for each $k \in \mathbb{N}$ there exists $k_l > k$ such that

$$c_{k_l} \ge \varepsilon_0.
\tag{6.15}$$

However, exactly as we have done with $\{c_k\}$ in the lines above, we may obtain a subsequence of $\{c_{k_l}\}$ which converges to 0. This contradicts (6.15).

Therefore

$$c_k \to 0, \text{ as } k \to \infty.$$

From this we may infer that

$$\Lambda_k \to \lambda, \text{ uniformly in } [-m, b], \forall m \in \mathbb{N} \text{ such that } -m < b.$$

The proof is complete.

Theorem 6.4.8 *Let M be a space time manifold. Assume that $\lambda : (-\infty, b] \to M$ is a causal future directed past inextensible curve.*

Under such hypotheses,

$$\lambda(s) \in \overline{I^+(\lambda)}, \forall s \in (-\infty, b].$$

Proof 6.3 Let $s \in (-\infty, b]$ and choose $s_1 < s$.

Thus, $\lambda|_{[s_1,s]}$ is a causal future directed curve such that denoting $p = \lambda(s_1)$ and $q = \lambda(s)$, we have that

$$q \in \overline{I^+(p)} \subset \overline{I^+(\lambda)}, \forall s \in (-\infty, b].$$

The proof is complete.

Theorem 6.4.9 *Let M be a normal space time manifold. Assume* $\lambda : (-\infty, c] \to M$ *is a causal future directed past inextensible curve which passes through a point* $p \in M$.

Under such hypotheses, for each $q \in I^+(p)$ *there exists a continuous and piece-wise* C^1 *class time-like future directed past inextensible curve* $\gamma : (-\infty, b] \to M$, *such that*

$$\gamma \subset I^+(\lambda)$$

and

$$\gamma(b) = q.$$

Proof 6.4 Let $\hat{\lambda} : [a, b] \to I^+(\lambda)$ be a time-like future directed curve such that

$$\hat{\lambda}(a) = p$$

and

$$\hat{\lambda}(b) = q.$$

We claim that $\hat{\lambda} \subset I^+(\lambda)$.

Indeed, let $s \in (a, b]$. Denoting $q_1 = \hat{\lambda}(s)$ we have

$$\hat{\lambda}(s) = q_1 \in I^+(p) \subset I^+(\lambda), \ \forall s \in (a, b].$$

So the concerning claim holds.

Let $\lambda_1 : (-\infty, b] \to M$ be the curve defined by

$$\lambda_1(s) = \begin{cases} \lambda(s), & \text{if } s \in (-\infty, a] \\ \hat{\lambda}(s), & \text{if } a \le s \le b \end{cases} \tag{6.16}$$

Since the graph of λ is connected and M is normal, there exists $r > 0$ such that if $\tilde{p}, \tilde{q} \in I^+(\lambda)$ and

$$0 < \|\tilde{p} - \tilde{q}\| < r,$$

then renaming $\tilde{p}, \tilde{q}$ if necessary, there exists a time-like future directed curve $\tilde{\lambda} : [c, d] \to I^+(\lambda)$ such that

$$\tilde{\lambda}(c) = \tilde{p}$$

and

$$\tilde{\lambda}(d) = \tilde{q}.$$

Let $\{s_n\} \subset (-\infty, b]$ be a real sequence such that $s_1 = b$, $s_n > s_{n+1}$, $\forall n \in \mathbb{N}$,

$$\lim_{n \to \infty} s_n = -\infty.$$

and also such that

$$\|\lambda_1(s_{n+1}) - \lambda_1(s_n)\| < \frac{r}{3}, \ \forall n \in \mathbb{N}.$$

Define $p_1 = q$. Since

$$\lambda_1 \subset \overline{I^+(\lambda)}$$

and $I^+(\lambda)$ is open, for each $n > 1$ we may select $p_n \in I^+(\lambda)$ such that

$$0 < \|p_n - \lambda_1(s_n)\| < \frac{r}{3}.$$

Observe that in such a case,

$$
\begin{aligned}
\|p_{n+1} - p_n\| &= \|p_{n+1} - \lambda_1(s_{n+1}) + \lambda_1(s_{n+1}) - \lambda_1(s_n) + \lambda_1(s_n) - p_n\| \\
&\leq \|p_{n+1} - \lambda_1(s_{n+1})\| + \|\lambda_1(s_{n+1}) - \lambda_1(s_n)\| + \|\lambda_1(s_n) - p_n\| \\
&< \frac{r}{3} + \frac{r}{3} + \frac{r}{3} \\
&= r, \ \forall n \in \mathbb{N}.
\end{aligned}
\tag{6.17}
$$

Moreover, $\{p_n\}$ may be chosen such that

$$0 < d(p_n, \lambda) < \frac{C}{1 + \sqrt{n}}, \ \forall n \in \mathbb{N},$$

for some appropriate constant $C > 0$.

Thus, from (6.17) and from the fact that M is normal, concerning such $r > 0$, we may obtain a smooth time-like future directed curve

$$\tilde{\lambda}_n : [s_{n+1}, s_n] \to I^+(\lambda)$$

such that

$$\tilde{\lambda}_n(s_{n+1}) = p_{n+1},$$

and

$$\tilde{\lambda}_n(s_n) = p_n.$$

Therefore, we may define $\gamma : (-\infty, b] \to M$ such that

$$\gamma = \{\tilde{\lambda}_n : [s_{n+1}, s_n] \to I^+(\lambda) \ : \ n \in \mathbb{N}\},$$

which is a continuous and piece-wise C^1 class time-like future directed past inextensible curve such that

$$\gamma(b) = q,$$

and

$$\gamma \subset I^+(\lambda).$$

The proof is complete.

Definition 6.4.10 *Let M be a space time manifold. We say that M is strongly causal if for each $p \in M$ and each neighborhood U of p, there exists a neighborhood V of p such that $V \subset U$ and no causal curve intersects V more than one time.*

Theorem 6.4.11 *Let M be a space-time manifold strongly causal. Let $K \subset M$ be a compact set. Under such hypotheses, each causal curve λ contained in K must have past and future final points.*

Proof 6.5 Let $\lambda : [-\infty, +\infty] \to M$ be a causal curve contained in K. Let $\{s_j\} \subset \mathbb{R}$ be such that $s_j < s_{j+1}$ and

$$\lim_{j \to \infty} s_j = +\infty.$$

Observe that

$$\{\lambda(s_j) = p_j\} \subset K$$

and K is compact. Hence, there exists a subsequence

$$\{p_{j_k}\}$$

and $p \in K$ such that

$$p_{j_k} \to p, \text{ as } k \to \infty.$$

Suppose, to obtain contradiction, we may obtain an open set U such that $p \in U$ and such that for each $s_0 \in \mathbb{R}$ there exists $s > s_0$ such that $\lambda(s) \notin U$. Thus, we have the same for all $V \subset U$ such that $p \in V$. Fixing an arbitrary $V \subset U$ with $p \in V$, we have that λ enters and leaves V more than one time, because each time λ enters V it does not remain completely in V. Since $V \subset U$ has been arbitrary, this contradicts the strong causality of M.

Thus, p is a future final point for λ. Similarly we may prove that λ has a past final point.

This completes the proof.

6.5 Dependence domains and hyperbolicity

Definition 6.5.1 *Let S be a closed and achronal set. We define the domain of future dependence of S, denoted by $D^+(S)$, by*

$$D^+(S) = \{p \in M : \text{ each piece-wise smooth causal future directed past}$$
$$\text{inextensible curve which passes through p intercepts } S\}. \quad (6.18)$$

Observe that

$$S \subset D^+(S) \subset J^+(S),$$

and since S is achronal, we have that

$$D^+(S) \cap I^-(S) = \emptyset.$$

The domain of past dependence of S, denoted by $D^-(S)$ is defined similarly. We also define

$$D(S) = D^+(S) \cup D^-(S),$$

Finally, an achronal set Σ for which $D(\Sigma) = M$ is said to be a Cauchy surface for M.

Observe that, in such a case,

$$\partial \Sigma = \emptyset.$$

Finally, a space-time manifold which has a Cauchy surface is said to be globally hyperbolic.

Theorem 6.5.2 *Let M be a normal space-time manifold and let $S \subset M$ be a set closed in M. Under such hypotheses, $p \in \overline{D^+(S)}$ if, and only if, each time-like future directed past inextensible curve which passes through p intercepts S.*

Proof 6.6   Suppose there exists a time-like future directed past inextensible curve which does not intercept S.

Hence, there exists a set U open in M such that $p \in U$ with such a propriety.

Thus, $U \cap D^+(S) = \emptyset$, so that $p \notin \overline{D^+(S)}$.

Reciprocally, suppose each time-like future direct past inextensible curve which passes through p intercepts S.

Thus, either $p \in S \subset D^+(S) \subset \overline{D^+(S)}$, and in such a case the proof would be finished, or $p \in I^+(S) \setminus S$.

In this latter case, let $q \in I^-(p) \cap I^+(S)$.

Suppose, to obtain contradiction, that $q \notin D^+(S)$.

Thus, there exists a causal future directed past inextensible curve λ which passes through q and does not intercept S.

Note that

$$\lambda \subset \overline{I^+(\lambda)} \setminus S,$$

so that, in such a case, since M is normal, similarly as in the proof of Theorem 6.4.9, we may obtain a piece-wise smooth time-like future directed past inextensible curve γ such that

$$\gamma \subset I^+(\lambda),$$

also such that $\gamma \cap S = \emptyset$ and γ passes through p, which contradicts the hypotheses in question.

Hence, if $q \in I^-(p) \cap I^+(p)$, then $q \in D^+(S)$, so that

$$I^-(p) \cap I^+(S) \subset D^+(S).$$

Since each neighborhood of $p \in I^+(S)$ intercepts

$$I^-(p) \cap I^+(S) \subset D^+(S),$$

we have that $p \in \overline{D^+(S)}$.

The proof is complete.

Theorem 6.5.3 *Let M be a space-time manifold. Let $S \subset M$ and let $\lambda : [a,b] \to M$ be a C^1 class future directed curve such that $\lambda(s) \in \partial I^+(S)$, $\forall s \in [a,b]$.*
Under such hypotheses, λ is a null geodesics, that is,

$$\frac{d\lambda(s)}{ds} \cdot \frac{d\lambda(s)}{ds} = 0, \ \forall s \in [a,b].$$

Proof 6.7   Suppose, to obtain contradiction, that there exists $s_0 \in (a,b)$ such that

$$\frac{d\lambda(s_0)}{ds} \cdot \frac{d\lambda(s_0)}{ds} < 0.$$

By continuity, there exists $\delta > 0$ such that

$$\frac{d\lambda(s)}{ds} \cdot \frac{d\lambda(s)}{ds} < 0, \ \forall s \in (s_0 - \delta, s_0 + \delta).$$

Define

$$p_1 = \lambda(s_0),$$

and

$$p_2 = \lambda\left(s_0 + \frac{\delta}{2}\right).$$

Thus, $p_2 \in I^+(p_1)$, and $p_1, p_2 \in \partial I^+(S)$, which contradicts $\partial I^+(S)$ to be achronal.

Hence,

$$\frac{d\lambda(s)}{ds} \cdot \frac{d\lambda(s)}{ds} = 0, \ \forall s \in [a,b].$$

The proof is complete.

Theorem 6.5.4 *Let M be a space-time manifold. Let $S \subset M$ and suppose $\lambda : [a,b] \to \overline{I^+(S)}$ is a future directed null geodesics.*
Under such hypotheses,

$$\lambda(s) \in \partial I^+(S), \ \forall s \in [a,b].$$

Proof 6.8   Since λ is a future directed null geodesics, we have that

$$\frac{d\lambda(s)}{ds} \cdot \frac{d\lambda(s)}{ds} = 0, \ \forall s \in [a,b].$$

From this, since $\lambda(s) \in \overline{I^+(S)}$, we get

$$\lambda(s) \in \partial I^+(S), \ \forall s \in [a,b].$$

The proof is complete.

Theorem 6.5.5 *Let M be a normal space-time manifold. Let $\Sigma \subset M$ be a Cauchy surface and let $\lambda : [-\infty, +\infty] \to M$ be a causal inextensible curve.*
Under such hypotheses, λ intercepts Σ, $I^+(\Sigma)$ and $I^-(\Sigma)$.

Proof 6.9 Suppose, to obtain contradiction, that λ does not intercept $I^-(\Sigma)$. Similarly as in the proof of Theorem 6.4.9, we may obtain a time-like past inextensible curve such that

$$\gamma \subset I^+(\lambda) \subset I^+(\Sigma \cup I^+(\Sigma)) = I^+(\Sigma).$$

Extending γ to the future indefinitely (if possible), such a curve cannot intercept Σ, because in such a case Σ would not be achronal, which is a contradiction.

However, since each causal inextensible curve must intercept Σ, we have got a final contradiction (that is, such a γ does not exists).

From this we may infer that λ intercepts $I^-(\Sigma)$.

Similarly, we may show that λ intercepts $I^+(\Sigma)$.

The proof is complete.

6.6 Existence of minimizer for the previous general functional

In this section, under some condition properly specified, we prove the existence of a minimizer for the general functional presented in the previous sections. We start with the following remark.

Remark 6.5 Considering the position field given by

$$\mathbf{r} : D \equiv [0, T] \times D_1 \to \mathbb{R}^{N+1},$$

and fixing a small $\varepsilon > 0$, define

$$
\begin{aligned}
U \;=\; & \{\tilde{u} = (\mathbf{r}, \phi, \mathbf{n}) \in C^2(\overline{D}; \mathbb{R}^{N+1}) \times C^1(\overline{D}; \mathbb{C}) \times C^1(\overline{D}; \mathbb{R}^{N+1}) \\
& \text{such that } |\phi|^2 \geq \varepsilon \text{ in } D, \; \forall k \in \{0, \ldots, m\}, \\
& \mathbf{r}(0, \mathbf{u}) = \hat{\mathbf{r}}_0, \text{ in } D_1, ; \mathbf{r}(T, \mathbf{u}) = \hat{\mathbf{r}}_1, \text{ in } D_1, \\
& \mathbf{r}(t, \mathbf{u}) = \hat{\mathbf{r}}_2, \text{ on } \partial D_1 \times [0, T], \phi(0, \mathbf{u}) = \hat{\phi}_0, \text{ in } D_1, \; \phi(cT, \mathbf{u}) = \hat{\phi}_1, \text{ in } D_1, \\
& \phi(ct, \mathbf{u}) = \hat{\phi}_2, \text{ on } \partial D_1 \times [0, T]\},
\end{aligned}
\tag{6.19}
$$

$$
\begin{aligned}
\tilde{U} \;=\; & \{(\mathbf{r}, \phi, \mathbf{n}) \in W^{2,2}(D; \mathbb{R}^{N+1}) \times W^{1,2}(D; \mathbb{C}) \times W^{1,2}(D; \mathbb{R}^{N+1}) \\
& \text{such that } \mathbf{r}(0, \mathbf{u}) = \hat{\mathbf{r}}_0, \text{ in } D_1, \; \mathbf{r}(T, \mathbf{u}) = \hat{\mathbf{r}}_1, \text{ in } D_1, \\
& \mathbf{r}(t, \mathbf{u}) = \hat{\mathbf{r}}_2, \text{ on } \partial D_1 \times [0, T], \\
& \phi(0, \mathbf{u}) = \hat{\phi}_0, \text{ in } D_1, \; \phi(cT, \mathbf{u}) = \hat{\phi}_1, \text{ in } D_1, \\
& \phi(ct, \mathbf{u}) = \hat{\phi}_2, \text{ on } \partial D_1 \times [0, T]\},
\end{aligned}
\tag{6.20}
$$

$$U_1 = \left\{ \tilde{u} \in U : \int_{D_1} |\phi(ct, \mathbf{u})|^2 \sqrt{-g}\, d\mathbf{u} = 1, \text{ on } [0, T] \right\}, \quad (6.21)$$

$$U_2 = \left\{ \tilde{u} \in U : \frac{\partial \mathbf{r}(\mathbf{u})}{\partial u_j} \cdot \mathbf{n}(\mathbf{u}) = 0, \text{ in } D \right\}, \quad (6.22)$$

and

$$U_3 = \{ \tilde{u} \in U : \mathbf{n}(\mathbf{u}) \cdot \mathbf{n}(\mathbf{u}) = 1, \text{ in } D \}. \quad (6.23)$$

Finally, we also define,

$$A = U_1 \cap U_2 \cap U_3,$$

and

$$A_1 = \tilde{U} \cap U_1 \cap U_2 \cap U_3.$$

■

With such definitions in mind, we state and prove the following existence theorem.

Theorem 6.6.1 *For $5 \leq m \leq 8$ and $m < N$, let $J_K : U \to \mathbb{R}$ be defined by*

$$\begin{aligned}
J_K(\tilde{u}) &= J\ (\tilde{u}) + \frac{K}{2} \int_0^T \left(\int_{D_1} |\phi|^2 \sqrt{-g}\, d\mathbf{u} - 1 \right)^2 dt \\
&\quad \frac{K}{2} \sum_{j=0}^{m} \int_D \left(\frac{\partial \mathbf{r}(\mathbf{u})}{\partial u_j} \cdot \mathbf{n}(\mathbf{u}) \right)^2 \sqrt{-g}\, d\mathbf{u}\, dt \\
&\quad + \frac{K}{2} \int_D (\mathbf{n}(\mathbf{u}) \cdot \mathbf{n}(\mathbf{u}) - 1)^2 \sqrt{-g}\, d\mathbf{u}\, dt, \quad (6.24)
\end{aligned}$$

where

$$\begin{aligned}
J(\tilde{u}) &= \frac{1}{2} \int_D |\phi|^2 g^{jk} b_{jl} b_k^l \sqrt{-g}\, d\mathbf{u}\, dt \\
&\quad + \frac{1}{2} \int_D g^{jk} \frac{\partial \phi}{\partial u_j} \frac{\partial \phi}{\partial u_k} \sqrt{-g}\, d\mathbf{u}\, dt \\
&\quad + \frac{1}{4} \int_D \left(\frac{\partial \phi}{\partial u_l} \phi^* + \frac{\partial \phi^*}{\partial u_l} \phi \right) \Gamma_{jk}^l g^{jk} \sqrt{-g}\, d\mathbf{u}\, dt, \quad (6.25)
\end{aligned}$$

where $K \in \mathbb{N}$ is a large constant.
   Let $\{\tilde{v}_n^K\}$ be a minimizing sequence for J_K such that

$$\alpha \leq J_K(\tilde{v}_n^K) < \alpha + \frac{1}{n},$$

where

$$\alpha = \inf_{\tilde{v} \in U} J_K(\tilde{u}).$$

Suppose such a sequence is such that

1. *There exists $c_0 > 0$ such that*

$$(g^{jk})_n^K y_j y_k \geq c_0 y_j y_j, \; \forall y = \{y_j\} \in \mathbb{R}^2, \; \forall n \in \mathbb{N}, \; in \; D.$$

2. *There exists $c_1 > 0$ such that*

$$|\phi_n^K|^2 (g^{jk})_n^K \mathbf{z}_j \cdot (g_l)_n^K (g^{ls})_n^K \mathbf{z}_s (g_k)_n^K \geq c_1 \mathbf{z}_j \cdot \mathbf{z}_j, \; \forall \{\mathbf{z}_j\} \in \mathbb{R}^{(N+1)(m+1)}, \; in \; D, \; \forall n \in \mathbb{N},$$

so that

$$|\phi_n^K|^2 (g^{jk})_n^K (b_{jl})_n^K (b_k^l)_n^K \geq c_1 \frac{\partial \mathbf{n}_n^K}{\partial u_i} \cdot \frac{\partial \mathbf{n}_n^K}{\partial u_i}, \; in \; D, \; \forall n \in \mathbb{N}.$$

3. *There exists $\{(c_2)_{ij}\}$ such that $(c_2)_{ij} > 0$, $\forall i, j \in \{0, \dots, m\}$ so that*

$$|\phi_n^K|^2 (g^{jk})_n^K (b_{jl})_n^K (b_k^l)_n^K \geq (c_2)_{ij} \left| \frac{\partial^2 \mathbf{r}_n^K}{\partial u_i \partial u_j} \right|^2, \; in \; D, \; \forall n \in \mathbb{N}.$$

4.

$$\|(\mathbf{g}_k)_n^K\|_{C^{1,\nu}(\overline{D})} \leq \hat{K}, \; \forall k \in \{0, \dots, m\}, \; \forall n \in \mathbb{N},$$

for some $\hat{K} \in \mathbb{R}^+$ and some $0 < \nu < 1$.

Under such hypotheses, there exists $\tilde{u}_0^K \in \tilde{U}$ such that

$$J_K(\tilde{u}_0^K) = \inf_{\tilde{u} \in U} J_K(\tilde{\mathbf{u}}).$$

Finally, there exists a subsequence $\{K_j\}$ of $\mathbb{N}$ and $\tilde{u}_0 \in A_1$ such that

$$J(\tilde{\mathbf{u}}_0) = \lim_{j \to \infty} J_{K_j}(\tilde{\mathbf{u}}_0^{K_j}) = \inf_{\tilde{u} \in A} J(\tilde{u}).$$

Proof 6.10 From the hypotheses we may infer that there exists $K_1 \in \mathbb{R}^+$ such that

1.

$$\left\| \frac{\partial^2 \mathbf{r}_n^K}{\partial u_i \partial u_j} \right\|_2 \leq K_1, \; \forall n \in \mathbb{N}, \; \forall i, j \in \{0, \dots, m\}.$$

2.

$$\|\phi_n^K\|_2 \leq K_1, \; \forall n \in \mathbb{N}.$$

3.

$$\left\| \frac{\partial \phi_n^K}{\partial u_j} \right\|_2, \; \forall n \in \mathbb{N}, \; \forall j \in \{0, \dots, m\}.$$

4.

$$\left\| \frac{\partial \mathbf{n}_n^K}{\partial u_j} \right\|_2 \leq K_1, \ \forall n \in \mathbb{N}, \ \forall i, j \in \{0, \ldots, m\}.$$

Observe that J_K is lower semi-continuous and bounded below so that, from the generalized Ekeland variational principle, we may find a sequence $\{\tilde{\mathbf{u}}\} \subset U$ such that

$$\alpha \leq J_K(\tilde{\mathbf{u}})_n^K) < \alpha + \frac{1}{n},$$

$$\|\tilde{\mathbf{u}}_n^K - \tilde{\mathbf{u}}_n^K\|_U \leq \frac{1}{\sqrt{n}},$$

and

$$\|\delta J_K(\tilde{\mathbf{u}}_n^K)\|_U \leq \frac{1}{\sqrt{n}}, \ \forall n \in \mathbb{N}.$$

Considering the expressions for the second variation of J_K, also from the generalized Ekeland variational principle, we may assume that there exists $\varepsilon_1 > 0$ such that there exists $K_0 \in \mathbb{N}$ such that if $K > K_0$, then there exists $n_0 \in \mathbb{N}$ such that, if $n > n_0$, then

$$\int_{D_1} |\phi_n^K|^2 (\sqrt{-g})_n^K \, d\mathbf{u} \geq \varepsilon_1,$$

and

$$\mathbf{n}_n^K \cdot \mathbf{n}^K \geq \varepsilon_1, \ \text{in } D.$$

From such results and from the variation of J_K in $\mathbf{n}$ we obtain that

$$\frac{\partial}{\partial u_j} \left((a_{ij}^l)_n^K \frac{\partial \mathbf{n}_n^K}{\partial u_i} \right) = (f_l)_n^K, \ \text{in } D,$$

for appropriate positive definite

$$\{(a_{ij})_n^K\}$$

of C^1 class and $\{(f_l)_n^K\} \in L^2$, $\forall l \in \{1, \ldots, N+1\}$, $i, j \in \{0, \ldots, m\}$.

From the study in Chapter 11 of book [14] of regularity of the Laplace equation solution, we have that

$$\mathbf{n}_n^K \in W^{2,2}$$

and since $\{(a_{ij})_n^K\}$ and $\{(f_l)_n^K\}$ are uniformly bounded in C^1 and L^2, respectively, there exists $K_3 \in \mathbb{R}^+$ such that

$$\|\mathbf{n}_n^K\|_{2,2} \leq K_3, \ \forall n \in \mathbb{N}.$$

With a similar reasoning, we may also obtain that

$$\|\phi_n^K\|_{2,2} \leq K_4, \ \forall l \in \mathbb{N},$$

for some $K_4 \in \mathbb{N}$.

From such results and the Rellich-Kondrachov theorem may obtain a subsequence $\{n_l\}$ of $\mathbb{N}$ and $\tilde{\mathbf{u}}_0^K \in \tilde{U}$ such that

1.

$$\phi_{n_l}^K \rightharpoonup \phi_0^K, \text{ as } l \to \infty, \text{ weakly in } W^{2,2},$$

2.

$$\phi_{n_l}^K \to \phi_0^K, \text{ as } l \to \infty, \text{ weakly in } W^{1,q},$$

3.

$$\mathbf{r}_{n_l}^K \rightharpoonup \mathbf{r}_0^K, \text{ as } l \to \infty, \text{ weakly in } W^{2,2},$$

4.

$$\mathbf{r}_{n_l}^K \to \mathbf{r}_0^K, \text{ as } l \to \infty, \text{ weakly in } W^{1,q},$$

5.

$$\mathbf{n}_{n_l}^K \rightharpoonup \mathbf{n}_0^K, \text{ as } l \to \infty, \text{ weakly in } W^{2,2},$$

6.

$$\mathbf{n}_{n_l}^K \to \mathbf{n}_0^K, \text{ as } l \to \infty, \text{ weakly in } W^{1,q},$$

$\forall 1 \le \frac{2m}{m-4} \equiv p^*$.

Moreover, we highlight that, up to a subsequence not relabeled,

$$|(\sqrt{-g})_{n_l}^K - (\sqrt{-g})_0^K|^4 \to 0, \text{ as } l \to \infty, \text{ a.e. in } D,$$

and

$$\left\| |(\sqrt{-g})_{n_l}^K - (\sqrt{-g})_0^K|^4 \right\|_\infty < \hat{K}_1, \forall l \in \mathbb{N},$$

for some appropriate $\hat{K}_1 \in \mathbb{R}^+$ so that, from the Lebesgue dominated convergence theorem, we get

$$\left\| (\sqrt{-g})_{n_l}^K - (\sqrt{-g})_0^K \right\|_4 \to 0, \text{ as } l \to \infty.$$

Thus,

$$\left| \int_D \frac{\partial \phi_{n_l}^K}{\partial u_j} \frac{\partial (\phi^*)_{n_l}^K}{\partial u_j} (g^{jk})_{n_l}^K (\sqrt{-g})_{n_l}^K \, d\mathbf{u}dt \right.$$

$$\left. - \int_D \frac{\partial \phi_0^K}{\partial u_j} \frac{\partial (\phi^*)_0^K}{\partial u_j} (g^{jk})_0^K (\sqrt{-g})_0^K \, d\mathbf{u}dt \right|$$

$$\le \int_D \left| \frac{\partial \phi_{n_l}^K}{\partial u_j} \frac{\partial (\phi^*)_{n_l}^K}{\partial u_j} (g^{jk})_{n_l}^K (\sqrt{-g})_{n_l}^K \, d\mathbf{u}dt \right.$$

$$\left. - \frac{\partial \phi_0^K}{\partial u_j} \frac{\partial (\phi^*)_0^K}{\partial u_j} (g^{jk})_0^K (\sqrt{-g})_0^K \right| \, d\mathbf{u}dt. \tag{6.26}$$

From this we obtain

$$\left| \int_D \frac{\partial \phi_{n_l}^K}{\partial u_j} \frac{\partial (\phi^*)_{n_l}^K}{\partial u_j} (g^{jk})_{n_l}^K (\sqrt{-g})_{n_l}^K \, d\mathbf{u}dt \right.$$

$$\left. - \int_D \frac{\partial \phi_0^K}{\partial u_j} \frac{\partial (\phi^*)_0^K}{\partial u_j} (g^{jk})_0^K (\sqrt{-g})_0^K \, d\mathbf{u}dt \right|$$

$$\leq \int_D \left| \frac{\partial \phi_{n_l}^K}{\partial u_j} \frac{\partial (\phi^*)_{n_l}^K}{\partial u_j} (g^{jk})_{n_l}^K (\sqrt{-g})_{n_l}^K \, d\mathbf{u}dt \right.$$

$$\left. - \frac{\partial \phi_0^K}{\partial u_j} \frac{\partial (\phi^*)_{n_l}^K}{\partial u_j} (g^{jk})_{n_l}^K (\sqrt{-g})_{n_l}^K \right| d\mathbf{u}dt$$

$$\int_D \left| \frac{\partial \phi_0^K}{\partial u_j} \frac{\partial (\phi^*)_{n_l}^K}{\partial u_j} (g^{jk})_{n_l}^K (\sqrt{-g})_{n_l}^K \, d\mathbf{u}dt \right.$$

$$\left. - \frac{\partial \phi_0^K}{\partial u_j} \frac{\partial (\phi^*)_0^K}{\partial u_j} (g^{jk})_{n_l}^K (\sqrt{-g})_{n_l}^K \right| d\mathbf{u}dt$$

$$\int_D \left| \frac{\partial \phi_0^K}{\partial u_j} \frac{\partial (\phi^*)_0^K}{\partial u_j} (g^{jk})_{n_l}^K (\sqrt{-g})_{n_l}^K \, d\mathbf{u}dt \right.$$

$$\left. - \frac{\partial \phi_0^K}{\partial u_j} \frac{\partial (\phi^*)_0^K}{\partial u_j} (g^{jk})_0^K (\sqrt{-g})_{n_l}^K \right| d\mathbf{u}dt$$

$$\int_D \left| \frac{\partial \phi_0^K}{\partial u_j} \frac{\partial (\phi^*)_0^K}{\partial u_j} (g^{jk})_0^K (\sqrt{-g})_{n_l}^K \, d\mathbf{u}dt \right.$$

$$\left. - \frac{\partial \phi_0^K}{\partial u_j} \frac{\partial (\phi^*)_0^K}{\partial u_j} (g^{jk})_0^K (\sqrt{-g})_0^K \right| d\mathbf{u}dt \tag{6.27}$$

Therefore, from this we have

$$\left| \int_D \frac{\partial \phi_{n_l}^K}{\partial u_j} \frac{\partial (\phi^*)_{n_l}^K}{\partial u_j} (g^{jk})_{n_l}^K (\sqrt{-g})_{n_l}^K \, d\mathbf{u}dt \right.$$

$$\left. - \int_D \frac{\partial \phi_0^K}{\partial u_j} \frac{\partial (\phi^*)_0^K}{\partial u_j} (g^{jk})_0^K (\sqrt{-g})_0^K \, d\mathbf{u}dt \right|$$

$$\leq K_1 \|\phi_{n_l}^K - \phi_0\|_{1,4} \, \|(\phi^*)_{n_l}\|_{1,4} \, \|(g^{jk})_{n_l}^K\|_4 \, \|(\sqrt{-g})_{n_l}^K\|_4$$

$$+ K_1 \|\phi_0^K\|_{1,4} \, \|(\phi^*)_{n_l}^K - (\phi^*)_0\|_{1,4} \, \|(g^{jk})_{n_l}^K\|_4 \, \|(\sqrt{-g})_{n_l}^K\|_4$$

$$+ K_1 \|\phi_0^K\|_{1,4}^2 \, \|(g^{jk})_{n_l}^K - (g^{jk})_0\|_{1,4} \, \|(\sqrt{-g})_{n_l}^K\|_4$$

$$+ K_1 \|\phi_0^K\|_{1,4}^2 \, \|(g^{jk})_0^K\|_{1,4}^2 \, \|(g^{jk})_{n_l}^K - (g^{jk})_0\|_4$$

$$\rightarrow \quad 0 \text{ as } l \rightarrow \infty. \tag{6.28}$$

With similar reasoning for the remaining functional parts, we may obtain

$$J_K(\tilde{\mathbf{u}}_{n_l}^K) \rightarrow J(\tilde{\mathbf{u}}_0^K) = \min_{\tilde{\mathbf{u}} \in \tilde{U}} J_K(\tilde{\mathbf{u}}).$$

At this point we emphasize that, through the Euler-Lagrange equations, the hypotheses and results so far obtained, we may infer that

$$\int_{D_1} |\phi_0^K|^2 \sqrt{-g_0^K} \, d\mathbf{u} - 1 = \mathcal{O}(1/K), \text{ on } [0, T],$$

$$\int_D \left(\frac{\partial \mathbf{r}_0^K}{\partial u_j} \cdot \mathbf{n}_0^K \right)^2 \sqrt{-g_0^K} \, d\mathbf{u} dt = \mathcal{O}(1/K), \, \forall j \in \{0, \dots, m\}$$

and

$$\int_D \left(\mathbf{n}_0^K \cdot \mathbf{n}_0^K - 1 \right)^2 \sqrt{-g_0^K} \, d\mathbf{u} dt = \mathcal{O}(1/K).$$

Finally, observe that the previous estimates are valid uniformly for the sequence $\{\tilde{\mathbf{u}}_0^K\}$ (the concerning constants do not depend on K) so that there exists $\tilde{\mathbf{u}}_0 \in \tilde{U}$ such that, up to a not relabeled subsequence

1.
$$\phi_0^K \rightharpoonup \phi_0, \text{ as } l \to \infty, \text{ weakly in } W^{2,2},$$

2.
$$\phi_0^K \to \phi_0, \text{ as } l \to \infty, \text{ weakly in } W^{1,q},$$

3.
$$\mathbf{r}_0^K \rightharpoonup \mathbf{r}_0, \text{ as } l \to \infty, \text{ weakly in } W^{2,2},$$

4.
$$\mathbf{r}_0^K \to \mathbf{r}_0, \text{ as } l \to \infty, \text{ weakly in } W^{1,q},$$

5.
$$\mathbf{n}_0^K \rightharpoonup \mathbf{n}_0, \text{ as } l \to \infty, \text{ weakly in } W^{2,2},$$

6.
$$\mathbf{n}_0^K \to \mathbf{n}_0, \text{ as } l \to \infty, \text{ weakly in } W^{1,q},$$

Moreover, from the previous estimates and concerning limits, we have that

$$\int_{D_1} |\phi_0|^2 \sqrt{-g_0} \, d\mathbf{u} - 1 = 0, \text{ on } [0, T],$$

$$\int_D \left(\frac{\partial \mathbf{r}_0}{\partial u_j} \cdot \mathbf{n}_0 \right)^2 \sqrt{-g_0} \, d\mathbf{u} dt = 0, \, \forall j \in \{0, \dots m\}$$

and

$$\int_D (\mathbf{n}_0 \cdot \mathbf{n}_0 - 1)^2 \sqrt{-g} \, d\mathbf{u} dt = 0.$$

From this we get

$$\tilde{\mathbf{u}}_0 \in A_1$$

so that

$$J(\tilde{\mathbf{u}}_0) = \lim_{K \to \infty} J_K(\tilde{\mathbf{u}}_0^K) = \inf_{\mathbf{u} \in A_1} J(\mathbf{u}).$$

The proof is complete.

6.7   Conclusion

In this chapter, we have obtained a variational formulation for relativistic mechanics based on standard tools of differential geometry. The novelty here is that the main manifold has its range in a space of dimension $n + 1$. In such a formulation, the concept of normal field plays a fundamental role.

In the second part of this chapter, we presented some formalism concerning the causal structure in a general space-time manifold defined locally by a function

$$(\mathbf{r} \circ \hat{\mathbf{u}}) : \Omega \times (-\infty, +\infty) \to \mathbb{R}^{n+1}.$$

It is worth highlighting that the main reference for this second part is the book [52].

Finally, in the last section, we developed an existence result for the main manifold variational formulation.

Chapter 7

A New Interpretation for the Bohr Atomic Model

7.1 Introduction

In this chapter, we propose again a variational formulation for the Klein-Gordon relativistic equation obtained through an extension of the classical mechanics approach to a more general context.

We introduce an energy part aiming to minimize and control, in a specific appropriate sense to be described in the next sections, the curvature field distribution along the concerned mechanical system.

Regarding the references, this work is based on the book [22] and the articles [25, 24]. Indeed, in the next sections we present some results similar to those presented in [22] and [24]. In the third section we cover in detail one of the main results, namely, the establishment of the Klein-Gordon relativistic equation obtained from the respective variational formulation.

At this point, we remark that details on the Sobolev Spaces involved may be found in [1, 13]. For standard references in quantum mechanics, we refer to [40, 37, 12] and the non-standard [11].

Finally, we emphasize that this article is not about Bohmian mechanics, even though the David Bohm's work has been always inspiring.

7.2 The Newtonian approach and a concerning extension

In this section, specifically for a free particle context, we shall obtain a close relationship between classical and quantum mechanics.

Let $\Omega \subset \mathbb{R}^3$ be an open, bounded and connected set with a regular (Lipschitzian) boundary denoted by $\partial \Omega$, on which we define a position field, in a free volume context, denoted by $\mathbf{r} : \Omega \times [0,T] \to \mathbb{R}^3$, where $[0,T]$ is a time interval.

Suppose also an associated density distribution scalar field is given by $(\rho \circ \mathbf{r})$: $\Omega \times [0,T] \to [0,+\infty)$, so that the kinetics energy for such a system, denoted by $J : U \times V \to \mathbb{R}$, is defined as

$$J(\mathbf{r},\rho) = \frac{1}{2} \int_0^T \int_\Omega \rho(\mathbf{r}(\mathbf{x},t)) \frac{\partial \mathbf{r}(\mathbf{x},t)}{\partial t} \cdot \frac{\partial \mathbf{r}(\mathbf{x},t)}{\partial t} \sqrt{g} \, d\mathbf{x} dt,$$

subject to

$$\int_\Omega \rho(\mathbf{r}(\mathbf{x},t)) \sqrt{g} \, d\mathbf{x} = m, \text{ on } [0,T],$$

where m is the total system mass, t denotes time and $d\mathbf{x} = dx_1 \, dx_2 \, dx_3$.

Here,

$$U = \{\mathbf{r} \in W^{1,2}(\Omega \times [0,T]) : \mathbf{r}(\mathbf{x},0) = \mathbf{r}_0(\mathbf{x})$$
$$\text{and } \mathbf{r}(\mathbf{x},T) = \mathbf{r}_1(\mathbf{x}), \text{ in } \Omega\}, \tag{7.1}$$

and

$$V = \{\rho(\mathbf{r}) \in L^2([0,T]; W^{1,2}(\Omega)) : \mathbf{r} \in U\}.$$

Also

$$\mathbf{g}_k = \frac{\partial \mathbf{r}(\mathbf{x},t)}{\partial x_k},$$

where we assume

$$\{\mathbf{g}_k, \, k \in \{1,2,3\}\}$$

to be a linearly independent set in $\Omega \times [0,T]$,

$$g_{jk} = \mathbf{g}_j \cdot \mathbf{g}_k,$$

$$\{g^{ij}\} = \{g_{ij}\}^{-1},$$

and

$$g = \det\{g_{jk}\}.$$

For such a standard Newtonian formulation, the kinetics energy takes into account just the tangential field given by the time derivative

$$\frac{\partial \mathbf{r}(\mathbf{x},t)}{\partial t}.$$

At this point, the idea is to complement such an energy with a new term, denoted by $\hat{R}$, which would consider also the control of curvature distribution along the mechanical system.

So, with such statements in mind, we redefine the concerning energy, denoting it again by $J : U \times V \times V_1 \to \mathbb{R}$, as

$$J(\mathbf{r},\rho) = -\frac{1}{2} \int_0^T \int_\Omega \rho(\mathbf{r}(\mathbf{x},t)) \frac{\partial \mathbf{r}(\mathbf{x},t)}{\partial t} \cdot \frac{\partial \mathbf{r}(\mathbf{x},t)}{\partial t} \sqrt{g} \, d\mathbf{x} dt$$

$$+ \frac{\gamma}{2} \int_0^T \int_\Omega \hat{R} \sqrt{g} \, d\mathbf{x} dt, \tag{7.2}$$

subject to

$$\int_\Omega \rho(\mathbf{r}(\mathbf{x},t)) \sqrt{g} \, d\mathbf{x} = m, \text{ on } [0,T],$$

where

$$\hat{R} = \sum_{i,j,k,l=1}^3 g^{ij} g^{kl} \frac{\partial}{\partial x_i} \left(\sqrt{\frac{\rho(\mathbf{x},t)}{m}} \frac{\partial \mathbf{r}(\mathbf{x},t)}{\partial x_j} \right) \cdot \frac{\partial}{\partial x_k} \left(\sqrt{\frac{\rho(\mathbf{x},t)}{m}} \frac{\partial \mathbf{r}(\mathbf{x},t)}{\partial x_l} \right),$$

and $\gamma > 0$ is a constant to be specified.

Thus, defining a complex function ϕ such that

$$|\phi| = \sqrt{\frac{\rho}{m}}$$

and observing that the Christoffel symbols Γ_{ij}^s are such that

$$\frac{\partial^2 \mathbf{r}(\mathbf{x},t)}{\partial x_i \partial x_j} = \sum_{s=1}^3 \Gamma_{ij}^s \frac{\partial \mathbf{r}(\mathbf{x},t)}{\partial x_s}, \ \forall i,j \in \{1,2,3\},$$

we have

$$\frac{\partial}{\partial x_i} \left(\phi \frac{\partial \mathbf{r}(\mathbf{x},t)}{\partial x_j} \right)$$

$$= \frac{\partial \phi}{\partial x_i} \frac{\partial \mathbf{r}(\mathbf{x},t)}{\partial x_j} + \phi \frac{\partial^2 \mathbf{r}(\mathbf{x},t)}{\partial x_i \partial x_j}$$

$$= \frac{\partial \phi}{\partial x_i} \frac{\partial \mathbf{r}(\mathbf{x},t)}{\partial x_j} + \phi \sum_{s=1}^3 \Gamma_{ij}^s \frac{\partial \mathbf{r}(\mathbf{x},t)}{\partial x_s}. \tag{7.3}$$

Therefore,

$$\left(\frac{\partial}{\partial x_i} \left(\phi \frac{\partial \mathbf{r}(\mathbf{x},t)}{\partial x_j} \right) \right) \cdot \left(\frac{\partial}{\partial x_k} \left(\phi^* \frac{\partial \mathbf{r}(\mathbf{x},t)}{\partial x_l} \right) \right)$$

$$= \left(\frac{\partial \phi}{\partial x_i} \frac{\partial \mathbf{r}(\mathbf{x},t)}{\partial x_j} + \phi \sum_{s=1}^3 \Gamma_{ij}^s \frac{\partial \mathbf{r}(\mathbf{x},t)}{\partial x_s} \right) \cdot \left(\frac{\partial \phi^*}{\partial x_k} \frac{\partial \mathbf{r}(\mathbf{x},t)}{\partial x_l} + \phi^* \sum_{p=1}^3 \Gamma_{kl}^p \frac{\partial \mathbf{r}(\mathbf{x},t)}{\partial x_p} \right)$$

$$= \sum_{s,p=1}^3 \left(g_{jl} \frac{\partial \phi}{\partial x_i} \frac{\partial \phi^*}{\partial x_k} + \phi \frac{\partial \phi^*}{\partial x_k} \Gamma_{ij}^s g_{sl} \right.$$

$$\left. + \phi^* \frac{\partial \phi}{\partial x_i} \Gamma_{kl}^p g_{pj} + |\phi|^2 \Gamma_{ij}^s \Gamma_{kl}^p g_{sp} \right). \tag{7.4}$$

From this, we may write,

$$\hat{R} = \sum_{i,j,k,l,p,s=1}^{3} g^{ij} g^{kl} \left(g_{jl} \frac{\partial \phi}{\partial x_i} \frac{\partial \phi^*}{\partial x_k} + \phi \frac{\partial \phi^*}{\partial x_k} \Gamma_{ij}^{s} g_{sl} \right.$$

$$\left. + \phi^* \frac{\partial \phi}{\partial x_i} \Gamma_{kl}^{p} g_{pj} + |\phi|^2 \Gamma_{ij}^{s} \Gamma_{kl}^{p} g_{sp} \right). \tag{7.5}$$

Already including the Lagrange multipliers concerning the restrictions, the final expression for the energy, denoted by $J : U \times V \to \mathbb{R}$, would be given by

$$J(\mathbf{r},\phi,E) = -\frac{1}{2} \int_0^T \int_\Omega m |\phi(\mathbf{r}(\mathbf{x},t))|^2 \frac{\partial \mathbf{r}(\mathbf{x},t)}{\partial t} \cdot \frac{\partial \mathbf{r}(\mathbf{x},t)}{\partial t} \sqrt{g}\, d\mathbf{x} dt$$

$$+ \frac{\gamma}{2} \int_0^T \int_\Omega \hat{R} \sqrt{g}\, d\mathbf{x} dt$$

$$- m \int_0^T E(t) \left(\int_\Omega |\phi(\mathbf{r})|^2 \sqrt{g}\, d\mathbf{x} - 1 \right) dt, \tag{7.6}$$

where,

$$U = \{ \mathbf{r} \in W^{1,2}(\Omega \times [0,T]) : \mathbf{r}(\mathbf{x},0) = \mathbf{r}_0(\mathbf{x})$$

$$\text{and } \mathbf{r}(\mathbf{x},T) = \mathbf{r}_1(\mathbf{x}), \text{ in } \Omega \}, \tag{7.7}$$

Finally, in particular for the special case in which

$$\mathbf{r}(\mathbf{x},t) \approx \mathbf{x},$$

we get

$$\frac{\partial \mathbf{r}(\mathbf{x},t)}{\partial t} \approx 0,$$

$\mathbf{g}_k \approx \mathbf{e}_k$, where

$$\{\mathbf{e}_1, \mathbf{e}_2, \mathbf{e}_3\}$$

is the canonical basis of $\mathbb{R}^3$.

Therefore, in such a case,

$$\frac{\gamma}{2} \int_0^T \int_\Omega \hat{R} \sqrt{g}\, d\mathbf{x} dt \approx \frac{\gamma T}{2} \sum_{k=1}^{3} \int_\Omega \frac{\partial \phi}{\partial x_k} \frac{\partial \phi^*}{\partial x_k}\, d\mathbf{x}.$$

Hence, with such last results we may infer that

$$J(\mathbf{r},\phi,E)/T \approx \tilde{J}(\phi,E)$$

$$= \frac{\gamma}{2} \sum_{k=1}^{3} \int_\Omega \frac{\partial \phi}{\partial x_k} \frac{\partial \phi^*}{\partial x_k}\, d\mathbf{x}$$

$$- E \left(\int_\Omega |\phi|^2 d\mathbf{x} - 1 \right). \tag{7.8}$$

This last energy is just the standard Schrödinger one in a free particle context.

7.3 A brief note on the relativistic context, the Klein-Gordon equation

Of particular interest is the case in which $\mathbf{x} = (x_1, x_2, x_3) \in \mathbb{R}^3$ and point-wise,

$$\mathbf{r}(\mathbf{x}, t) = (ct, X_1(t, \mathbf{x}), X_2(t, \mathbf{x}), X_3(t, \mathbf{x})),$$

where

$$M = \{\mathbf{r}(\mathbf{x}, t) \ : \ (\mathbf{x}, t) \in \Omega \times [0, T]\},$$

for an appropriate $\Omega \subset \mathbb{R}^3$.

Also, denoting $d\mathbf{x} = dx_1 dx_2 dx_3$, the mass differential would be given by

$$dm = \frac{\rho(\mathbf{r})}{\sqrt{1 - v^2/c^2}} \sqrt{-g} \, d\mathbf{x} = \frac{|R(\mathbf{r})|^2}{\sqrt{1 - v^2/c^2}} \sqrt{-g} \, d\mathbf{x},$$

and the semi-classical kinetics energy differential would be expressed by

$$
\begin{aligned}
dE_c &= \frac{\partial \mathbf{r}(t, \mathbf{x})}{\partial t} \cdot \frac{\partial \mathbf{r}(t, \mathbf{x})}{\partial t} \, dm \\
&= -\left(\frac{d\bar{t}}{dt}\right)^2 dm \\
&= -(c^2 - v^2) \, dm,
\end{aligned}
\tag{7.9}
$$

so that

$$dE_c = -c^2 (\sqrt{1 - v^2/c^2}) |R(\mathbf{r})|^2 \sqrt{-g} \, d\mathbf{x},$$

where

$$d\bar{t}^2 = c^2 dt^2 - dX_1(t, \mathbf{x})^2 - dX_2(t, \mathbf{x})^2 - dX_3(t, \mathbf{x})^2.$$

Thus, the concerning energy is expressed by,

$$
\begin{aligned}
J_1(\mathbf{r}, R) &= -\int_0^T \int_\Omega dE_c \, dt + \frac{\gamma}{2} \int_0^T \int_\Omega \hat{R} \sqrt{-g} \, d\mathbf{x} \, dt \\
&= c^2 \int_0^T \int_\Omega |R(\mathbf{r})|^2 \sqrt{1 - v^2/c^2} \sqrt{-g} \, d\mathbf{x} \, dt \\
&\quad + \frac{\gamma}{2} \int_0^T \int_\Omega \hat{R} \sqrt{-g} \, d\mathbf{x} \, dt,
\end{aligned}
\tag{7.10}
$$

subject to

$$R(\mathbf{r}(\mathbf{x}, 0)) = R_0(\mathbf{x})$$
$$R(\mathbf{r}(\mathbf{x}, T)) = R_1(\mathbf{x})$$

and

$$R(\mathbf{r}(\mathbf{x}, t)) = 0, \text{ on } \partial\Omega \times [0, T],$$

$$\int_\Omega |R(\mathbf{r})|^2 \sqrt{-g} \, d\mathbf{x} = m, \text{ on } [0, T].$$

Here, we have denoted

$$x_0 = ct,$$

$$(x_0, \mathbf{x}) = (x_0, x_1, x_2, x_3),$$

$$\mathbf{g}_k = \frac{\partial \mathbf{r}(t, \mathbf{x})}{\partial x_k},$$

where we assume

$$\{\mathbf{g}_k, \ k \in \{0, 1, 2, 3\}\}$$

to be a linearly independent set in $\Omega \times [0, T]$,

$$g = det\{g_{ij}\},$$

$$g_{ij} = \mathbf{g}_i \cdot \mathbf{g}_j,$$

$$\{g^{ij}\} = \{g_{ij}\}^{-1},$$

where such a product is given by

$$\mathbf{y} \cdot \mathbf{z} = -y_0 z_0 + \sum_{i=1}^{3} y_i z_i, \ \forall \mathbf{y} = (y_0, y_1, y_2, y_3), \ \mathbf{z} = (z_0, z_1, z_2, z_3) \in \mathbb{R}^4.$$

Moreover,

$$\hat{R} = \sum_{i,j,k,l=0}^{3} g^{ij} g^{kl} \frac{\partial}{\partial x_i} \left(\frac{R(\mathbf{x}, t)}{\sqrt{m}} \frac{\partial \mathbf{r}(\mathbf{x}, t)}{\partial x_j} \right) \cdot \frac{\partial}{\partial x_k} \left(\frac{R(\mathbf{x}, t)}{\sqrt{m}} \frac{\partial \mathbf{r}(\mathbf{x}, t)}{\partial x_l} \right).$$

Therefore, defining $\phi \in W^{1,2}(\Omega \times [0, T]; \mathbb{C})$ as

$$\phi(\mathbf{x}, t) = \frac{R(\mathbf{r}(\mathbf{x}, t))}{\sqrt{m}},$$

and recalling that the Christoffel symbols Γ_{ij}^s are such that

$$\frac{\partial^2 \mathbf{r}(\mathbf{x}, t)}{\partial x_i \partial x_j} = \sum_{s=0}^{3} \Gamma_{ij}^s \frac{\partial \mathbf{r}(\mathbf{x}, t)}{\partial x_s}, \ \forall i, j \in \{0, 1, 2, 3\},$$

similarly as in the last section, we may obtain

$$\hat{R} = \sum_{i,j,k,l,p,s=0}^{3} g^{ij} g^{kl} \left(g_{jl} \frac{\partial \phi}{\partial x_i} \frac{\partial \phi^*}{\partial x_k} + \phi \frac{\partial \phi^*}{\partial x_k} \Gamma_{ij}^s g_{sl} \right.$$

$$\left. + \phi^* \frac{\partial \phi}{\partial x_i} \Gamma_{kl}^p g_{pj} + |\phi|^2 \Gamma_{ij}^s \Gamma_{kl}^p g_{sp} \right). \tag{7.11}$$

Finally, we would also have

$$v = \sqrt{\left(\frac{\partial X_1}{\partial t} \right)^2 + \left(\frac{\partial X_2}{\partial t} \right)^2 + \left(\frac{\partial X_3}{\partial t} \right)^2}.$$

In particular for the special case in which

$$\mathbf{r}(\mathbf{x},t) \approx (ct,\mathbf{x}),$$

so that

$$\frac{\partial \mathbf{r}(\mathbf{x},t)}{\partial t} \approx (c,0,0,0),$$

we would obtain

$$\mathbf{g}_0 \approx (1,0,0,0),\ \mathbf{g}_1 \approx (0,1,0,0),\ \mathbf{g}_2 \approx (0,0,1,0)\ \text{and}\ \mathbf{g}_3 \approx (0,0,0,1) \in \mathbb{R}^4.$$

so that

$$\frac{\gamma}{2} \int_0^T \int_\Omega \hat{R}\, \sqrt{g}\, d\mathbf{x}dt \ \approx\ \frac{\gamma}{2} \int_0^T \int_\Omega \left(-\frac{1}{c^2} \frac{\partial \phi(\mathbf{x},t)}{\partial t} \frac{\partial \phi^*(\mathbf{x},t)}{\partial t} \right.$$
$$\left. + \sum_{k=1}^3 \frac{\partial \phi(\mathbf{x},t)}{\partial x_k} \frac{\partial \phi^*(\mathbf{x},t)}{\partial x_k} \right) d\mathbf{x}dt, \qquad (7.12)$$

and

$$c^2 \int_0^T \int_\Omega |R(\mathbf{r})|^2 \sqrt{1 - v^2/c^2}\, \sqrt{-g}\, d\mathbf{x}\, dt \approx mc^2 \int_0^T \int_\Omega |\phi(\mathbf{x},t)|^2\, d\mathbf{x}dt.$$

Hence, with such last results we may infer that

$$J_1(\mathbf{r},\phi,E) \ \approx\ \frac{\gamma}{2} \left(\int_0^T \int_\Omega -\frac{1}{c^2} \frac{\partial \phi(\mathbf{x},t)}{\partial t} \frac{\partial \phi^*(\mathbf{x},t)}{\partial t}\, d\mathbf{x}dt \right.$$
$$\left. + \sum_{k=1}^3 \int_\Omega \int_0^T \frac{\partial \phi(\mathbf{x},t)}{\partial x_k} \frac{\partial \phi^*(\mathbf{x},t)}{\partial x_k}\, d\mathbf{x}dt \right)$$
$$+ mc^2 \int_0^T \int_\Omega |\phi(\mathbf{x},t)|^2\, d\mathbf{x}dt$$
$$- m \int_0^T E(t) \left(\int_\Omega |\phi(\mathbf{x},t)|^2 d\mathbf{x} - 1 \right) dt. \qquad (7.13)$$

The Euler Lagrange equations for such an energy are given by

$$\frac{\gamma}{2} \left(\frac{1}{c^2} \frac{\partial^2 \phi(\mathbf{x},t)}{\partial t^2} - \sum_{k=1}^3 \frac{\partial^2 \phi(\mathbf{x},t)}{\partial x_k^2} \right)$$
$$+ mc^2 \phi(\mathbf{x},t) - E_1(t)\phi(\mathbf{x},t) = 0,\ \text{in}\ \Omega, \qquad (7.14)$$

where,

$$\phi(\mathbf{x},0) = \phi_0(\mathbf{x}),\ \text{in}\ \Omega,$$
$$\phi(\mathbf{x},T) = \phi_1(\mathbf{x}),\ \text{in}\ \Omega,$$
$$\phi(\mathbf{x},t) = 0,\ \text{on}\ \partial\Omega \times [0,T]$$

and $E_1(t) = mE(t)$.

Equation (7.14) is the relativistic Klein-Gordon one.

For $E_1(t) = E_1 \in \mathbb{R}$ (not time dependent), at this point we suggest a solution (and implicitly related time boundary conditions) $\phi(\mathbf{x}, t) = e^{-\frac{iE_1 t}{\hbar}}\phi_2(\mathbf{x})$, where

$$\phi_2(\mathbf{x}) = 0, \text{ on } \partial\Omega.$$

Therefore, replacing this solution into equation (7.14), we would obtain

$$\left(\frac{\gamma}{2}\left(-\frac{E_1^2}{c^2\hbar^2}\phi_2(\mathbf{x}) - \sum_{k=1}^{3}\frac{\partial^2\phi_2(\mathbf{x})}{\partial x_k^2} \right) + mc^2\phi_2(\mathbf{x}) - E_1\phi_2(\mathbf{x}) \right) e^{-\frac{iE_1 t}{\hbar}} = 0,$$

in Ω.

Denoting

$$E_2 = -\frac{\gamma E_1^2}{2c^2\hbar^2} + mc^2 - E_1,$$

the final eigenvalue problem would stand for

$$-\frac{\gamma}{2}\sum_{k=1}^{3}\frac{\partial^2\phi_2(\mathbf{x})}{\partial x_k^2} + E_2\phi_2(\mathbf{x}) = 0, \text{ in } \Omega$$

where E_1 is such that

$$\int_\Omega |\phi_2(\mathbf{x})|^2\, d\mathbf{x} = 1.$$

Moreover, from (7.14), such a solution $\phi(\mathbf{x}, t) = e^{-\frac{iE_1 t}{\hbar}}\phi_2(\mathbf{x})$ is also such that

$$\frac{\gamma}{2}\left(\frac{1}{c^2}\frac{\partial^2\phi(\mathbf{x}, t)}{\partial t^2} - \sum_{k=1}^{3}\frac{\partial^2\phi(\mathbf{x}, t)}{\partial x_k^2} \right)$$

$$+ mc^2\phi(\mathbf{x}, t) = i\hbar\frac{\partial\phi(\mathbf{x}, t)}{\partial t}, \text{ in } \Omega. \tag{7.15}$$

At this point, we recall that in quantum mechanics,

$$\gamma = \hbar^2/m.$$

Finally, we remark that this last equation (7.15) is a kind of relativistic Schrödinger-Klein-Gordon equation.

7.4 A second model and the respective energy expression

In a free volume context, denote again by $\mathbf{r} : \Omega \times [0, T] \to \mathbb{R}^3$ a position field, where $[0, T]$ is a time interval.

Suppose also an associated density distribution scalar field is given by $(\rho \circ \mathbf{r})$: $\Omega \times [0, T] \to [0, +\infty)$, so that the kinetics energy for such a system, denoted by $J : U \times V \to \mathbb{R}$, is defined as

$$J(\mathbf{r}, \rho) = \frac{1}{2} \int_0^T \int_\Omega \rho(\mathbf{r}(\mathbf{x}, t)) \frac{\partial \mathbf{r}(\mathbf{x}, t)}{\partial t} \cdot \frac{\partial \mathbf{r}(\mathbf{x}, t)}{\partial t} \sqrt{g} \, d\mathbf{x} dt,$$

subject to

$$\int_\Omega \rho(\mathbf{r}(\mathbf{x}, t)) \sqrt{g} \, d\mathbf{x} = m, \text{ on } [0, T].$$

We recall that for such a standard Newtonian formulation, the kinetics energy takes into account just the tangential field given by the time derivative

$$\frac{\partial \mathbf{r}(\mathbf{x}, t)}{\partial t}.$$

Now the new idea is to complement such an energy with a new term which would consider also the variation of a normal field $\mathbf{n}$ and concerning distribution of curvature, such that

$$\mathbf{n} \cdot \frac{\partial \mathbf{r}(\mathbf{x}, t)}{\partial t} = 0, \text{ in } \Omega \times [0, T].$$

So, with such statements in mind, we redefine the concerning energy, denoting it again by $J : U \times V \times V_1 \to \mathbb{R}$, as

$$
\begin{aligned}
J(\mathbf{r}, \mathbf{n}, \rho) \ = \ & -\frac{1}{2} \int_0^T \int_\Omega \rho(\mathbf{r}(\mathbf{x}, t)) \frac{\partial \mathbf{r}(\mathbf{x}, t)}{\partial t} \cdot \frac{\partial \mathbf{r}(\mathbf{x}, t)}{\partial t} \sqrt{g} \, d\mathbf{x} dt \\
& + \frac{\gamma}{2} \int_0^T \int_\Omega \hat{R} \sqrt{g} \, d\mathbf{x} dt,
\end{aligned}
\tag{7.16}
$$

where $\gamma > 0$ is an appropriate constant,

$$\hat{R} = g^{ij} \hat{R}_{ij},$$

$$\hat{R}_{jk} = \hat{R}^i_{jik},$$

$$\hat{R}^i_{jkl} = b^l_i \, b_{jk},$$

$$b_{ij} = -\frac{1}{\sqrt{m}} \frac{\partial \left(\sqrt{\rho(\mathbf{r})} \mathbf{n}(\mathbf{r}) \right)}{\partial x_j} \cdot \mathbf{g}_i,$$

$$b^i_j = g^{il} b_{lj},$$

and,

$$\{g^{ij}\} = \{g_{ij}\}^{-1},$$

$\forall i, j, k, l \in \{1, 2, 3\}$.

subject to

$$\mathbf{n}(\mathbf{r}) \cdot \mathbf{n}(\mathbf{r}) = 1, \text{ in } \Omega \times [0, T],$$

$$\mathbf{n}(\mathbf{r}) \cdot \frac{\partial \mathbf{r}}{\partial t} = 0, \text{ in } \Omega \times [0, T],$$

and

$$\int_\Omega \rho(\mathbf{r}(\mathbf{x},t))\sqrt{g}\,d\mathbf{x} = m, \text{ on } [0,T].$$

Here

$$V_1 = \{\mathbf{n}(\mathbf{r}) \in L^2(\Omega \times [0,T]) : \mathbf{r} \in U\}.$$

Thus, defining ϕ such that

$$|\phi| = \sqrt{\frac{\rho}{m}}$$

and already including the Lagrange multipliers concerning the restrictions, the final expression for the energy, denoted by $J : U \times V \times V_1 \times V_2 \times [V_3]^2 \to \mathbb{R}$, would be given by

$$
\begin{aligned}
J(\mathbf{r},\mathbf{n},\phi,E,\lambda_1,\lambda_2) =\ & -\frac{1}{2}\int_0^T \int_\Omega m|\phi(\mathbf{r}(\mathbf{x},t))|^2 \frac{\partial \mathbf{r}(\mathbf{x},t)}{\partial t}\cdot \frac{\partial \mathbf{r}(\mathbf{x},t)}{\partial t}\sqrt{g}\,d\mathbf{x}dt \\
& +\frac{\gamma}{2}\int_0^T \int_\Omega \hat{R}\sqrt{g}\,d\mathbf{x}dt \\
& -m\int_0^T E(t)\left(\int_\Omega |\phi(\mathbf{r})|^2 \sqrt{g}\,d\mathbf{x} - 1\right)dt \\
& +\langle \lambda_1, \mathbf{n}\cdot\mathbf{n} - 1\rangle_{L^2} \\
& +\left\langle \lambda_2, \mathbf{n}\cdot \frac{\partial \mathbf{r}}{\partial t}\right\rangle_{L^2},
\end{aligned}
\tag{7.17}
$$

where,

$$
\begin{aligned}
U =\ & \{\mathbf{r} \in W^{1,2}(\Omega \times [0,T]) : \mathbf{r}(\mathbf{x},0) = \mathbf{r}_0(\mathbf{x}) \\
& \text{and } \mathbf{r}(\mathbf{x},T) = \mathbf{r}_1(\mathbf{x}), \text{ in } \Omega\},
\end{aligned}
\tag{7.18}
$$

$$V = \{\phi(\mathbf{r}) \in L^2([0,T]; W^{1,2}(\Omega;\mathbb{C})) : \mathbf{r} \in U\},$$

$$V_1 = \{\mathbf{n}(\mathbf{r}) \in L^2(\Omega \times [0,T]) : \mathbf{r} \in U\},$$

$$V_2 = L^2([0,T]),$$

$$V_3 = L^2(\Omega \times [0,T]).$$

Moreover,

$$\hat{R} = g^{ij}\hat{R}_{ij},$$

$$\hat{R}_{jk} = \hat{R}^i_{jik},$$

$$\hat{R}^i_{jkl} = b^l_i\, b^*_{jk},$$

$$b_{ij} = -\frac{\partial(\phi(\mathbf{r})\mathbf{n}(\mathbf{r}))}{\partial x_j}\cdot \mathbf{g}_i,$$

$$b^i_j = g^{il} b_{lj},$$

$\forall i,j,k,l \in \{1,2,3\}.$

Finally, in particular for the special case in which

$$\mathbf{r}(\mathbf{x},t) \approx \mathbf{x},$$

so that

$$\frac{\partial \mathbf{r}(\mathbf{x},t)}{\partial t} \approx 0,$$

and

$$\mathbf{n} \cdot \frac{\partial \mathbf{r}}{\partial t} \approx 0,$$

we may set

$$\mathbf{n} = \mathbf{c},$$

where $\mathbf{c} \in \mathbb{R}^3$ is a constant such that

$$\mathbf{c} \cdot \mathbf{c} = 1,$$

and obtain

$$\mathbf{g}_k \approx \mathbf{e}_k,$$

where

$$\{\mathbf{e}_1, \mathbf{e}_2, \mathbf{e}_3\}$$

is the canonical basis of $\mathbb{R}^3$.

Therefore, in such a case,

$$\frac{\gamma}{2} \int_0^T \int_\Omega \hat{R} \sqrt{g}\, d\mathbf{x}dt \approx \frac{\gamma T}{2} \sum_{k=1}^3 \int_\Omega \frac{\partial \phi}{\partial x_k} \frac{\partial \phi^*}{\partial x_k}\, d\mathbf{x}.$$

Hence, we would also obtain

$$
\begin{aligned}
J(\mathbf{r}, \mathbf{n}, \phi, E, \lambda_1, \lambda_2)/T &\approx \tilde{J}(\phi, E) \\
&= \frac{\gamma}{2} \sum_{k=1}^3 \int_\Omega \frac{\partial \phi}{\partial x_k} \frac{\partial \phi^*}{\partial x_k}\, d\mathbf{x} \\
&\quad - E\left(\int_\Omega |\phi|^2 d\mathbf{x} - 1\right).
\end{aligned}
\tag{7.19}
$$

This last energy is just the standard Schrödinger one in a free particle context.

7.5 A brief note on the relativistic context for such a second model

We recall to have denoted by c the speed of light and

$$d\bar{t}^2 = c^2 dt^2 - dX_1^2 - dX_2^2 - dX_3^2.$$

In a relativistic free particle context, the Hilbert variational formulation could be extended, for a motion in a pseudo Riemannian relativistic C^1 class manifold M, where locally

$$M = \{\mathbf{r}(\mathbf{u}) \; : \; \mathbf{u} \in \Omega\},$$

$$\mathbf{u} = (u_1, u_2, u_3, u_4) \in \mathbb{R}^4,$$

and

$$\mathbf{r} : \Omega \subset \mathbb{R}^4 \to \mathbb{R}^4$$

point-wise stands for,

$$\mathbf{r}(\mathbf{u}) = (ct(\mathbf{u}), X_1(\mathbf{u}), X_2(\mathbf{u}), X_3(\mathbf{u})),$$

to a functional J_1 where denoting $\rho(\mathbf{r}) = |R(\mathbf{r})|^2$, the mass differential is given by

$$dm = \frac{\rho(\mathbf{r})}{\sqrt{1 - v^2/c^2}} \sqrt{|g|} \, d\mathbf{u} = \frac{|R(\mathbf{r})|^2}{\sqrt{1 - v^2/c^2}} \sqrt{|g|} \, d\mathbf{u},$$

the semi-classical kinetics energy differential is given by

$$
\begin{aligned}
dE_c &= \frac{\partial \mathbf{r}(\mathbf{u})}{\partial t} \cdot \frac{\partial \mathbf{r}(\mathbf{u})}{\partial t} \, dm \\
&= -\left(\frac{d\bar{t}}{dt}\right)^2 dm \\
&= -(c^2 - v^2) \, dm,
\end{aligned}
\tag{7.20}
$$

so that

$$dE_c = -c^2 \left(\sqrt{1 - v^2/c^2}\right) |R(\mathbf{r})|^2 \sqrt{|g|} \, d\mathbf{u},$$

and

$$
\begin{aligned}
J_1(\mathbf{r}, R, \mathbf{n}) &= -\int_\Omega dE_c + \frac{\gamma}{2} \int_\Omega \hat{R} \sqrt{|g|} \, d\mathbf{u} \\
&= c^2 \int_\Omega |R(\mathbf{r})|^2 \sqrt{1 - v^2/c^2} \, \sqrt{|g|} \, d\mathbf{u} \\
&\quad + \frac{\gamma}{2} \int_\Omega \hat{R} \sqrt{|g|} \, d\mathbf{u},
\end{aligned}
\tag{7.21}
$$

subject to

$$\int_\Omega |R(\mathbf{r})|^2 \sqrt{|g|} \, d\mathbf{u} = m,$$

where m is the particle mass at rest.

Moreover,

$$\mathbf{n}(\mathbf{r}) \cdot \frac{\partial \mathbf{r}}{\partial \bar{t}} = 0, \text{ in } \Omega,$$

where

$$\frac{\partial \mathbf{r}}{\partial \bar{t}} = \frac{\partial \mathbf{r}}{\partial t} \frac{\partial t}{\partial \bar{t}}$$

$$= \frac{\frac{\partial \mathbf{r}}{\partial t}}{\frac{\partial \bar{t}}{\partial t}}$$

$$= \frac{\partial \mathbf{r}}{c \partial t} \frac{1}{\sqrt{1 - v^2/c^2}}, \qquad (7.22)$$

and

$$\mathbf{n}(\mathbf{r}) \cdot \mathbf{n}(\mathbf{r}) = 1, \text{ in } \Omega.$$

Where γ is an appropriate positive constant to be specified.
Also,

$$\mathbf{g}_k = \frac{\partial \mathbf{r}(\mathbf{u})}{\partial u_k},$$

$$g = det\{g_{ij}\},$$

$$g_{ij} = \mathbf{g}_i \cdot \mathbf{g}_j,$$

where here, in this subsection, such a product is given by

$$\mathbf{y} \cdot \mathbf{z} = -y_0 z_0 + \sum_{i=1}^{3} y_i z_i, \ \forall \mathbf{y} = (y_0, y_1, y_2, y_3), \ \mathbf{z} = (z_0, z_1, z_2, z_3) \in \mathbb{R}^4,$$

$$\hat{R} = g^{ij} \hat{R}_{ij},$$

$$\hat{R}_{jk} = \hat{R}^i_{jik},$$

$$\hat{R}^i_{jkl} = b^l_i b^*_{jk},$$

$$b_{ij} = -\frac{1}{\sqrt{m}} \frac{\partial (R(\mathbf{r})\mathbf{n}(\mathbf{r}))}{\partial u_j} \cdot \mathbf{g}_i,$$

$$b^i_j = g^{il} b_{lj},$$

and,

$$\{g^{ij}\} = \{g_{ij}\}^{-1},$$

$\forall i, j, k, l \in \{1, 2, 3, 4\}$.

Finally,

$$v = \sqrt{\left(\frac{\partial X_1}{\partial t}\right)^2 + \left(\frac{\partial X_2}{\partial t}\right)^2 + \left(\frac{\partial X_3}{\partial t}\right)^2},$$

where,

$$\frac{\partial X_k(\mathbf{u})}{\partial t} = \frac{\partial X_k(\mathbf{u})}{\partial u_j} \cdot \frac{\partial u_j}{\partial t}$$

$$= \sum_{j=1}^{4} \frac{\frac{\partial X_k(\mathbf{u})}{\partial u_j}}{\frac{\partial t(\mathbf{u})}{\partial u_j}}, \ \forall k \in \{1, 2, 3\}. \qquad (7.23)$$

Here, the Einstein sum convention holds.

Remark 7.1   The role of the variable **u** concerns the idea of establishing a relation between t, X_1, X_2 and X_3. The dimension of M may vary with the problem in question.

■

7.6   A brief note on the case including electro-magnetic effects

In this section, we address in a specific special relativistic context, the inclusion of electromagnetic effects.

7.6.1   About a specific Lorentz transformation

In this section, we assume the particle/volume motion is such that we are in a special relativity approximate context.

Consider the specific Lorentz transformation defined by a matrix $\{a_{jk}\}$, where the coordinates of the cartesian systems

$$(\mathbf{x}, t) = (x_1, x_2, x_3, t) = (x_1, x_2, x_3, x_4/(i_m c))$$

and

$$(\mathbf{x}', t') = (x_1', x_2', x_3', t') = (x_1', x_2', x_3', x_4'/(i_m c))$$

are related by the equations,

$$x_j' = \sum_{k=1}^{4} a_{jk} x_k, \ \forall j \in \{1, 2, 3, 4\},$$

where

$$x_4 = i_m ct \ \text{ and } x_4' = i_m ct'.$$

More specifically, we consider the case in which $\{a_{jk}\}$ is generated by a motion of the origin $0'$ of the system $(\mathbf{x}', t')$ with velocity $\mathbf{v} = (v_1, v_2, v_3)$ in relation to the origin 0 of the system $(\mathbf{x}, t)$. In such a motion, we assume the axis $0'x_j'$ keeps parallel to $0x_j$, $\forall j \in \{1, 2, 3\}$.

So, indeed, $\{a_{jk}\}$ is such that

$$x_j' = x_j + \left(\frac{1}{\sqrt{1 - \frac{v^2}{c^2}}} - 1 \right) \frac{\mathbf{x} \cdot \mathbf{v}}{v^2} v_j - \frac{tv_j}{\sqrt{1 - \frac{v^2}{c^2}}},$$

$\forall j \in \{1, 2, 3\}$, and

$$x_4'/(i_m c) = t' = \frac{1}{\sqrt{1 - \frac{v^2}{c^2}}} \left(t - \frac{\mathbf{x} \cdot \mathbf{v}}{c^2} \right),$$

where, as above indicated

$$\mathbf{x} = (x_1, x_2, x_3),$$
$$\mathbf{x}' = (x'_1, x'_2, x'_3),$$

and,

$$\mathbf{v} = (v_1, v_2, v_3).$$

7.6.2 Describing the self interaction energy and obtaining a final variational formulation

Considering the model for an electronic field with position field given by $\mathbf{r}(\mathbf{x},t)$ over the set $\Omega \times [0,T]$, where $\Omega \subset \mathbb{R}^3$, and a mass/charge density given by $\rho(\mathbf{r}) = |R(\mathbf{r})|^2$, we shall define the self interaction electric field differential, as indicated in the next lines.

First, denote in this section $dx = dx_1\, dx_2\, dx_3$,

$$\mathbf{r}(x,t) = (ct, X_1(x,t), X_2(x,t), X_3(x,t)),$$

$$\mathbf{r}_1(x,t) = (X_1(x,t), X_2(x,t), X_3(x,t)),$$

$$\Delta \mathbf{r}_1(x,\tilde{x},t) = \mathbf{r}_1(x,t) - \mathbf{r}_1(\tilde{x},t),$$

and

$$dq(\tilde{x},t) = K_1 |R(\mathbf{r}(\tilde{x},t))|^2 \sqrt{g(\mathbf{r}(\tilde{x},t))} d\tilde{x}.$$

Denote

$$\mathbf{v} = \frac{\partial \mathbf{r}_1(\tilde{x},t)}{\partial t} - \frac{\partial \mathbf{r}_1(x,t)}{\partial t},$$

and define

$$\Delta t(x,\tilde{x},t) = -\frac{1}{\sqrt{1 - \frac{v^2}{c^2}}} \frac{\Delta \mathbf{r}_1(x,\tilde{x},t)) \cdot \mathbf{v}}{c^2}.$$

Define also

$$\mathbf{v}' = \frac{\partial \mathbf{r}_1(\tilde{x}, t - \Delta t(x,\tilde{x},t))}{\partial t} - \frac{\partial \mathbf{r}_1(x, t - \Delta t(x,\tilde{x},t))}{\partial t},$$

and,

$$\Delta \hat{\mathbf{r}}_1(x,\tilde{x},t) = \Delta \mathbf{r}_1(x,\tilde{x}, t - \Delta t(x,\tilde{x},t))$$
$$+ \left(\frac{1}{\sqrt{1 - \frac{(v')^2}{c^2}}} - 1 \right) \frac{(\Delta \mathbf{r}_1(x,\tilde{x}, t - \Delta t(x,\tilde{x},t)) \cdot \mathbf{v}')\mathbf{v}'}{c^2}, \quad (7.24)$$

where

$$\Delta \mathbf{r}_1(x,\tilde{x}, t - \Delta t(x,\tilde{x},t)) = \mathbf{r}_1(x, t - \Delta t(x,\tilde{x},t)) - \mathbf{r}_1(\tilde{x}, t - \Delta t(x,\tilde{x},t)).$$

Thus, the electric field generated by $dq(\tilde{x}, t - \Delta t(x, \tilde{x}, t))$ at $\mathbf{r}(x, t)$ is given by

$$
\begin{aligned}
d\mathbf{E}'(x, \tilde{x}, t) &= Kdq(\tilde{x}, t - \Delta t(x, \tilde{x}, t)) \frac{\Delta \hat{\mathbf{r}}_1(x, \tilde{x}, t)}{|\Delta \hat{\mathbf{r}}_1(x, \tilde{x}, t)|^3} \\
&= KK_1 |R(\mathbf{r}(\tilde{x}, t - \Delta t(x, \tilde{x}, t)))|^2 \frac{\Delta \hat{\mathbf{r}}_1(x, \tilde{x}, t)}{|\Delta \hat{\mathbf{r}}_1(x, \tilde{x}, t)|^3} \sqrt{g(\mathbf{r}(\tilde{x}, t - \Delta t(x, \tilde{x}, t)))} d\tilde{x}.
\end{aligned}
$$

Now, we define

$$
dF_{E'} = \begin{pmatrix}
0 & 0 & 0 & -i_m dE_1' \\
0 & 0 & 0 & -i_m dE_2' \\
0 & 0 & 0 & -i_m dE_3' \\
i_m dE_1' & i_m dE_2' & i_m dE_3' & 0
\end{pmatrix} \tag{7.25}
$$

and define dF_E through the relations

$$
(dF_{E'})_{jk} = \hat{a}_{jl}\hat{a}_{kp}(dF_E)_{lp},
$$

where $\{\hat{a}_{jk}\}$ is such that

$$
x_j' = \hat{a}_{jk}x_k,
$$

or more specifically,

$$
x_j' = x_j + \left(\frac{1}{\sqrt{1 - \frac{v^2}{c^2}}} - 1 \right) \frac{\mathbf{x} \cdot \mathbf{v}}{v^2} v_j - \frac{tv_j}{\sqrt{1 - \frac{v^2}{c^2}}},
$$

$\forall j \in \{1, 2, 3\}$, and

$$
x_4' / (i_m c) = t' = \frac{1}{\sqrt{1 - \frac{v^2}{c^2}}} \left(t - \frac{\mathbf{x} \cdot \mathbf{v}}{c^2} \right),
$$

where here,

$$
\mathbf{v} = \frac{\partial \mathbf{r}_1(x, t)}{\partial t}.
$$

At this point we assume a functional $W(R(\mathbf{r}), \mathbf{r})$, which corresponds to the self-interacting energy, is such that

$$
\delta_R W(R(\mathbf{r}), \mathbf{r}) = |R(\mathbf{r})| \int_\Omega K \frac{dq(\tilde{x}, t - \Delta t(x, \tilde{x}, t))}{|\Delta \hat{\mathbf{r}}_1(x, \tilde{x}, t)|}
$$

and,

$$
\delta_\mathbf{r} W(R(\mathbf{r}), \mathbf{r}) = \delta_R W(R(\mathbf{r}), \mathbf{r}) \frac{\partial R(\mathbf{r})}{\partial \mathbf{r}} + |R(\mathbf{r}(x, t))|^2 \int_\Omega \sum_{k=1}^3 d\tilde{E}_k(x, \tilde{x}, t) e_k
$$

where $\{e_1, e_2, e_3\}$ is the canonical basis of $\mathbb{R}^3$ and we have denoted

$$dF_E(x,\tilde{x},t) = \begin{pmatrix} 0 & d\tilde{B}_3 & -d\tilde{B}_2 & -i_m d\tilde{E}_1 \\ -d\tilde{B}_3 & 0 & d\tilde{B}_1 & -i_m d\tilde{E}_2 \\ d\tilde{B}_2 & -d\tilde{B}_1 & 0 & -i_m d\tilde{E}_3 \\ i_m d\tilde{E}_1 & i_m d\tilde{E}_2 & i_m d\tilde{E}_3 & 0, \end{pmatrix} \tag{7.26}$$

$$F_E(x,t) = \int_\Omega dF_E(x,\tilde{x},t) = \begin{pmatrix} 0 & \tilde{B}_3 & -\tilde{B}_2 & -i_m \tilde{E}_1 \\ -\tilde{B}_3 & 0 & \tilde{B}_1 & -i_m \tilde{E}_2 \\ \tilde{B}_2 & -\tilde{B}_1 & 0 & -i_m \tilde{E}_3 \\ i_m \tilde{E}_1 & i_m \tilde{E}_2 & i_m \tilde{E}_3 & 0 \end{pmatrix} \tag{7.27}$$

where these last integrations are in $\tilde{x}$.
Also,

$$\tilde{\mathbf{E}} = (\tilde{E}_1, \tilde{E}_2, \tilde{E}_3),$$

and

$$\tilde{\mathbf{B}} = (\tilde{B}_1, \tilde{B}_2, \tilde{B}_3).$$

At this point, up to a concerning Lorentz transformation, we assume that it is possible to express the total electric field $\mathbf{E}$ by

$$\mathbf{E} = \tilde{\mathbf{E}} + \nabla\Phi + \frac{1}{c}\frac{\partial \mathbf{A}}{\partial t},$$

for appropriate functions $\Phi \in W^{1,2}(\Omega \times [0,T])$ and $\mathbf{A} \in W^{1,2}(\Omega \times [0,T]; \mathbb{R}^3)$.
Also

$$\mathbf{B} = \mathbf{B}_0 + \tilde{\mathbf{B}} - \operatorname{curl} \mathbf{A},$$

where $\mathbf{A} = (A_1, A_2, A_3)$ is a magnetic potential.

From the standard literature, we also define $A_4 = i_m \Phi$ and assume the Lorentz condition,

$$\operatorname{div} \mathbf{A} + \frac{1}{c}\frac{\partial \Phi}{\partial t} = 0, \text{ in } \Omega \times [0,T].$$

So, the system energy may be written as

$$
\begin{aligned}
& J(R,\mathbf{r},\mathbf{n},\Phi,\mathbf{A},E,\lambda) \\
&= c^2 \int_0^T \int_\Omega |R(\mathbf{r})|^2 \sqrt{1 - \frac{v^2}{c^2}} \sqrt{-g}\, dx\, dt \\
&\quad + \frac{1}{c} \int_0^T \int_\Omega K_1 |R(\mathbf{r})|^2 \frac{\partial \mathbf{r}_1}{\partial t} \cdot \mathbf{A} \sqrt{-g}\, dx\, dt \\
&\quad + \frac{\gamma}{2} \int_0^T \int_\Omega \hat{R}\sqrt{-g}\, dx\, dt \\
&\quad + W(R(\mathbf{r}),\mathbf{r}) \\
&\quad + K_1 \int_0^T \int_\Omega \Phi(\mathbf{r})|R(\mathbf{r})|^2 \sqrt{-g}\, dx\, dt \\
&\quad + \frac{1}{8\pi}\left(\|\mathbf{B}\|_{\Omega\times[0,T],2}^2 + \|\mathbf{E}\|_{\Omega\times[0,T],2}^2 \right) \\
&\quad - \frac{1}{2} \int_0^T E(t)\left(\int_\Omega |R(\mathbf{r})|^2 \sqrt{-g}\, dx - m_e \right) dt \\
&\quad + \left\langle \lambda_1, \mathbf{n}\cdot \frac{\partial \mathbf{r}}{\partial \bar{t}} \right\rangle_{L^2} + \langle \lambda_2, \mathbf{n}\cdot\mathbf{n} - 1\rangle_{L^2} \\
&\quad + \left\langle \lambda_3, \operatorname{div}\mathbf{A} + \frac{1}{c}\frac{\partial \Phi}{\partial t} \right\rangle_{L^2}.
\end{aligned}
\tag{7.28}
$$

Also,

$$
\mathbf{g}_k = \frac{\partial \mathbf{r}(x,t)}{\partial \hat{x}_k},
$$

where here $\{\hat{x}_k\} = (x_1,x_2,x_3,ct)$. Moreover,

$$
g = det\{g_{ij}\},
$$

$$
g_{ij} = \mathbf{g}_i \cdot \mathbf{g}_j,
$$

where here again, such a product is given by

$$
\mathbf{y}\cdot\mathbf{z} = -y_4 z_4 + \sum_{i=1}^3 y_i z_i,\ \forall \mathbf{y} = (y_1,y_2,y_3,y_4),\ \mathbf{z} = (z_1,z_2,z_3,z_4) \in \mathbb{R}^4,
$$

$$
\hat{R} = g^{ij}\hat{R}_{ij},
$$

$$
\hat{R}_{jk} = \hat{R}^i_{jik},
$$

$$
\hat{R}^i_{jkl} = b^l_i\, b^*_{jk},
$$

$$
b_{ij} = -\frac{1}{\sqrt{m}}\left(\frac{\partial\, (R(\mathbf{r})\mathbf{n}(\mathbf{r}))}{\partial \hat{x}_j} - i_m A_j R(\mathbf{r})\mathbf{n}(\mathbf{r}) \right) \cdot \mathbf{g}_i.
$$

Furthermore,

$$b^i_j = g^{il} b_{lj},$$

and,

$$\{g^{ij}\} = \{g_{ij}\}^{-1},$$

$\forall i, j, k \in \{1, 2, 3, 4\}$.

Finally, here we would also have

$$v = \sqrt{\left(\frac{\partial X_1}{\partial t}\right)^2 + \left(\frac{\partial X_2}{\partial t}\right)^2 + \left(\frac{\partial X_3}{\partial t}\right)^2},$$

It is worth mentioning that $\lambda_1, \lambda_2, \lambda_3$ are appropriate Lagrange multipliers concerning the respective constraints.

Remark 7.2   It is possible that in some cases we cannot find $W(R(\mathbf{r}), \mathbf{r})$, which corresponds to the self-interacting energy, such that

$$\delta_R W(R(\mathbf{r}), \mathbf{r}) = |R(\mathbf{r})| \int_\Omega K \frac{dq(\tilde{x}, t - \Delta t(x, \tilde{x}, t))}{|\Delta \hat{\mathbf{r}}_1(x, \tilde{x}, t)|}$$

and,

$$\delta_{\mathbf{r}} W(R(\mathbf{r}), \mathbf{r}) = \delta_R W(R(\mathbf{r}), \mathbf{r}) \frac{\partial R(\mathbf{r})}{\partial \mathbf{r}} + |R(\mathbf{r}(x, t))|^2 \int_\Omega \sum_{k=1}^3 d\tilde{E}_k(x, \tilde{x}, t) e_k.$$

In such a case, such last equations may be just approximately satisfied, so that we could define an optimization problem corresponding to find a critical point of the functional J plus a positive constant multiplied by the L^2 norms of analytical expressions corresponding to these last two equations.   ■

7.7   A new interpretation of the Bohr atomic model

This section develops a new interpretation of Bohr atomic model through classical and quantum mechanics.

In a second step, we consider as a generalization of such a model, the issue of an interacting system composed of a large amount of same type atoms.

At this point, we start to describe such a model.

Let $\Omega = B_{R_0}(\mathbf{0}) \subset \mathbb{R}^3$ be an open ball with center at $\mathbf{0} \in \mathbb{R}^3$. Let $[0, T]$ be a time interval. For $n \in \mathbb{N}$, consider a system with $\sum_{l=0}^{n-1}(2l + 1)$ electrons and the same number of protons, where the protons are supposed to be at rest at $\mathbf{x} = \mathbf{0} \in \mathbb{R}^3$. Moreover, the electrons are distributed in n layers $l \in \{0, \ldots, n-1\}$, each layer l with $2l + 1$ electrons.

We denote the position field for the electron $j \in -l,\dots,0,\dots,l$ at the layer l, by $\mathbf{r}_j^l : \Omega \times [0,T] \to \mathbb{R}^3$, where

$$
\begin{aligned}
\mathbf{r}_j^l(\mathbf{x},t) \;=\; & R_j^l(r)\Big(\sin((w_1)_j^l(\mathbf{x})t + \theta_j^l)\cos((w_2)_j^l(\mathbf{x})t + \phi_j^l)\mathbf{i} \\
& + \sin((w_1)_j^l(\mathbf{x})t + \theta_j^l)\sin((w_2)_j^l(\mathbf{x})t + \phi_j^l)\mathbf{j} \\
& + \cos((w_1)_j^l(\mathbf{x})t + \theta_j^l)\mathbf{k} \Big).
\end{aligned}
\tag{7.29}
$$

We also recall that $\mathbf{x} \in \mathbb{R}^3$ in spherical coordinates corresponds to (r,θ,ϕ) and $\{\mathbf{i},\mathbf{j},\mathbf{k}\}$ is the canonical basis of $\mathbb{R}^3$.

Moreover, the density scalar field for such a same electron is denoted by $m_e|\varphi_j^l|^2 : \Omega \to \mathbb{R}$, where

$$
\varphi_j^l(\mathbf{x}) = \hat{\varphi}_j^l(r)(L^-)^{(l+j)}[(\sin\theta)^l e^{i\phi l}],
$$

i denotes the imaginary unit,

$$
L_x = -\frac{\hbar}{i}\left(\sin\phi \frac{\partial}{\partial\theta} + \cot\theta\cos\phi\frac{\partial}{\partial\phi}\right),
$$

$$
L_y = \frac{\hbar}{i}\left(\cos\phi \frac{\partial}{\partial\theta} - \cot\theta\sin\phi\frac{\partial}{\partial\phi}\right)
$$

and

$$
L^- = \frac{1}{\hbar}(L_x - iL_y),
$$

and where $(L^-)^0 = I_d$ (identity operator).

Remark 7.3

In principle, we would expect $\mathbf{r}_j^l$ to be an injective function, so that

$$
\tilde{\varphi}_j^l(\mathbf{r}_j^l(\mathbf{x},t)) = \varphi_j^l(\mathbf{x},t) = \varphi_j^l((\mathbf{r}_j^l)^{-1}(\mathbf{r}_j^l(\mathbf{x},t)))
$$

is well defined.

This may not be the case for the motion indicated in (7.29). Thus, such a concerning motion suggests us a new interpretation of the Bohr atomic model and related wave particle duality for the electrons in the atom in question. ∎

We also define,

$$
U = \{\varphi = \{\varphi_j^l\} \in W^{1,2}(\Omega;\mathbb{C}^Z) \;:\; \varphi_j^l = 0 \text{ on } \partial\Omega\},
$$

and

$$
V = \{\mathbf{r} = \{\mathbf{r}_j^l\} \in W^{1,2}(\Omega \times [0,T];\mathbb{R}^{3Z}) \;:\; R_j^l(0) = 0, \text{ and } R_j^l(R_0) = R_0\}.
$$

For such a system, we consider the following types of energy.

1. Kinetics energy, denoted by E_c, where

$$E_c = \frac{1}{2}\sum_{l=0}^{n-1}\sum_{j=-l}^{l}\int_0^T\int_\Omega m_e|\varphi_j^l(\mathbf{x})|^2\frac{\partial \mathbf{r}_j^l(\mathbf{x},t)}{\partial t}\cdot\frac{\partial \mathbf{r}_j^l(\mathbf{x},t)}{\partial t}\,d\mathbf{x}dt,$$

where m_e denotes the mass of a single electron and

$$d\mathbf{x} = dx_1dx_2dx_3.$$

2. A regularizing part for the position field, denoted by E_r, where

$$E_r = \frac{1}{2}\sum_{l=0}^{n-1}\sum_{j=-l}^{l}\sum_{k=1}^{3}\int_0^T\int_\Omega A^l|\varphi_j^l(\mathbf{x})|^2\frac{\partial \mathbf{r}_j^l(\mathbf{x},t)}{\partial x_k}\cdot\frac{\partial \mathbf{r}_j^l(\mathbf{x},t)}{\partial x_k}\,d\mathbf{x}dt,$$

with $A^l > 0$ to be specified, $\forall l \in \{0,\dots,n-1\}$.

3. Coulomb electronic interaction (classical), denoted by E_{int}, where in a first approximation, we consider only the interaction for the same layer electrons, neglecting the interactions between different layer electrons. Thus,

$$E_{int} = \frac{1}{4}\sum_{l=0}^{n-1}\sum_{j=-l}^{l}\sum_{k=-l}^{l}Ke^2\int_0^T\int_\Omega\int_\Omega\frac{|\varphi_j^l(\mathbf{x})|^2|\varphi_k^l(\tilde{\mathbf{x}})|^2}{|\mathbf{r}_j^l(\mathbf{x},t)-\mathbf{r}_k^l(\tilde{\mathbf{x}},t)|}\,d\mathbf{x}d\tilde{\mathbf{x}}dt,$$

where e is the charge of a single electron and $K > 0$ is a an appropriate constant to be specified.

4. Coulomb interaction of each electron with the heavier nucleus, denoted by E_{int}^p, where

$$E_{in}^p = \frac{1}{2}\sum_{l=0}^{n-1}\sum_{j=-l}^{l}Ke^2Z\int_0^T\int_\Omega\frac{|\varphi_j^l(\mathbf{x})|^2}{|\mathbf{r}_j^l(\mathbf{x},t)|}\,d\mathbf{x}dt.$$

5. Energy related to the presence of external potentials V_j^l, denoted by E_p, where

$$E_p = \frac{1}{2}\sum_{l=0}^{n-1}\sum_{j=-l}^{l}\int_0^T\int_\Omega V_j^l(\mathbf{x})|\varphi_j^l(\mathbf{x})|^2\,d\mathbf{x}dt.$$

6. A regularizing and curvature distribution control term for the scalar density field (quantum part), denoted by E_q, where

$$E_q = \frac{1}{2}\sum_{l=0}^{n-1}\sum_{j=-l}^{l}\sum_{k=1}^{3}\gamma_j^l\int_0^T\int_\Omega\frac{\partial(\varphi_j^l(\mathbf{x})\mathbf{n}_j^l(\mathbf{x},t))}{\partial x_k}\cdot\frac{\partial(\varphi_j^l(\mathbf{x})\mathbf{n}_j^l(\mathbf{x},t))}{\partial x_k}\,d\mathbf{x}dt,$$

where the normal field $\mathbf{n}_j^l$ may be given by

$$
\begin{aligned}
\mathbf{n}_j^l(\mathbf{x},t) \;=\; & \sin((w_1)_j^l(\mathbf{x})t + \theta_j^l)\cos((w_2)_j^l(\mathbf{x})t + \phi_j^l)\mathbf{i} \\
& + \sin((w_1)_j^l(\mathbf{x})t + \theta_j^l)\sin((w_2)_j^l(\mathbf{x})t + \phi_j^l)\mathbf{j} \\
& + \cos((w_1)_j^l(\mathbf{x})t + \theta_j^l)\mathbf{k},
\end{aligned} \tag{7.30}
$$

so that,

$$
\mathbf{n}_j^l(\mathbf{x},t) \cdot \frac{\partial \mathbf{r}_j^l(\mathbf{x},t)}{\partial t} = 0, \text{ in } \Omega \times [0,T],
$$

$\forall j \in \{-l,\ldots,0,\ldots,l\},\ \forall l \in \{0,\ldots,n-1\}.$

7. Constraints: The system is subject to the following constraints,

$$
\int_\Omega |\varphi_j^l(\mathbf{x})|^2\, dx = 1,\ \forall l \in \{0,\ldots,n-1\},\ j \in \{-l,\ldots,0,\ldots,l\}.
$$

Hence, the total system energy is given by the functional $J : U \times V \times \mathbb{R}^Z \to \mathbb{R}$ where already including the Lagrange multipliers, we have

$$
\begin{aligned}
J(\varphi,\mathbf{r},E) \;=\; & -E_c + E_r + E_{in} - E_{in}^p + E_p + E_q \\
& - \frac{1}{2}\sum_{l=0}^{n-1}\sum_{j=-l}^{l} E_j^l T \left(\int_\Omega |\varphi_j^l(\mathbf{x})|^2\, dx - 1 \right).
\end{aligned} \tag{7.31}
$$

Summarizing,

$$
\begin{aligned}
& J(\varphi,\mathbf{r},E) \\
=\; & -\frac{1}{2}\sum_{l=0}^{n-1}\sum_{j=-l}^{l}\int_0^T\int_\Omega m_e|\varphi_j^l(\mathbf{x})|^2 \frac{\partial \mathbf{r}_j^l(\mathbf{x},t)}{\partial t}\cdot\frac{\partial \mathbf{r}_j^l(\mathbf{x},t)}{\partial t}\, d\mathbf{x}dt \\
& + \frac{1}{2}\sum_{l=0}^{n-1}\sum_{j=-l}^{l}\sum_{k=1}^{3}\int_0^T\int_\Omega A^l|\varphi_j^l(\mathbf{x})|^2\frac{\partial \mathbf{r}_j^l(\mathbf{x},t)}{\partial x_k}\cdot\frac{\partial \mathbf{r}_j^l(\mathbf{x},t)}{\partial x_k}\, d\mathbf{x}dt \\
& + \frac{1}{4}\sum_{l=0}^{n-1}\sum_{j=-l}^{l}\sum_{k=-l}^{l} Ke^2\int_0^T\int_\Omega\int_\Omega \frac{|\varphi_j^l(\mathbf{x})|^2|\varphi_k^l(\tilde{\mathbf{x}})|^2}{|\mathbf{r}_j^l(\mathbf{x},t)-\mathbf{r}_k^l(\tilde{\mathbf{x}},t)|}\, d\mathbf{x}d\tilde{\mathbf{x}}dt \\
& - \frac{1}{2}\sum_{l=0}^{n-1}\sum_{j=-l}^{l} Ke^2 Z\int_0^T\int_\Omega \frac{|\varphi_j^l(\mathbf{x})|^2}{|\mathbf{r}_j^l(\mathbf{x},t)|}\, d\mathbf{x}dt \\
& + \frac{1}{2}\sum_{l=0}^{n-1}\sum_{j=-l}^{l}\sum_{k=1}^{3}\gamma_j^l\int_0^T\int_\Omega \frac{\partial(\varphi_j^l(\mathbf{x})\mathbf{n}_j^l(\mathbf{x},t))}{\partial x_k}\cdot\frac{\partial(\varphi_j^l(\mathbf{x})\mathbf{n}_j^l(\mathbf{x},t))}{\partial x_k}\, d\mathbf{x}dt \\
& - \frac{1}{2}\sum_{l=0}^{n-1}\sum_{j=-l}^{l} E_j^l T\left(\int_\Omega |\varphi_j^l(\mathbf{x})|^2\, dx - 1 \right).
\end{aligned} \tag{7.32}
$$

With such statements and definitions in mind, we define the control problem of finding $\{\theta_j^l, \phi_j^l\} \in \mathbb{R}^Z \times \mathbb{R}^Z$, which minimizes

$$J_1(\varphi, \mathbf{r}, \{E\})$$

where

$$
\begin{aligned}
& J_1(\varphi, \mathbf{r}, \{E\}) \\
&= \frac{1}{4} \sum_{l=0}^{n-1} \sum_{j=-l}^{l} \sum_{k=-l}^{l} Ke^2 \int_0^T \int_\Omega \int_\Omega \frac{|\varphi_j^l(\mathbf{x})|^2 |\varphi_k^l(\tilde{\mathbf{x}})|^2}{|\mathbf{r}_j^l(\mathbf{x},t) - \mathbf{r}_k^l(\tilde{\mathbf{x}},t)|} \, d\mathbf{x} d\tilde{\mathbf{x}} dt, \quad (7.33)
\end{aligned}
$$

subject to

1.

$$
\begin{aligned}
& m_e |\varphi_j^l(\mathbf{x})|^2 \frac{\partial^2 \mathbf{r}_j^l(\mathbf{x},t)}{\partial t^2} \\
& -A^l \sum_{k=1}^{3} \frac{\partial}{\partial x_k} \left(|\varphi_j^l(\mathbf{x})|^2 \frac{\partial \mathbf{r}_j^l(\mathbf{x},t))}{\partial x_k} \right) \\
& -\sum_{k=-l}^{l} |\varphi_j^l(\mathbf{x})|^2 \int_\Omega \frac{Ke^2 |\varphi_k^l(\tilde{\mathbf{x}})|^2 (\mathbf{r}_j^l(\mathbf{x},t) - \mathbf{r}_k^l(\tilde{\mathbf{x}},t))}{|\mathbf{r}_j^l(\mathbf{x},t) - \mathbf{r}_k^l(\tilde{\mathbf{x}},t)|^3} \, d\tilde{\mathbf{x}} \\
& +KZe^2 \frac{|\varphi_j^l(\mathbf{x})|^2}{|\mathbf{r}_j^l(\mathbf{x},t)|^3} \mathbf{r}_j^l(\mathbf{x},t) \\
&= \mathbf{0}, \text{ in } \Omega \times [0,T], \quad (7.34)
\end{aligned}
$$

2.

$$
\begin{aligned}
& -m_e \hat{\varphi}_j^l(r) \frac{1}{T} \int_0^T \frac{\partial \mathbf{r}_j^l(\mathbf{x},t)}{\partial t} \cdot \frac{\partial \mathbf{r}_j^l(\mathbf{x},t)}{\partial t} \, dt + \\
& -\gamma_j^l \frac{1}{r^2} \frac{\partial}{\partial r} \left(r^2 \frac{\partial \hat{\varphi}_j^l(r)}{\partial r} \right) \\
& +\gamma_j^l \frac{l(l+1)}{r^2} \hat{\varphi}_j^l(r) \\
& +\gamma_j^l \hat{\varphi}_j^l(r) \frac{1}{T} \int_0^T \sum_{k=1}^{3} \left(\frac{\partial \mathbf{n}_j^l}{\partial x_k} \cdot \frac{\partial \mathbf{n}_j^l}{\partial x_k} \right) dt \\
& +\sum_{k=-l}^{l} \hat{\varphi}_j^l(r) \frac{1}{T} \int_0^T \int_\Omega \frac{Ke^2 |\varphi_k^l(\tilde{\mathbf{x}})|^2}{|\mathbf{r}_j^l(\mathbf{x},t) - \mathbf{r}_k^l(\tilde{\mathbf{x}},t)|} \, d\tilde{\mathbf{x}} dt \\
& -KZe^2 \frac{\hat{\varphi}_j^l(r)}{R_j^l(r)} \\
& -E_j^l \hat{\varphi}_j^l(r) = 0, \text{ in } [0, R_0], \quad (7.35)
\end{aligned}
$$

and up to a normalizing constant for $\varphi(\mathbf{x})$,

3.
$$\int_0^{R_0} |\hat{\phi}_j^l(r)|^2 r^2 \, dr = 1,$$

$$\forall j \in \{-l,\dots,0,\dots,l\}, \ \forall l \in \{0,\dots,n-1\}.$$

7.8 A system with a large number of interacting atoms

Now consider a system with a large number N of interacting same type atoms, each one with $Z = \sum_{l=0}^{n-1}(2l+1)$ electrons and the same number of protons.

Consider also the problem of finding the N nucleus positions, each one composed of Z protons, in an open, bounded, connected set $\Omega \subset \mathbb{R}^3$ with a Lipschitzian boundary denoted by $\partial\Omega$.

We define the position field for the electron j, in the layer l at the atom k, where the nucleus is located at $\mathbf{x}_k \in \Omega$, denoted by $\mathbf{r}_j^l(\cdot,\mathbf{x}_k,\cdot) : \Omega \times [0,T] \to \mathbb{R}^3$, as

$$\mathbf{r}_j^l(\mathbf{x},\mathbf{x}_k,t) = \mathbf{x}_k + \mathbf{R}_j^l(\mathbf{x},\mathbf{x}_k)e^{iw_j^l(\mathbf{x},\mathbf{x}_k)t} \tag{7.36}$$

Also, the respective density scalar field is denoted by $\varphi = \{\varphi_j^l : \Omega \to \mathbb{C}\}$, Here,

$$\varphi(\cdot,\mathbf{x}_k) \in U, \forall k \in \{1,\dots,N\}$$

and

$$\mathbf{r}(\cdot,\mathbf{x}_k,\cdot) \in V, \ \forall k \in \{1,\dots,N\}.$$

With such statements in mind, we consider the control problem of finding $\{\mathbf{x}_k\}_{k=1}^N$ and $\{w_j^l(\mathbf{x},\mathbf{x}_k)\}$ which minimizes $J_2 + J_3 + J_4$, where

$$J_2 = \sum_{l_1=0}^{n-1}\sum_{l_2=0}^{n-1}\sum_{j=-l_1}^{l_1}\sum_{k=-l_2}^{l_2}\sum_{k_1=1}^{N}\sum_{k_2=1}^{N} Ke^2 \times$$
$$\times \int_0^T \int_\Omega \int_\Omega \frac{|\varphi_j^{l_1}(\mathbf{x},\mathbf{x}_{k_1})|^2 |\varphi_k^{l_2}(\tilde{\mathbf{x}},\mathbf{x}_{k_2})|^2}{|\mathbf{r}_j^{l_1}(\mathbf{x},\mathbf{x}_{k_1},t) - \mathbf{r}_k^{l_2}(\tilde{\mathbf{x}},\mathbf{x}_{k_2},t)|} \, dx d\tilde{x} dt \tag{7.37}$$

$$J_3 = \sum_{l=0}^{n-1}\sum_{j=-l}^{l}\sum_{k=1}^{N}\sum_{k_1=1}^{N} Ke^2 Z \int_0^T \int_\Omega \frac{|\varphi_j^l(\mathbf{x},\mathbf{x}_k)|^2}{|\mathbf{r}_j^l(\mathbf{x},\mathbf{x}_k,t) - \mathbf{x}_{k_1}|} \, dx dt, \tag{7.38}$$

and

$$J_4 = \sum_{k=1}^{N}\sum_{k_1=1}^{N} \frac{Ke^2 Z^2}{|\mathbf{x}_k - \mathbf{x}_{k_1}|},$$

subject to

1.

$$
m_e |\varphi_j^l(\mathbf{x}, \mathbf{x}_k)|^2 \frac{\partial^2 \mathbf{r}_j^l(\mathbf{x}, \mathbf{x}_k, t)}{\partial t^2}
$$

$$
- \sum_{s=1}^{3} A^l \frac{\partial}{\partial x_s} \left(|\varphi(\mathbf{x}, \mathbf{x}_k)|^2 \frac{\partial \mathbf{r}_j^l(\mathbf{x}, \mathbf{x}_k, t)}{\partial x_s} \right)
$$

$$
- \sum_{l_1=0}^{n-1} \sum_{p=-l_1}^{l_1} \sum_{k_1=1}^{N} Ke^2 |\varphi_j^l(\mathbf{x}, \mathbf{x}_k)|^2 \int_{\Omega} \frac{|\varphi_p^{l_1}(\tilde{\mathbf{x}}, \mathbf{x}_{k_1})|^2 (\mathbf{r}_j^l(\mathbf{x}, \mathbf{x}_k, t) - \mathbf{r}_p^{l_1}(\tilde{\mathbf{x}}, \mathbf{x}_{k_1}, t))}{|\mathbf{r}_j^l(\mathbf{x}, \mathbf{x}_k, t) - \mathbf{r}_p^{l_1}(\tilde{\mathbf{x}}, \mathbf{x}_{k_1}, t)|^3} d\tilde{\mathbf{x}}
$$

$$
+ Ke^2 Z \sum_{k_1=1}^{N} \frac{|\varphi_j^l(\mathbf{x}, \mathbf{x}_k)|^2 (\mathbf{r}_j^l(\mathbf{x}, \mathbf{x}_k, t) - \mathbf{x}_{k_1})}{|\mathbf{r}_j^l(\mathbf{x}, \mathbf{x}_k, t) - \mathbf{x}_{k_1}|^3}
$$

$$
= \quad \mathbf{0}, \text{ in } \Omega. \tag{7.39}
$$

2.

$$
\mathbf{n}_j^l(\mathbf{x}, \mathbf{x}_k, t) \cdot \frac{\partial \mathbf{r}_j^l(\mathbf{x}, \mathbf{x}_k, t)}{\partial t} = 0, \text{ in } \Omega \times [0, T],
$$

3.

$$
\mathbf{n}_j^l(\mathbf{x}, \mathbf{x}_k, t) \cdot \mathbf{n}_j^l(\mathbf{x}, \mathbf{x}_k, t) = 1, \text{ in } \Omega \times [0, T],
$$

4.

$$
- \sum_{s=1}^{3} \gamma_j^l \frac{\partial^2 \varphi_j^l(\mathbf{x}, \mathbf{x}_k)}{\partial x_s^2}
$$

$$
+ \gamma_j^l \varphi_j^l(\mathbf{x}, \mathbf{x}_k) \frac{1}{T} \int_0^T \sum_{s=1}^{3} \left(\frac{\partial \mathbf{n}_j^l(\mathbf{x}, \mathbf{x}_k, t)}{\partial x_s} \cdot \frac{\partial \mathbf{n}_j^l(\mathbf{x}, \mathbf{x}_k, t)}{\partial x_s} \right) dt
$$

$$
+ \varphi_j^l(\mathbf{x}, \mathbf{x}_k) \sum_{l_1=0}^{n-1} \sum_{k=-l_1}^{l_1} \sum_{k_1=1}^{N} \frac{1}{T} \int_{\Omega} \frac{|\varphi_k^{l_1}(\tilde{\mathbf{x}}, \mathbf{x}_{k_1})|^2}{|\mathbf{r}_j^l(\mathbf{x}, \mathbf{x}_k) - \mathbf{r}_k^{l_1}(\tilde{\mathbf{x}}, \mathbf{x}_{k_1}, t)|} d\tilde{\mathbf{x}} dt
$$

$$
- \sum_{k_1=1}^{N} Ke^2 Z \frac{\varphi_j^l(\mathbf{x}, \mathbf{x}_k)}{|\mathbf{r}(\mathbf{x}, \mathbf{x}_k, t) - \mathbf{x}_{k_1}|}
$$

$$
+ \sum_{s=1}^{3} A^l \varphi_j^l(\mathbf{x}, \mathbf{x}_k) \frac{1}{T} \int_0^T \frac{\partial \mathbf{r}_j^l(\mathbf{x}, \mathbf{x}_k, t)}{\partial x_s} \cdot \frac{\partial \mathbf{r}_j^l(\mathbf{x}, \mathbf{x}_k, t)}{\partial x_s} dt
$$

$$
- m_e \varphi_j^l(\mathbf{x}, \mathbf{x}_k) \frac{1}{T} \int_0^T \frac{\partial \mathbf{r}_j^l(\mathbf{x}, \mathbf{x}_k, t)}{\partial t} \cdot \frac{\partial \mathbf{r}_j^l(\mathbf{x}, \mathbf{x}_k, t)}{\partial t} dt
$$

$$
- E_{jk}^l \varphi_j^l(\mathbf{x}, \mathbf{x}_k) = 0, \text{ in } \Omega. \tag{7.40}
$$

5.

$$
\int_{\Omega} |\varphi_j^l(\mathbf{x}, \mathbf{x}_k)|^2 dx = 1, \forall l \in \{0, \ldots, n-1\}, \ j \in \{-l, \ldots, 0, \ldots, l\}, \ k \in \{1, \ldots, N\}.
$$

7.8.1 A proposal for the case in which N is very large

As N is very large, we shall propose a limit density scalar field $\varphi_j^l(\mathbf{x}, \mathbf{y})$, that is

$$\varphi_j^l : \Omega \times \Omega \to \mathbb{C}.$$

Also, we shall propose, as the position vector field, $\mathbf{r}_j^l(\mathbf{x}, \mathbf{y}, t)$, that is $\mathbf{r}_j^l :$ $\Omega \times \Omega \times [0, T] \to \mathbb{R}^3$, where

$$\mathbf{r}_j^l(\mathbf{x}, \mathbf{y}, t) \;=\; \mathbf{y} + \mathbf{R}_j^l(\mathbf{x}, \mathbf{y})e^{iw_j^l(\mathbf{x}, \mathbf{y})t}. \tag{7.41}$$

We assume $\varphi = \{\varphi_j^l\} \in U$, where

$$U = \{\varphi = \{\varphi_j^l\} \in W^{1,2}(\Omega \times \Omega; \mathbb{C}^Z) \;:\; \varphi_j^l = 0 \text{ on } \partial(\Omega \times \Omega)\}$$

and $\mathbf{r} = \{\mathbf{r}_j^l\} \in V$, where here

$$V = \{\mathbf{r} = \{\mathbf{r}_j^l\} \in W^{1,2}(\Omega \times \Omega \times [0, T]; \mathbb{R}^{3Z}) \;:\; \mathbf{R}_j^l = \mathbf{0} \text{ on } \partial(\Omega \times \Omega)\}.$$

For the protons, we specify the density scalar field $\varphi_p : \Omega \times \Omega \to \mathbb{C}$, and the respective position field $\mathbf{r}_p(\mathbf{x}, \mathbf{y}) = \mathbf{y}$. Moreover, $\varphi_p \in U_p$, where

$$U_p = \{\varphi_p \in W^{1,2}(\Omega \times \Omega; \mathbb{C}) \;:\; \varphi_p(\mathbf{x}, \mathbf{y}) = 0, \text{ in } \partial(\Omega \times \Omega)\}.$$

In the distributional sense, we should approximately expect to obtain

$$\varphi_p(\mathbf{x}, \mathbf{y}) = \delta(\mathbf{x} - \mathbf{y}), \text{ in } (\Omega \times \Omega)^0$$

where $(\Omega \times \Omega)^0$ denotes the interior of $\Omega \times \Omega$. Also, $\delta(\mathbf{x} - \mathbf{y})$ denotes a standard Dirac delta.

With such statements in mind, we consider the control problem of finding $\{w_j^l\}$ which minimizes $J_2 + J_3 + J_4$, where

$$J_2 \;=\; \sum_{l_1=0}^{n-1} \sum_{l_2=0}^{n-1} \sum_{j=-l_1}^{l_1} \sum_{k=-l_2}^{l_2} Ke^2 \times$$

$$\times \int_0^T \int_\Omega \int_\Omega \int_\Omega \int_\Omega \frac{|\varphi_j^{l_1}(\mathbf{x}, \mathbf{y})|^2 |\varphi_k^{l_2}(\tilde{\mathbf{x}}, \tilde{\mathbf{y}})|^2}{|\mathbf{r}_j^{l_1}(\mathbf{x}, \mathbf{y}, t) - \mathbf{r}_k^{l_2}(\tilde{\mathbf{x}}, \tilde{\mathbf{y}}, t)|}\, d\mathbf{x}d\tilde{\mathbf{x}}d\mathbf{y}d\tilde{\mathbf{y}}dt \tag{7.42}$$

$$J_3 \;=\; \sum_{l=0}^{n-1} \sum_{j=-l}^{l} \sum_{k=1}^{N} Ke^2 Z \int_0^T \int_\Omega \int_\Omega \int_\Omega \frac{|\varphi_j^l(\mathbf{x}, \mathbf{y})|^2}{|\mathbf{r}_j^l(\mathbf{x}, \mathbf{y}, t) - \tilde{\mathbf{y}}|}\, d\mathbf{x}d\mathbf{y}d\tilde{\mathbf{y}}dt, \tag{7.43}$$

and

$$J_4 = T \int_\Omega \int_\Omega \int_\Omega \int_\Omega \frac{Ke^2 Z^2 |\varphi_p(\mathbf{x}, \mathbf{y})|^2 |\varphi_p(\tilde{\mathbf{x}}, \tilde{\mathbf{y}})|^2}{|\mathbf{y} - \tilde{\mathbf{y}}|}\, d\mathbf{x}d\tilde{\mathbf{x}}d\mathbf{y}d\tilde{\mathbf{y}},$$

subject to

1.

$$
m_e |\varphi_j^l(\mathbf{x},\mathbf{y})|^2 \frac{\partial^2 \mathbf{r}_j^l(\mathbf{x},\mathbf{y},t)}{\partial t^2}
$$

$$
-A^l \sum_{s=1}^{3} \left(\frac{\partial}{\partial x_s} \left(|\varphi(\mathbf{x},\mathbf{y})|^2 \frac{\partial \mathbf{r}_j^l(\mathbf{x},\mathbf{y},t)}{\partial x_s} \right) \right.
$$

$$
+ \frac{\partial}{\partial y_s} \left(|\varphi(\mathbf{x},\mathbf{y})|^2 \frac{\partial \mathbf{r}_j^l(\mathbf{x},\mathbf{y},t)}{\partial y_s} \right) \bigg)
$$

$$
- \sum_{l_1=0}^{n-1} \sum_{s=-l_1}^{l_1} Ke^2 |\varphi_j^l(\mathbf{x},\mathbf{y})|^2 \int_\Omega \int_\Omega \frac{|\varphi_s^{l_1}(\tilde{\mathbf{x}},\tilde{\mathbf{y}})|^2 (\mathbf{r}_j^l(\mathbf{x},\mathbf{y},t) - \mathbf{r}_s^{l_1}(\tilde{\mathbf{x}},\tilde{\mathbf{y}},t))}{|\mathbf{r}_j^l(\mathbf{x},\mathbf{y},t) - \mathbf{r}_s^{l_1}(\tilde{\mathbf{x}},\tilde{\mathbf{y}},t)|^3} \, d\tilde{\mathbf{x}} d\tilde{\mathbf{y}}
$$

$$
+ Ke^2 Z |\varphi_j^l(\mathbf{x},\mathbf{y})|^2 \int_\Omega \frac{(\mathbf{r}_j^l(\mathbf{x},\mathbf{y},t) - \tilde{\mathbf{y}})}{|\mathbf{r}_j^l(\mathbf{x},\mathbf{y},t) - \tilde{\mathbf{y}}|^3} \, d\tilde{\mathbf{y}}
$$

$$
= \mathbf{0}, \text{ in } \Omega \times \Omega \times [0,T]. \tag{7.44}
$$

2.

$$
\mathbf{n}_j^l(\mathbf{x},\mathbf{y},t) \cdot \frac{\partial \mathbf{r}_j^l(\mathbf{x},\mathbf{y},t)}{\partial t} = 0, \text{ in } \Omega \times \Omega \times [0,T],
$$

3.

$$
\mathbf{n}_j^l(\mathbf{x},\mathbf{y},t) \cdot \mathbf{n}_j^l(\mathbf{x},\mathbf{y},t) = 1, \text{ in } \Omega \times \Omega \times [0,T],
$$

4.

$$
-\gamma_j^l \sum_{s=1}^{3} \left(\frac{\partial^2 \varphi_j^l(\mathbf{x},\mathbf{y})}{\partial x_s^2} + \frac{\partial^2 \varphi_j^l(\mathbf{x},\mathbf{y})}{\partial y_s^2} \right)
$$

$$
+ \gamma_j^l \varphi_j^l(\mathbf{x},\mathbf{y}) \frac{1}{T} \int_0^T \sum_{s=1}^{3} \left(\left(\frac{\partial \mathbf{n}_j^l(\mathbf{x},\mathbf{y},t)}{\partial x_s} \cdot \frac{\partial \mathbf{n}_j^l(\mathbf{x},\mathbf{y},t)}{\partial x_s} \right) \right.
$$

$$
+ \left(\frac{\partial \mathbf{n}_j^l(\mathbf{x},\mathbf{y},t)}{\partial y_s} \cdot \frac{\partial \mathbf{n}_j^l(\mathbf{x},\mathbf{y},t)}{\partial y_s} \right) \bigg) dt
$$

$$
+ \varphi_j^l(\mathbf{x},\mathbf{y}) \sum_{l_1=0}^{n-1} \sum_{k=-l_1}^{l_1} \frac{1}{T} \int_0^T \int_\Omega \int_\Omega \frac{|\varphi_k^{l_1}(\tilde{\mathbf{x}},\tilde{\mathbf{y}})|^2}{|\mathbf{r}_j^l(\mathbf{x},\mathbf{y},t) - \mathbf{r}_k^{l_1}(\tilde{\mathbf{x}},\tilde{\mathbf{y}},t)|} \, d\tilde{\mathbf{x}} d\tilde{\mathbf{y}} dt
$$

$$
- \sum_{k_1=1}^{N} Ke^2 Z \, \varphi_j^l(\mathbf{x},\mathbf{y}) \int_\Omega \int_\Omega \frac{|\varphi_p(\tilde{\mathbf{x}},\tilde{\mathbf{y}})|^2}{|\mathbf{r}_j^l(\mathbf{x},\mathbf{y},t) - \tilde{\mathbf{y}}|} \, d\tilde{\mathbf{x}} d\tilde{\mathbf{y}}
$$

$$
+ A^l \sum_{s=1}^{3} \left(\varphi_j^l(\mathbf{x},\mathbf{y}) \frac{1}{T} \int_0^T \left(\frac{\partial \mathbf{r}_j^l(\mathbf{x},\mathbf{y},t)}{\partial x_s} \cdot \frac{\partial \mathbf{r}_j^l(\mathbf{x},\mathbf{y},t)}{\partial x_s} \right. \right.
$$

$$
+ \frac{\partial \mathbf{r}_j^l(\mathbf{x},\mathbf{y},t)}{\partial y_s} \cdot \frac{\partial \mathbf{r}_j^l(\mathbf{x},\mathbf{y},t)}{\partial y_s} \bigg) dt \bigg)
$$

$$
- m_e \varphi_j^l(\mathbf{x},\mathbf{y}) \frac{1}{T} \int_0^T \frac{\partial \mathbf{r}_j^l(\mathbf{x},\mathbf{y},t)}{\partial t} \cdot \frac{\partial \mathbf{r}_j^l(\mathbf{x},\mathbf{y},t)}{\partial t} \, dt
$$

$$
- E_j^l(\mathbf{y}) \varphi_j^l(\mathbf{x},\mathbf{y}) = 0, \text{ in } \Omega \times \Omega, \tag{7.45}
$$

5.

$$\int_\Omega |\varphi_j^l(\mathbf{x},\mathbf{y})|^2 \, dx = 1, \ \forall l \in \{0,\dots,n-1\}, \ j \in \{-l,\dots,0,\dots,l\}, \ \mathbf{y} \in \Omega,$$

6.

$$-\gamma_p \sum_{s=1}^{3} \left(\frac{\partial^2 \varphi_p(\mathbf{x},\mathbf{y})}{\partial x_s^2} + \frac{\partial^2 \varphi_p(\mathbf{x},\mathbf{y})}{\partial y_s^2} \right)$$

$$+\varphi_p(\mathbf{x},\mathbf{y}) \sum_{l=0}^{n-1} \frac{Ke^2 Z}{T} \int_0^T \int_\Omega \int_\Omega \frac{|\varphi_j^l(\tilde{\mathbf{x}},\tilde{\mathbf{y}})|^2}{|\mathbf{y} - \mathbf{r}_j^l(\tilde{\mathbf{x}},\tilde{\mathbf{y}},t)|} \, d\tilde{\mathbf{x}} d\tilde{\mathbf{y}} dt$$

$$-Ke^2 Z^2 \, \varphi_p(\mathbf{x},\mathbf{y}) \int_\Omega \int_\Omega \frac{|\varphi_p(\tilde{\mathbf{x}},\tilde{\mathbf{y}})|^2}{|\mathbf{y} - \tilde{\mathbf{y}}|} \, d\tilde{\mathbf{x}} d\tilde{\mathbf{y}}$$

$$-E_p(\mathbf{y})\varphi_p(\mathbf{x},\mathbf{y}) = 0, \text{ in } \Omega \times \Omega. \tag{7.46}$$

7.

$$\int_\Omega |\varphi_p(\mathbf{x},\mathbf{y})|^2 \, dx = 1, \ \forall \mathbf{y} \in \Omega.$$

7.9 About the definition of Temperature

Consider a system with $N = \sum_{j=1}^{N_0} N_j$ particles and suppose that N_j particles have a state in a set of C_j states. Thus the total number of possible states for the concerning N_j particles is given by

$$\Delta\Gamma_j = \frac{(C_j)^{N_j}}{N_j!}.$$

Here we have considered permutations are equivalent states.
Define thus,

$$S_j = \ln \Delta\Gamma_j$$

and the Entropy of the system, denoted by S, by

$$S = A \sum_{j=1}^{N_0} S_j,$$

for some normalizing constant $A > 0$.
Thus,

$$S = A \sum_{j=1}^{N_0} \ln \left(\frac{(C_j)^{N_j}}{N_j!} \right),$$

so that

$$S = A \sum_{j=1}^{N_0} N_j \ln C_j - \ln(N_j!).$$

If N_j is large enough, we may use the approximation

$$\ln(N_j!) \approx N_j \ln \left(\frac{N_j}{e} \right).$$

Thus, in particular, if $C_j = 1$, $\forall j \in \{1, \ldots, N_0\}$, mathematically we obtain

$$S = \sum_{j=1}^{N_0} S_j \approx - \sum_{j=1}^{N_0} N_j \ln(N_j/e).$$

At this point, with such a results in mind, we define the local density $\hat{N}_j(x,t)$ by

$$\hat{N}_j(x,t) = \frac{|\phi_j(x,t)|^2}{|\phi(x,t)|^2} N,$$

where

$$|\phi(x,t)|^2 = \sum_{j=1}^{N_0} |\phi_j(x,t)|^2,$$

and where $\phi_j : \Omega \to \mathbb{C}$ corresponds to the wave function with energy E_j.

We also define the Entropy scalar field

$$S(x,t) = A \sum_{j=1}^{N_0} S_j(x,t),$$

where

$$
\begin{aligned}
S_j(x,t) &= -A\hat{N}_j(x,t) \ln \left(\frac{\hat{N}_j(x,t)}{e} \right) \\
&= -A \frac{|\phi_j(x,t)|^2}{|\phi(x,t)|^2} N \ln \left(\frac{|\phi_j(x,t)|^2}{|\phi(x,t)|^2} \frac{N}{e} \right).
\end{aligned}
\tag{7.47}
$$

Now observe that the position field for each particle is given by

$$\hat{\mathbf{r}}_j(x,t) = \mathbf{x} + \mathbf{r}_j(x,t),$$

so that the total energy is given by

$$
\begin{aligned}
E &= \sum_{j=1}^{N_0} \frac{-1}{2} \int_0^T \int_\Omega m_e |\phi_j(x,t)|^2 \frac{\partial \mathbf{r}_j(x,t)}{\partial t} \cdot \frac{\partial \mathbf{r}_j(x,t)}{\partial t} \, dx \, dt \\
&\quad \sum_{j=1}^{N_0} \frac{\gamma_j}{2} \int_0^T \int_\Omega \frac{\partial (\phi_j \mathbf{n}_j)}{\partial x_k} \cdot \frac{\partial (\phi_j \mathbf{n}_j)}{\partial x_k} \, dx \, dt.
\end{aligned}
\tag{7.48}
$$

At this point, we define the scalar temperature field, denoted by $T(x,t)$, as the one such that

$$\frac{\partial S}{\partial E} = \frac{1}{T(x,t)},$$

so that, we define

$$T(x,t) = \sum_{j=1}^{N_0} \frac{\frac{\partial E}{\partial \phi_j}}{\frac{\partial S}{\partial \phi_j}}, \tag{7.49}$$

and thus,

$$
T(x,t)
$$
$$
= \sum_{j=1}^{N_0} \frac{\left(-m_e \phi_j \frac{\partial \mathbf{r}_j(x,t)}{\partial t} \cdot \frac{\partial \mathbf{r}_j(x,t)}{\partial t} - \gamma_j \sum_{k=1}^{3} \frac{\partial^2 (\phi_j \mathbf{n}_j)}{\partial x_k^2} \cdot \mathbf{n}_j \right)}{-2\frac{\phi_j N}{|\phi|^2} \left(\ln \left(\frac{|\phi_j|^2}{|\phi|^2} \frac{N}{e} \right) + 1 \right) A} \tag{7.50}
$$

This last formula show us that the temperature is directly proportional to the energy (the internal one, related to the electronic motion) and inversely proportional to the entropy.

7.10   A note on the Entropy concept

First define, for a wave function in a non-relativistic free particle context, for a motion developing on a time interval $[0, T]$,

$$W(E) = \frac{1}{T} \int \int_{\Omega_E} |\phi(x)|^2 \, dx \, dt$$

where

$$\Omega_E = \{ (x,t) \in \Omega \times [0,T] \; : \; E(x,t) \leq E \}.$$

At this point we define the entropy S by,

$$S(E) = -\int_{E_0}^{E} W(\hat{E}) \ln(W(\hat{E})) \, d\hat{E}, \tag{7.51}$$

where $E(x,t)$ will be specified in the next lines.

We define also the temperature $\hat{T} = \hat{T}(E)$ through the relation

$$\frac{dS(E)}{dE} = \frac{1}{\hat{T}} = -W(E) \ln W(E),$$

where we must emphasize the dependence $\hat{T} = \hat{T}(E)$.

In a free particle context, we assume

$$
\begin{aligned}
E = E(x,t) &= m_e |\phi(x)|^2 \frac{\partial \mathbf{r}(x,t)}{\partial t} \cdot \frac{\partial \mathbf{r}(x,t)}{\partial t} \\
&= \hat{E}_q - \mu |\phi(x)|^2, \tag{7.52}
\end{aligned}
$$

where here μ is such that

$$\int_\Omega |\phi(x)|^2 \, dx = 1.$$

Also,

$$
\begin{aligned}
\hat{E}_q(x,t) &= -\gamma \sum_{k=1}^{3} \frac{\partial^2(\phi \mathbf{n})}{\partial x_k^2} \cdot (\phi \mathbf{n}) \\
&= E_1(x,t)\mathbf{n} \cdot \mathbf{n} \\
&= E_1(x,t),
\end{aligned}
\tag{7.53}
$$

where $E_1(x,t)$ is the Lagrange multiplier such that

$$\mathbf{n} \cdot \mathbf{n} = 1, \text{ in } \Omega \times [0,T].$$

Summarizing,

$$E(x,t) = E_1(x,t) - \mu |\phi(x)|^2.$$

Finally,

$$
\begin{aligned}
dS(E) &= -W(E)\ln(W(E)) \, dE \\
&= \frac{dE}{\hat{T}}.
\end{aligned}
\tag{7.54}
$$

Hence,

$$dS(E(x,t)) = \frac{dE_1(x,t)}{\hat{T}} - \frac{d(\mu |\phi(x)|^2)}{\hat{T}}.$$

7.11   About modeling a chemical reaction

Let us consider a volume $\Omega \subset \mathbb{R}^3$ and a possible chemical reaction in Ω, in which α_1 units of mass of a solid substance type 1 reacts with α_2 units of mass of a liquid of type 2 to produce 1 (one) unit of mass of a gaseous substance of type 3, that is

$$\alpha_1 + \alpha_2 = 1.$$

For such a system, we define

$$w_i(x,t) = \left(\frac{\rho_i(x,t)}{\sum_{k=1}^{3} \rho_k(x,t)} \right)^{\beta_i(x,t)},$$

where $\rho_k(x,t)$ denotes the point wise density of substance of type k, and

$$\beta_k(x,t) = \beta_k(P(x,t), \hat{T}(x,t), \rho(x,t))$$

must be obtained experimentally.

We define also, for such a system,

$$S(w) = -\sum_{k=1}^{3} \int_0^T \int_\Omega w_k(x,t) \ln(w_k(x,t)) \, dx \, dt.$$

Remark 7.11.1 *The functions* $\beta_k(\hat{T}, P, \rho)$ *must be obtained such that the direction of the chemical reaction is properly modeled for the concerning point-wise values of* $\hat{T}(x,t), P(x,t), \rho(x,t)$.

7.11.1 About the variational formulation modeling such a chemical reaction

We define the problem of finding a critical point of a functional $J(\mathbf{r}, \phi, E, \lambda, \mathbf{n}, \hat{T}, \mathbf{u})$, that is the problem of finding a solution for the equation,

$$\delta J(\mathbf{r}, \phi, E, \lambda, \mathbf{n}, \hat{T}, \mathbf{u}) = \mathbf{0},$$

where $J : U \times V_1 \times V_2 \times V_3 \times V_4 \times V_5 \times V_6 \to \mathbb{R}$ will be specified in the next lines. In this model, we consider the substance s composed of atoms of type s with Z_s electrons, protons and neutrons, where $Z_s = \sum_{l=0}^{n_s}(2l+1)$ and n_s is the number of electronic layers for each atom of type s. We assume each layer l initially has $2l+1$ electrons and the possible molecular arrangements are obtained from the system motion and behavior, locally and as a whole.

We also define,

$$U = \left\{ \mathbf{r} = \{(\mathbf{r}_s)_j^l\} \in W^{1,2}\left(\Omega \times \Omega \times [0,T]; \mathbb{R}^{\sum_{s=1}^3 3Z_s}\right) : (\mathbf{r}_s)_j^l = \mathbf{x} \text{ on } \Gamma_0, \text{ on } [0,T]\right\},$$

$$V_1 = \left\{ \phi = \{(\phi_s)_j^l, (\phi_p)_s\} \in W^{1,2}\left(\Omega \times \Omega \times [0,T]; \mathbb{C}^{\left(\sum_{s=1}^3 Z_s\right)+3}\right) \right.$$
$$\left. : (\phi_s)_j^l = (\phi_p)_s = 0, \text{ on } \Gamma_1 \text{ on } [0,T]\right\}, \qquad (7.55)$$

where

$$\Gamma_0, \Gamma_1 \subset \partial(\Omega \times \Omega),$$

$$V_2 = \mathbb{R}^{\left(\sum_{s=1}^3 Z_s\right)+3},$$

$$V_3 = L^2\left(\Omega \times [0,T]; \mathbb{R}^{5+\left(\sum_{s=1}^3 2Z_s\right)}\right),$$

$$V_4 = L_2([0,T]; \mathbb{R}^3),$$

$$V_5 = W^{2,2}(\Omega \times [0,T]; \mathbb{R}^4)$$

and

$$V_6 = L^2\left(\Omega \times [0,T]; \mathbb{R}^{\sum_{s=1}^3 Z_s}\right).$$

Finally, the functional J is expressed as

$$J = J_1 + J_2 + J_3 + J_4 + J_5 + J_6 + J_7,$$

where

$$J_1(\mathbf{r}, \phi) = -\sum_{s=1}^{3}\sum_{l=0}^{n_s-1}\sum_{j=-l}^{l}\int_0^T\int_\Omega\int_\Omega |(\phi_s)_j^l|^2 \frac{\partial(\mathbf{r}_s)_j^l}{\partial t}\cdot\frac{\partial(\mathbf{r}_s)_j^l}{\partial t}\,d\mathbf{x}\,d\mathbf{y}\,dt$$

$$-\sum_{s=1}^{3}\int_0^T\int_\Omega\int_\Omega |(\phi_p)_s|^2 \frac{\partial(\mathbf{r}_p)_s}{\partial t}\cdot\frac{\partial(\mathbf{r}_p)_s}{\partial t}\,d\mathbf{x}\,d\mathbf{y}\,dt, \qquad (7.56)$$

$$J_2(\mathbf{r},\phi)$$

$$= \sum_{s=1}^{3}\sum_{q=1}^{3}\sum_{l=0}^{n_s-1}\sum_{l_1=0}^{n_q-1}\sum_{k=-l_1}^{l_1}$$

$$\left(Ke^2\int_0^T\int_\Omega\int_\Omega\int_\Omega\int_\Omega \frac{|(\phi_s)_j^l(\mathbf{x},\mathbf{y},t)|^2|(\phi_q)_k^{l_1}(\tilde{\mathbf{x}},\tilde{\mathbf{y}},t)|^2}{|(\mathbf{r}_s)_j^l(\mathbf{x},\mathbf{y},t)-(\mathbf{r}_q)_k^{l_1}(\tilde{\mathbf{x}},\tilde{\mathbf{y}},t)|}\,d\mathbf{x}\,d\mathbf{y}\,d\tilde{\mathbf{x}}\,d\tilde{\mathbf{y}}\,dt\right)$$

$$-\sum_{s=1}^{3}\sum_{q=1}^{3}\sum_{l=0}^{n_s-1}Ke^2\int_0^T\int_\Omega\int_\Omega\int_\Omega\int_\Omega \frac{|(\phi_s)_j^l(\mathbf{x},\mathbf{y},t)|^2|(\phi_p)_q(\tilde{\mathbf{x}},\tilde{\mathbf{y}},t)|^2}{|(\mathbf{r}_s)_j^l(\mathbf{x},\mathbf{y},t)-(\mathbf{r}_p)_q(\tilde{\mathbf{x}},\tilde{\mathbf{y}},t)|}\,d\mathbf{x}\,d\mathbf{y}\,d\tilde{\mathbf{x}}\,d\tilde{\mathbf{y}}\,dt$$

$$+\sum_{s=1}^{3}\sum_{q=1}^{3}Ke^2\int_0^T\int_\Omega\int_\Omega\int_\Omega\int_\Omega \frac{|(\phi_p)_s(\mathbf{x},\mathbf{y},t)|^2|(\phi_p)_q(\tilde{\mathbf{x}},\tilde{\mathbf{y}},t)|^2}{|(\mathbf{r}_p)_s(\mathbf{x},\mathbf{y},t)-(\mathbf{r}_p)_q(\tilde{\mathbf{x}},\tilde{\mathbf{y}},t)|}\,d\mathbf{x}\,d\mathbf{y}\,d\tilde{\mathbf{x}}\,d\tilde{\mathbf{y}}\,dt, \quad (7.57)$$

$$J_3(\mathbf{r},\phi,\mathbf{n})$$

$$= \sum_{s=1}^{3}\sum_{l=0}^{n_s-1}\sum_{j=-l^l}^{l}(\gamma_s)_j^l\int_0^T\int_\Omega\int_\Omega\sum_{k=1}^{3}\left(\frac{\partial[(\phi_s)_j^l(\mathbf{n}_s)_j^l]}{\partial x_k}\cdot\frac{\partial[(\phi_s)_j^l(\mathbf{n}_s)_j^l]}{\partial x_k}\right.$$

$$\left.+\frac{\partial[(\phi_s)_j^l(\mathbf{n}_s)_j^l]}{\partial y_k}\cdot\frac{\partial[(\phi_s)_j^l(\mathbf{n}_s)_j^l]}{\partial y_k}\right)d\mathbf{x}\,d\mathbf{y}\,dt$$

$$+\sum_{s=1}^{3}(\gamma_p)_s\int_0^T\int_\Omega\int_\Omega\sum_{k=1}^{3}\left(\frac{\partial[(\phi_s)_j^l(\mathbf{n}_s)_j^l]}{\partial x_k}\cdot\frac{\partial[(\phi_s)_j^l(\mathbf{n}_s)_j^l]}{\partial x_k}\right.$$

$$\left.+\frac{\partial[(\phi_s)_j^l(\mathbf{n}_s)_j^l]}{\partial y_k}\cdot\frac{\partial[(\phi_s)_j^l(\mathbf{n}_s)_j^l]}{\partial y_k}\right)d\mathbf{x}\,d\mathbf{y}\,dt. \qquad (7.58)$$

Also,

$$J_4(\mathbf{r},\phi,E,\lambda,\hat{T},\mathbf{u})$$

$$= \frac{1}{2}\int_0^T\int_\Omega H_{ijkl}(\rho_1)\varepsilon_{ij}(\hat{\mathbf{u}})\varepsilon_{kl}(\hat{\mathbf{u}})\,(1-\chi_{\rho_f})\,d\mathbf{x}\,dt$$

$$-\int_0^T\int_\Omega f_i\hat{u}_i\,(1-\chi_{\rho_f})\,d\mathbf{x}\,dt-\int_0^T\int_{\Gamma_t}\hat{f}_i\hat{u}_i\,(1-\chi_{\rho_f})\,d\Gamma\,dt$$

$$+\sum_{k=1}^{3}\int_0^T\int_\Omega \lambda_k\left(\frac{\partial(\rho u_k)}{\partial t}+\sum_{j=1}^{3}\frac{\partial[(\rho u_k)u_j]}{\partial x_j}-\sum_{j=1}^{3}\frac{\partial\tau_{jk}}{\partial x_j}+\frac{\partial P}{\partial x_k}-g_k\right)\chi_{\rho_f}\,d\mathbf{x}\,dt$$

$$+\int_0^T\int_\Omega \lambda_4\left(\frac{\partial\rho}{\partial t}+\sum_{j=1}^{3}\frac{\partial(\rho u_j)}{\partial x_j}\right)\chi_{\rho_f}\,d\mathbf{x}\,dt$$

$$+\int_0^T\int_\Omega \lambda_5\left(\frac{\partial E_n}{\partial t}+\sum_{j=1}^{3}\frac{\partial}{\partial x_j}\left(u_jE_n-\sum_{k=1}^{3}u_k\tau_{jk}-q_j\right)\right)\chi_{\rho_f}\,d\mathbf{x}\,dt, \qquad (7.59)$$

where $\Gamma_t \subset \partial\Omega$ and, for appropriate constants $\hat{K} > 0$, $R > 0$ and $K_B > 0$, we have

$$\mathbf{q} = \hat{K}\nabla\hat{T},$$

$$P = \rho_3 R\hat{T},$$

$$\rho_f = \rho_2 + \rho_3,$$

$$\tau_{ij} = -\frac{\rho}{m_f}K_B\hat{T}\delta_{ij} - \frac{2}{3}\mu(\rho,\hat{T},P)\sum_{k=1}^{3}\frac{\partial u_k}{\partial x_k}\delta_{ij} + \mu(\rho,\hat{T},P)\left(\frac{\partial u_i}{\partial x_j} + \frac{\partial u_j}{\partial x_i}\right),$$

and

$$E_n(x,t) = \begin{cases} \frac{3}{2}\frac{m_f}{\rho}K_B\hat{T} + \frac{1}{2}\rho_f|\mathbf{u}|^2, & \text{if } \rho_f(x,t) \neq 0, \\ 0, & \text{if } \rho_f(x,t) = 0. \end{cases} \qquad (7.60)$$

where

$$m_f(t) = \int_\Omega \rho(x,t)\chi_{\rho_f}\,d\mathbf{x}.$$

Moreover, we define the strain tensor (for the solid type 1), by

$$\varepsilon(\hat{\mathbf{u}}) = \{\varepsilon_{ij}(\hat{\mathbf{u}})\} = \left\{\frac{\hat{u}_{i,j} + \hat{u}_{j,i}}{2}\right\}.$$

On the other hand,

$$\{H_{ijkl}(\rho_1)\}$$

is a positive definite tensor which represents the stiffness matrix for the solid type 1, which is assumed to depend linearly on ρ_1. Also, $f \in L^2(\Omega;\mathbb{R}^3)$ and $\hat{f} \in L^2(\partial\Omega;\mathbb{R}^3)$ are external loads effectively acting on the solid type 1 only where $\chi_{\rho_f} = 0$.

Remark 7.11.2 *We assume the concerning tensor is such that*

$$H_{ijkl}(\rho_1(x,t)) \approx 0$$

if

$$\rho_1(x,t) \approx 0,$$

and expect, in an appropriate sense, at least approximately

$$\chi_{\rho_1} \approx 1 - \chi_{\rho_f}.$$

Here, generically,

$$\chi_{\rho_f}(x,t) = \begin{cases} 1, & \text{if } \rho_f(x,t) > 0, \\ 0, & \text{if } \rho_f(x,t) = 0. \end{cases} \qquad (7.61)$$

Furthermore, denoting the initial mass of the substance type s by $(m_0)_s$, we have

$$
\begin{aligned}
m_s(t) &= (m_0)_s - \int_0^t \int_{\partial\Omega} \rho_s(\mathbf{x},\hat{t})\mathbf{u}\cdot\mathbf{n}\,d\Gamma\,d\hat{t} \\
&\quad - \int_0^t \int_{\partial\Omega} \alpha_s \rho_3(\mathbf{x},\hat{t})\mathbf{u}\cdot\mathbf{n}\,d\Gamma\,d\hat{t}
\end{aligned}
\tag{7.62}
$$

$\forall s \in \{1,2\}$, where here $\mathbf{n} = (n_1, n_2, n_3)$ denotes the outward normal field to $\partial\Omega$ and $m_s(t)$ denotes the mass of substance type s at the time t. We emphasize to have assumed $(m_0)_3 = 0$ and the substance type 3 may only leave the system represented by Ω (not enter it).

Considering such assumptions and statements, we define

$$
\begin{aligned}
&J_5(\mathbf{r},\phi,E,\lambda,\mathbf{u}) \\
&= -\frac{1}{2}\sum_{s=1}^{3}\sum_{l=0}^{n_s-1}\sum_{j=-l}^{l}\int_0^T\int_{\Omega}(E_s)_j^l(\mathbf{y},t)\left(\int_{\Omega}|(\phi_s)_j^l(\mathbf{x},\mathbf{y},t)|^2\,d\mathbf{x} - (m_e)_s(t)\right)d\mathbf{y}\,dt \\
&\quad -\frac{1}{2}\sum_{s=1}^{3}\int_0^T\int_{\Omega}(E_s)_p(\mathbf{y},t)\left(\int_{\Omega}|(\phi_s)_p(\mathbf{x},\mathbf{y},t)|^2\,d\mathbf{x} - \frac{(m_p+m_n)}{m_e}(m_e)_s(t)Z_s\right)d\mathbf{y}\,dt \\
&\quad +\int_0^T\sum_{s=1}^{2}\lambda_{5+s}(t)\left(m_s(t)-(m_0)_s + \int_0^t\int_{\partial\Omega}\rho_s(\mathbf{x},\hat{t})\mathbf{u}\cdot\mathbf{n}\,d\Gamma\,d\hat{t}\right. \\
&\quad \left.+\int_0^t\int_{\partial\Omega}\alpha_s\rho_3(\mathbf{x},\hat{t})\mathbf{u}\cdot\mathbf{n}\,d\Gamma\,d\hat{t}\right)dt \\
&\quad +\int_0^T\lambda_9(t)\left(m_1(t)+m_2(t)+m_3(t)\right. \\
&\quad \left.-\left((m_0)_1+(m_0)_2-\sum_{s=1}^{3}\int_0^t\int_{\partial\Omega}\rho_s(\mathbf{x},\hat{t})\mathbf{u}\cdot\mathbf{n}\,d\Gamma\,d\hat{t}\right)\right)dt,
\end{aligned}
\tag{7.63}
$$

where $(m_e)_s(t)$ is such that

$$
Z_s(m_e)_s(t) + Z_s\frac{m_p+m_n}{m_e}(m_e)_s(t) = m_s(t), \; \forall s \in \{1,2,3\}.
$$

Here, m_e, m_p, m_n denotes the mass of a single electron, proton and neutron, respectively.

We also define

$$
\begin{aligned}
J_6(\mathbf{r},\lambda,\mathbf{n}) &= \sum_{s=1}^{3}\sum_{l=0}^{n_s-1}\sum_{j=-l}^{l}\int_0^T\int_{\Omega}(\lambda_7^s)_j^l((\mathbf{n}_s)_j^l\cdot(\mathbf{n}_s)_j^l-1)\,d\mathbf{x}dt \\
&\quad +\sum_{s=1}^{3}\sum_{l=0}^{n_s-1}\sum_{j=-l}^{l}\int_0^T\int_{\Omega}(\lambda_8^s)_j^l(\mathbf{n}_s)_j^l\cdot\frac{\partial(\mathbf{r}_s)_j^l}{\partial t}\,d\mathbf{x}dt
\end{aligned}
\tag{7.64}
$$

and

$$
J_7(\phi,\hat{T}) = -S(w),
$$

where

$$\rho_s(\mathbf{x},t) = \sum_{l=0}^{n_s-1} \sum_{j=-l}^{l} \int_\Omega |(\phi_s)_j^l(\mathbf{x},\mathbf{y},t)|^2 \, d\mathbf{y}$$
$$+ \int_\Omega |(\phi_p)_s(\mathbf{x},\mathbf{y},t)|^2 \, d\mathbf{y}, \tag{7.65}$$

and

$$m_s(t) = \int_\Omega \rho_s(\mathbf{x},t) \, d\mathbf{x}.$$

Moreover,

$$(\mathbf{r}_s)_j^l(\mathbf{x},\mathbf{y},t) = \mathbf{r}_s(\mathbf{x},t) + \hat{\mathbf{r}}_s(\mathbf{x},t) + (\tilde{\mathbf{r}}_s)_j^l(\mathbf{x},\mathbf{y},t) \approx \mathbf{x},$$

where

$$\mathbf{r}_s(\mathbf{x},t) = \mathbf{x}, \text{ in } \Omega.$$

Also,

$$\hat{\mathbf{r}}_s(\mathbf{x},t) = \mathbf{0}, \text{ if } s = 2,3,$$

and

$$\hat{\mathbf{r}}_1(\mathbf{x},t) = \hat{\mathbf{u}}(\mathbf{x},t) = (\hat{u}_1(\mathbf{x},t), \hat{u}_2(\mathbf{x},t), \hat{u}_3(\mathbf{x},t))$$

refers to the displacement field for the solid part, and

$$\mathbf{u} = (u_1, u_2, u_3)$$

is the velocity field for the fluid part.

Furthermore,

$$(\mathbf{r}_p)_s(\mathbf{x},\mathbf{y},t) = \mathbf{y} + (\hat{\mathbf{r}}_p)_s(\mathbf{x},\mathbf{y},t) \approx \mathbf{y}.$$

7.11.2   The final variational formulation

This previous variational formulation may be useful in a nano-technology context, for example.

However, since it is a multi-scale one, it is of difficult computation. So, with such statements in mind, we shall propose a final macroscopic version for such a model, which we shall denote by $\tilde{J}$.

Concerning an analogy relating the previous formulation, in the next lines we set,

$$(\mathbf{r}_s)_j^l(\mathbf{x},\mathbf{y},t) = \mathbf{x}, \text{ for } s = 2,3$$

which translates into

$$\mathbf{r}_s(\mathbf{x},t) = \mathbf{x}, \text{ for } s = 2,3,$$

and

$$\mathbf{r}_1(\mathbf{x},t) = \mathbf{x} + \hat{\mathbf{u}}(\mathbf{x},t) = \mathbf{x} + (\hat{u}_1(\mathbf{x},t), \hat{u}_2(\mathbf{x},t), \hat{u}_3(\mathbf{x},t)).$$

Moreover,

$$\mathbf{u} = (u_1, u_2, u_3)$$

is the velocity field for the fluid part.

Finally, we also set

$$(\mathbf{n_s})^l_j = \mathbf{k}_0$$

for an appropriate unit constant vector $\mathbf{k}_0 \in \mathbb{R}^3$.

Concerning the new proposed formulation, we define,

$$\tilde{J} = \tilde{J}_1 + \tilde{J}_2 + \tilde{J}_3 + \tilde{J}_4 + \tilde{J}_5,$$

where

$$\tilde{J}_1(\mathbf{r}, \phi) \;=\; -\sum_{s=1}^{3} \int_0^T \int_\Omega |\phi_1|^2 \frac{\partial \mathbf{r}_1}{\partial t} \cdot \frac{\partial \mathbf{r}_1}{\partial t} \, d\mathbf{x} \; dt, \tag{7.66}$$

$$\tilde{J}_2(\mathbf{r}, \phi)$$
$$= \sum_{s=1}^{3} \sum_{q=1}^{3} K \int_0^T \int_\Omega \int_\Omega \frac{|\phi_s(\mathbf{x},t)|^2 |\phi_q(\tilde{\mathbf{x}},t)|^2}{|\mathbf{r}_s(\mathbf{x},t) - \mathbf{r}_q(\tilde{\mathbf{x}},t)|} \, d\mathbf{x} \, d\tilde{\mathbf{x}} \, dt \tag{7.67}$$

$$\tilde{J}_3(\phi)$$
$$= \sum_{s=1}^{3} \gamma_s \int_0^T \int_\Omega \sum_{k=1}^{3} \left(\frac{\partial \phi_s}{\partial x_k} \frac{\partial \phi_s}{\partial x_k} \right) d\mathbf{x} \, dt. \tag{7.68}$$

Also,

$$\tilde{J}_4(\mathbf{r}, \phi, E, \lambda, \hat{T}, \mathbf{u})$$
$$= \frac{1}{2} \int_0^T \int_\Omega H_{ijkl}(\rho_1) \varepsilon_{ij}(\hat{\mathbf{u}}) \varepsilon_{kl}(\hat{\mathbf{u}}) (1 - \chi_{\rho_f}) \, d\mathbf{x} \, dt$$
$$- \int_0^T \int_\Omega f_i \hat{u}_i (1 - \chi_{\rho_f}) \, d\mathbf{x} \, dt - \int_0^T \int_{\Gamma_t} \hat{f}_i \hat{u}_i (1 - \chi_{\rho_f}) \, d\Gamma \, dt$$
$$+ \sum_{k=1}^{3} \int_0^T \int_\Omega \lambda_k \left(\frac{\partial(\rho u_k)}{\partial t} + \sum_{j=1}^{3} \frac{\partial[(\rho u_k)u_j]}{\partial x_j} - \sum_{j=1}^{3} \frac{\partial \tau_{jk}}{\partial x_j} + \frac{\partial P}{\partial x_k} - g_k \right) \chi_{\rho_f} \, d\mathbf{x} \, dt$$
$$+ \int_0^T \int_\Omega \lambda_4 \left(\frac{\partial \rho}{\partial t} + \sum_{j=1}^{3} \frac{\partial(\rho u_j)}{\partial x_j} \right) \chi_{\rho_f} \, d\mathbf{x} \, dt$$
$$+ \int_0^T \int_\Omega \lambda_5 \left(\frac{\partial E_n}{\partial t} + \sum_{j=1}^{3} \frac{\partial}{\partial x_j} \left(u_j E_n - \sum_{k=1}^{3} u_k \tau_{jk} - q_j \right) \right) \chi_{\rho_f} \, d\mathbf{x} \, dt, \tag{7.69}$$

where again

$$\mathbf{q} = \hat{K} \nabla \hat{T},$$

$$P = \rho_3 R \hat{T},$$

$$\rho_f = \rho_2 + \rho_3,$$

$$\tau_{ij} = -\frac{\rho}{m_f} K_B T \delta_{ij} - \frac{2}{3} \mu(\rho, \hat{T}, P) \sum_{k=1}^{3} \frac{\partial u_k}{\partial x_k} \delta_{ij} + \mu(\rho, \hat{T}, P) \left(\frac{\partial u_i}{\partial x_j} + \frac{\partial u_j}{\partial x_i} \right),$$

and

$$E_n(x,t) = \begin{cases} \frac{3}{2} \frac{m_f}{\rho} K_B \hat{T} + \frac{1}{2} \rho_f |\mathbf{u}|^2, & \text{if } \rho_f(x,t) \neq 0, \\ 0, & \text{if } \rho_f(x,t) = 0. \end{cases} \tag{7.70}$$

where

$$m_f(t) = \int_\Omega \rho(x,t) \chi_{\rho_f} \, d\mathbf{x}.$$

Here also again, generically

$$\chi_{\rho_f}(x,t) = \begin{cases} 1, & \text{if } \rho_f(x,t) > 0, \\ 0, & \text{if } \rho_f(x,t) = 0. \end{cases} \tag{7.71}$$

Finally,

$$\tilde{J}_5(\mathbf{r}, \phi, \lambda, \hat{T}, \mathbf{u})$$

$$= \int_0^T \sum_{s=1}^{2} \lambda_{5+s}(t) \left(m_s(t) - (m_0)_s + \int_0^t \int_{\partial\Omega} \rho_s(\mathbf{x}, \hat{t}) \mathbf{u} \cdot \mathbf{n} \, d\Gamma \, d\hat{t} \right.$$

$$+ \int_0^t \int_{\partial\Omega} \alpha_s \rho_3(\mathbf{x}, \hat{t}) \mathbf{u} \cdot \mathbf{n} \, d\Gamma \, d\hat{t} \Big) \, dt$$

$$+ \int_0^T \lambda_8(t) \left(m_1(t) + m_2(t) + m_3(t) \right.$$

$$- \left((m_0)_1 + (m_0)_2 - \sum_{s=1}^{3} \int_0^t \int_{\partial\Omega} \rho_s(\mathbf{x}, \hat{t}) \mathbf{u} \cdot \mathbf{n} \, d\Gamma \, d\hat{t} \right) \Big) \, dt$$

$$- S(w), \tag{7.72}$$

where

$$\rho_s(\mathbf{x},t) = |(\phi_s)(\mathbf{x},t)|^2, \tag{7.73}$$

$$m_s(t) = \int_\Omega \rho_s(\mathbf{x},t) \, d\mathbf{x}, \ \forall s \in \{1,2,3\}$$

and

$$\rho(\mathbf{x},t) = \sum_{s=1}^{3} \rho_s(\mathbf{x},t).$$

7.12   A note on the Spin operator

We finish this article with a result about the Spin operator in a relativistic context.

Consider a wave function $\phi(\mathbf{r})$ related to the scalar density field of a particle with position field given by

$$\mathbf{r} : \Omega \times [0,T] \to \mathbb{R}^3.$$

Observe that, in a special relativity context, the field of velocity

$$\frac{\partial \mathbf{r}(\mathbf{x},t)}{\partial t}$$

induces a Lorentz type transformation concerning an observer at $(0,0,0) \in \mathbb{R}^3$. So, the corresponding transform of the vector

$$\mathbf{r}(\mathbf{x},t) = (ct, X_1(\mathbf{x},t), X_2(\mathbf{x},t), X_3(\mathbf{x},t))$$

will be the vector

$$(ct', X_1', X_2', X_3'),$$

where

$$X_j' = \left(\frac{1}{\sqrt{1 - \frac{v^2}{c^2}}} - 1 \right) \left(\frac{\mathbf{r}_1 \cdot \mathbf{v}}{v^2} \right) v_j - \frac{tv_j}{\sqrt{1 - \frac{v^2}{c^2}}} + X_j(\mathbf{x},t),$$

$\forall j \in \{1,2,3\}$, and

$$t' = \frac{1}{\sqrt{1 - \frac{v^2}{c^2}}} \left(t - \frac{\mathbf{r}_1 \cdot \mathbf{v}}{c^2} \right),$$

where

$$\mathbf{r}_1(\mathbf{x},t) = (X_1(\mathbf{x},t), X_2(\mathbf{x},t), X_3(\mathbf{x},t)),$$

$$\mathbf{v} = \frac{\partial \mathbf{r}_1(\mathbf{x},t)}{\partial t},$$

and

$$v = \sqrt{\left(\frac{\partial X_1(\mathbf{x},t)}{\partial t} \right)^2 + \left(\frac{\partial X_2(\mathbf{x},t)}{\partial t} \right)^2 + \left(\frac{\partial X_3(\mathbf{x},t)}{\partial t} \right)^2}.$$

We assume there exists a function such that φ

$$\phi(\mathbf{r}(\mathbf{x},t)) = \varphi(X_1', X_2', X_3', t').$$

At this point, we shall define the angular momentum operator. First, we consider a rotation about the z axis, so that we define

$$\begin{aligned} \mathbf{r}_\varepsilon(\mathbf{x},t) &= (X_1(\mathbf{x},t), X_2(\mathbf{x},t), X_3(\mathbf{x},t)) + \varepsilon(-x_2, x_1, 0) \\ &= ((X_1)_\varepsilon, (X_2)_\varepsilon, (X_3)_\varepsilon), \end{aligned} \tag{7.74}$$

where

$$(X_1)_\varepsilon = X_1(\mathbf{x},t) - \varepsilon x_2,$$
$$(X_2)_\varepsilon = X_2(\mathbf{x},t) + \varepsilon x_1,$$
$$(X_3)_\varepsilon = X_3(\mathbf{x},t),$$

and also

$$t_{\varepsilon} = t.$$

In such a case, we have

$$\mathbf{v}_{\varepsilon} = \frac{\partial \mathbf{r}_{\varepsilon}(\mathbf{x},t)}{\partial t} = \mathbf{v},$$

so that

$$v_{\varepsilon} = v.$$

Hence, we define the angular momentum coordinate $J_z(\varphi(\mathbf{X}',t'))$, by

$$J_z(\varphi(\mathbf{X}',t')) = -i\hbar \frac{\partial \varphi(\mathbf{X}'_{\varepsilon},t'_{\varepsilon})}{\partial \varepsilon}\Big|_{\varepsilon=0}, \qquad (7.75)$$

where

$$\mathbf{X}'_{\varepsilon} = \left(\frac{1}{\sqrt{1-\frac{v^2}{c^2}}} - 1\right) \left(\frac{\mathbf{r}_{\varepsilon} \cdot \mathbf{v}}{v^2}\right)\mathbf{v} - \frac{t\mathbf{v}}{\sqrt{1-\frac{v^2}{c^2}}} + \mathbf{r}_{\varepsilon}(\mathbf{x},t),$$

and

$$t'_{\varepsilon} = \frac{1}{\sqrt{1-\frac{v^2}{c^2}}} \left(t - \frac{\mathbf{r}_{\varepsilon} \cdot \mathbf{v}}{c^2}\right),$$

so that

$$\begin{aligned}
J_z(\varphi(\mathbf{X}',t')) &= -i\hbar \sum_{j=1}^{3} \frac{\partial \varphi(\mathbf{X}'_{\varepsilon},t'_{\varepsilon})}{\partial X_j} \frac{\partial(X'_{\varepsilon})_j}{\partial \varepsilon}\Big|_{\varepsilon=0} \\
&\quad -i\hbar \frac{\partial \varphi(\mathbf{X}'_{\varepsilon},t'_{\varepsilon})}{\partial t} \frac{\partial t'_{\varepsilon}}{\partial \varepsilon}\Big|_{\varepsilon=0}.
\end{aligned} \qquad (7.76)$$

Observe that

$$\frac{\partial \mathbf{X}'_{\varepsilon}}{\partial \varepsilon} = \left(\frac{1}{\sqrt{1-\frac{v^2}{c^2}}} - 1\right) \left(\frac{(-x_2,x_1,0) \cdot \mathbf{v}}{v^2}\right)\mathbf{v} + (-x_2,x_1,0),$$

and

$$\frac{\partial t'_{\varepsilon}}{\partial \varepsilon} = -\frac{1}{\sqrt{1-\frac{v^2}{c^2}}} \frac{(-x_2,x_1,0) \cdot \mathbf{v}}{c^2}.$$

From such last results, we have

$$
\begin{aligned}
J_z(\varphi(\mathbf{X}',t')) &= -i\hbar \sum_{j=1}^{3} \frac{\partial \varphi(\mathbf{X}',t')}{\partial X_j} \left(\frac{1}{\sqrt{1-\frac{v^2}{c^2}}} - 1 \right) \left(\frac{(-x_2,x_1,0)\cdot\mathbf{v}}{v^2} \right) v_j \\
&\quad + i\hbar \frac{\partial \varphi(\mathbf{X}',t')}{\partial t} \frac{1}{\sqrt{1-\frac{v^2}{c^2}}} \frac{(-x_2,x_1,0)\cdot\mathbf{v}}{c^2} \\
&\quad - i\hbar \left(-x_2 \frac{\partial \varphi(\mathbf{X}',t')}{\partial X_1} + x_1 \frac{\partial \varphi(\mathbf{X}',t')}{\partial X_2} \right) \\
&= S_z(\varphi(\mathbf{X}',t')) + L_z(\varphi(\mathbf{X}',t')),
\end{aligned}
\tag{7.77}
$$

where

$$
L_z(\varphi(\mathbf{X}',t')) = -i\hbar \left(-x_2 \frac{\partial \varphi(\mathbf{X}',t')}{\partial X_1} + x_1 \frac{\partial \varphi(\mathbf{X}',t')}{\partial X_2} \right),
$$

and

$$
\begin{aligned}
S_z(\varphi(\mathbf{X}',t')) &= -i\hbar \sum_{j=1}^{3} \frac{\partial \varphi(\mathbf{X}',t')}{\partial X_j} \left(\frac{1}{\sqrt{1-\frac{v^2}{c^2}}} - 1 \right) \left(\frac{(-x_2,x_1,0)\cdot\mathbf{v}}{v^2} \right) v_j \\
&\quad + i\hbar \frac{\partial \varphi(\mathbf{X}',t')}{\partial t} \frac{1}{\sqrt{1-\frac{v^2}{c^2}}} \frac{(-x_2,x_1,0)\cdot\mathbf{v}}{c^2}.
\end{aligned}
\tag{7.78}
$$

Similarly we may obtain L_x, L_y, S_x, S_y.
Finally defining

$$
\mathbf{J} = \mathbf{S} + \mathbf{L},
$$

where

$$
\mathbf{L} = (L_x, L_y, L_z)
$$

and

$$
\mathbf{S} = (S_x, S_y, S_z),
$$

we call $\mathbf{L}$ the orbital angular momentum operator and $\mathbf{S}$ the spin one.

Chapter 8

Existence and Duality for Superconductivity and Related Models

8.1 Introduction

In this chapter, we present a theorem which represents a duality principle suitable for a large class of non-convex variational problems.

At this point, we refer to the exceptionally important article "A contribution to contact problems for a class of solids and structures" by W.R. Bielski and J.J. Telega, [8], published in 1985, as the first one to successfully apply and generalize the convex analysis approach to a model in non-convex and non-linear mechanics.

The present work is, in some sense, a kind of extension of this previous work [8] combined with a D.C. approach presented in [50] and others such as [7], which greatly influenced and inspired my work and recent book [13].

First, we recall that about the year 1950 Ginzburg and Landau introduced a theory to model the super-conducting behavior of some types of materials below a critical temperature T_c, which depends on the material in question. They postulated that free density energy may be written close to T_c as

$$F_s(T) = F_n(T) + \frac{\hbar}{4m} \int_{\Omega} |\nabla \phi|_2^2 \, dx + \frac{\alpha(T)}{4} \int_{\Omega} |\phi|^4 \, dx - \frac{\beta(T)}{2} \int_{\Omega} |\phi|^2 \, dx,$$

where ϕ is a complex parameter, $F_n(T)$ and $F_s(T)$ are the normal and super-conducting free energy densities, respectively (see [11] for details). Here, $\Omega \subset$

$\mathbb{R}^3$ denotes the super-conducting sample with a boundary denoted by $\partial\Omega = \Gamma$. The complex function $\phi \in W^{1,2}(\Omega;\mathbb{C})$ is intended to minimize $F_s(T)$ for a fixed temperature T.

Denoting $\alpha(T)$ and $\beta(T)$ simply by α and β, the corresponding Euler-Lagrange equations are given by:

$$\begin{cases} -\frac{\hbar}{2m}\nabla^2\phi + \alpha|\phi|^2\phi - \beta\phi = 0, & \text{in } \Omega \\[2mm] \frac{\partial\phi}{\partial\mathbf{n}} = 0, & \text{on } \partial\Omega. \end{cases} \tag{8.1}$$

This last system of equations is well known as the Ginzburg-Landau (G-L) one. In the physics literature, the G-L energy in which a magnetic potential here denoted by $\mathbf{A}$ is included, is also well known. The functional in question is given by:

$$\begin{aligned} J(\phi,\mathbf{A}) &= \frac{1}{8\pi}\int_{\mathbb{R}^3} |\operatorname{curl}\mathbf{A} - \mathbf{B}_0|_2^2 \, dx + \frac{\hbar^2}{4m}\int_\Omega \left|\nabla\phi - \frac{2ie}{\hbar c}\mathbf{A}\phi\right|_2^2 \, dx \\[2mm] &\quad + \frac{\alpha}{4}\int_\Omega |\phi|^4 \, dx - \frac{\beta}{2}\int_\Omega |\phi|^2 \, dx \end{aligned} \tag{8.2}$$

Considering its minimization on the space U, where

$$U = W^{1,2}(\Omega;\mathbb{C}) \times W^{1,2}(\mathbb{R}^3;\mathbb{R}^3),$$

through the physics notation, the corresponding Euler-Lagrange equations are:

$$\begin{cases} \frac{1}{2m}\left(-i\hbar\nabla - \frac{2e}{c}\mathbf{A}\right)^2\phi + \alpha|\phi|^2\phi - \beta\phi = 0, & \text{in } \Omega \\[2mm] \left(i\hbar\nabla\phi + \frac{2e}{c}\mathbf{A}\phi\right)\cdot\mathbf{n} = 0, & \text{on } \partial\Omega, \end{cases} \tag{8.3}$$

and

$$\begin{cases} \operatorname{curl}(\operatorname{curl}\mathbf{A}) = \operatorname{curl}\mathbf{B}_0 + \frac{4\pi}{c}\tilde{J}, & \text{in } \Omega \\[2mm] \operatorname{curl}(\operatorname{curl}\mathbf{A}) = \operatorname{curl}\mathbf{B}_0, & \text{in } \mathbb{R}^3 \setminus \overline{\Omega}, \end{cases} \tag{8.4}$$

where

$$\tilde{J} = -\frac{ie\hbar}{2m}(\phi^*\nabla\phi - \phi\nabla\phi^*) - \frac{2e^2}{mc}|\phi|^2\mathbf{A}.$$

and

$$\mathbf{B}_0 \in L^2(\mathbb{R}^3;\mathbb{R}^3)$$

is a known applied magnetic field.

At this point, we emphasize to denote generically

$$\langle g,h\rangle_{L^2} = \int_\Omega Re[g]Re[h] \, dx - \int_\Omega Im[g]Im[h] \, dx,$$

$\forall h, g \in L^2(\Omega; \mathbb{C})$, where $Re[a], Im[a]$ denote the real and imaginary parts of a, $\forall a \in \mathbb{C}$, respectively.

Moreover, existence of a global solution for a similar problem has been proved in section 8.2 of this chapter and in [22].

Finally, for the subsequent theoretical results, we assume a simplified atomic units context.

8.2   A global existence result for the full complex Ginzburg-Landau system

In this section, we present a global existence result for the complex Ginzburg-Landau system in superconductivity.

The main result is summarized by the following theorem.

Theorem 8.2.1 *Let $\Omega, \Omega_1 \subset \mathbb{R}^3$ be open, bounded and connected sets whose regular (Lipschistzian) boundaries are denoted by $\partial\Omega$ and $\partial\Omega_1$, respectively. Assume $\overline{\Omega} \subset \Omega_1$.*

Consider the functional $J : U \to \mathbb{R}$ defined by

$$
\begin{aligned}
J(\phi, \mathbf{A}) &= \frac{\gamma}{2} \int_\Omega |\nabla\phi - i\rho\mathbf{A}\phi|^2 \, dx \\
&+ \frac{\alpha}{2} \int_\Omega (|\phi|^2 - \beta)^2 \, dx + K_0 \| \, curl \, \mathbf{A} - \mathbf{B}_0\|_{2,\Omega_1}^2,
\end{aligned}
\tag{8.5}
$$

where $\alpha, \beta, \gamma, \rho, K_0$ are positive constants, i is the imaginary unit,

$$U = U_1 \times U_2,$$

$$U_1 = W^{1,2}(\Omega; \mathbb{C})$$

and

$$U_2 = W^{1,2}(\Omega_1; \mathbb{R}^3).$$

Also $\mathbf{B}_0 \in W^{1,2}(\Omega_1; \mathbb{R}^3)$ denotes an external magnetic field.

Under such hypotheses, there exists $(\phi_0, \mathbf{A}_0) \in U$ such that

$$J(\phi_0, \mathbf{A}_0) = \min_{(\phi, \mathbf{A}) \in U} J(\phi, \mathbf{A}).$$

Proof 8.1   Denote

$$\alpha_1 = \inf_{(\phi, \mathbf{A}) \in U} J(\phi, \mathbf{A}).$$

From the results in [22], fixing the gauge of London, we may find a minimizing sequence for J such that

$$\alpha_1 \leq J(\phi_n, \mathbf{A}_n) < \alpha_1 + \frac{1}{n},$$

$$\text{div } \mathbf{A}_n = 0, \text{ in } \Omega_1,$$

and

$$\mathbf{A}_n \cdot \mathbf{n} = 0, \text{ on } \partial\Omega_1, \ \forall n \in \mathbb{N}.$$

From div $\mathbf{A}_n = 0$ we get

$$\text{curl curl } \mathbf{A}_n = -\nabla^2 \mathbf{A}_n,$$

so that from this and

$$\mathbf{A}_n \cdot \mathbf{n} = 0 \text{ on } \partial\Omega_1,$$

we have that

$$\int_{\Omega_1} \text{curl } \mathbf{A} \cdot \text{curl } \mathbf{A} \, dx = \sum_{k=1}^{3} \int_{\Omega_1} \nabla(A_k)_n \nabla(A_k)_n \, dx, \ \forall n \in \mathbb{N}.$$

From the expression of J, for such a minimizing sequence, there exists a real $K_1 > 0$ such that

$$\|\nabla \mathbf{A}_n\|_{2,\Omega_1} = \|\text{ curl } \mathbf{A}_n\|_{2,\Omega_1} \leq K_1, \ \forall n \in \mathbb{N}.$$

From this and the boundary conditions in question, there exists a real $K_2 > 0$ such that

$$\|\mathbf{A}_n\|_{1,2,\Omega_1} \leq K_2, \ \forall n \in \mathbb{N}.$$

Thus, from the Rellich-Kondrachov theorem, there exists $\mathbf{A}_0 \in U_2$ such that, up to a not relabeled subsequence

$$\mathbf{A}_n \rightharpoonup \mathbf{A}_0, \text{ weakly in } W^{1,2}(\Omega_1, \mathbb{R}^3),$$

and

$$\mathbf{A}_n \to \mathbf{A}_0, \text{ strongly in } L^2(\Omega_1, \mathbb{R}^3) \text{ and } L^4(\Omega_1; \mathbb{R}^3).$$

From this and the generalized Hölder inequality, we obtain

$$
\begin{aligned}
\alpha_1 + \frac{1}{n} > J(\phi_n, \mathbf{A}_n) \ \geq \ & \frac{\gamma}{2} \int_\Omega |\nabla \phi_n|^2 \, dx \\
& - K_3 \|\mathbf{A}_n\|_{4,\Omega} \|\nabla \phi_n\|_{2,\Omega} \|\phi_n\|_{4,\Omega} + \frac{\gamma}{2} \int_\Omega |\mathbf{A}_n|^2 |\phi_n|^2 \, dx \\
& + \frac{\alpha}{2} \int_\Omega |\phi_n|^4 \, dx - \alpha\beta \int_\Omega |\phi_n|^2 \, dx \\
& K_0 \|\text{ curl } \mathbf{A}_n - \mathbf{B}_0\|_{2,\Omega_1}^2 + K_5 \\
\geq \ & \frac{\gamma}{2} \int_\Omega |\nabla \phi_n|^2 \, dx \\
& - K_4 \|\nabla \phi_n\|_{2,\Omega} \|\phi_n\|_{4,\Omega} + \frac{\gamma}{2} \int_\Omega |\mathbf{A}_n|^2 |\phi_n|^2 \, dx \\
& + \frac{\alpha}{2} \int_\Omega |\phi_n|^4 \, dx - \alpha\beta \int_\Omega |\phi_n|^2 \, dx \\
& K_0 \|\text{ curl } \mathbf{A}_n - \mathbf{B}_0\|_{2,\Omega_1}^2 + K_5, \ \forall n \in \mathbb{N}.
\end{aligned}
\tag{8.6}
$$

From this, there exists $K_6 > 0$ such that

$$\|\nabla \phi_n\|_{2,\Omega} \leq K_6,$$

and

$$\|\phi_n\|_{4,\Omega} \leq K_6, \; \forall n \in \mathbb{N}.$$

Thus, from the Rellich-Kondrachov theorem, there exists $\phi_0 \in U_1$ such that, up to a not relabeled subsequence,

$$\phi_n \rightharpoonup \phi_0, \text{ weakly in } W^{1,2}(\Omega; \mathbb{C}),$$

and

$$\phi_n \rightarrow \phi_0, \text{ strongly in } L^2(\Omega; \mathbb{C}) \text{ and } L^4(\Omega; \mathbb{C}).$$

From such results, we may obtain

$$\nabla \phi_n - i\rho \mathbf{A}_n \phi_n \rightharpoonup \nabla \phi_0 - i\rho \mathbf{A}_0 \phi_0, \text{ weakly in } L^2(\Omega; \mathbb{C}).$$

Therefore, from this, the convexity of J in $v = \nabla \phi - i\rho \mathbf{A} \phi$ and $w = \text{curl } \mathbf{A}$, and from

$$\phi_n \rightarrow \phi_0, \text{ strongly in } L^2(\Omega; \mathbb{C}) \text{ and } L^4(\Omega; \mathbb{C})$$

and

$$\text{curl } \mathbf{A}_n \rightharpoonup \text{curl } \mathbf{A}_0, \text{ weakly in } L^2(\Omega_1, \mathbb{R}^3),$$

we get

$$\alpha_1 = \inf_{(\phi, \mathbf{A}) \in U} J(\phi, \mathbf{A}) = \liminf_{n \rightarrow \infty} J(\phi_n, \mathbf{A}_n) \geq J(\phi_0, \mathbf{A}_0).$$

The proof is complete.

8.3 A brief initial description of our proposal for duality

Let $\Omega \subset \mathbb{R}^3$ be an open, bounded and connected set with a regular boundary denoted by $\partial \Omega$. Let $U = W_0^{1,2}(\Omega)$ and let $J : U \rightarrow \mathbb{R}$ be a functional defined by

$$J(u) = G_1(u) + F_1(u) - \langle u, f \rangle_{L^2},$$

where

$$G_1(u) = \frac{\gamma}{2} \int_\Omega \nabla u \cdot \nabla u \, dx,$$

and

$$F_1(u) = \frac{\alpha}{2} \int_\Omega (u^2 - \beta)^2 \, dx,$$

where α, β and γ are positive real constants and $f \in L^2(\Omega)$.

Observe that there exists $\eta \in \mathbb{R}$ such that

$$\eta = \inf_{u \in U} J(u) = J(u_0),$$

for some $u_0 \in U$. (We recall that the existence of a global minimizer may be proven by the direct method of the calculus of variations).

In our approach, we combine the ideas of J.J. Telega [7, 8] generalizing the approach in Ekeland and Temam [31] for establishing the dual functionals through the Legendre transform definition, with a D.C. approach for non-convex optimization developed by J.F. Toland, [50].

At this point we would define the functionals $F : U \to \mathbb{R}$ and $G : U \to \mathbb{R}$, where

$$F(u) = \frac{\gamma}{2} \int_\Omega \nabla u \cdot \nabla u \, dx + \frac{K}{2} \int_\Omega u^2 \, dx,$$

and

$$G(u,v) = -\frac{\alpha}{2} \int_\Omega (u^2 - \beta + v)^2 \, dx + \frac{K}{2} \int_\Omega u^2 \, dx + \langle u, f \rangle_{L^2},$$

so that

$$
\begin{aligned}
J(u) &= F(u) - G(u,0) \\
&= \frac{\gamma}{2} \int_\Omega \nabla u \cdot \nabla u \, dx + \frac{\alpha}{2} \int_\Omega (u^2 - \beta)^2 \, dx - \langle u, f \rangle_{L^2},
\end{aligned}
\tag{8.7}
$$

where such a functional, for a large $K > 0$, is represented as a difference of two convex functionals in a large domain region proportional to $K > 0$.

The second step is to define the corresponding dual functionals F^*, G^*, where

$$
\begin{aligned}
F^*(v_1^*) &= \sup_{u \in U} \{ \langle u, v_1^* \rangle_{L^2} - F(u) \} \\
&= \frac{1}{2} \int_\Omega \frac{(v_1^*)^2}{K - \gamma \nabla^2} \, dx
\end{aligned}
\tag{8.8}
$$

and

$$
\begin{aligned}
G^*(v_1^*, v_0^*) &= \sup_{u \in U} \inf_{v \in L^2} \{\langle u, v_1^* \rangle_{L^2} - \langle v, v_0^* \rangle_{L^2} - G(u,v)\} \\
&= \sup_{u \in U} \inf_{v \in L^2} \{\langle u, v_1^* \rangle_{L^2} - \langle v, v_0^* \rangle_{L^2} + \frac{\alpha}{2} \int_\Omega (u^2 - \beta + v)^2 \, dx \\
&\quad - \frac{K}{2} \int_\Omega u^2 \, dx - \langle u, f \rangle_{L^2} \} \\
&= \sup_{u \in U} \inf_{w \in L^2} \{\langle u, v_1^* \rangle_{L^2} - \langle w - u^2 + \beta, v_0^* \rangle_{L^2} + \frac{\alpha}{2} \int_\Omega w^2 \, dx \\
&\quad - \frac{K}{2} \int_\Omega u^2 \, dx - \langle u, f \rangle_{L^2} \} \\
&= \sup_{u \in U} \inf_{w \in L^2} \{\langle u, v_1^* \rangle_{L^2} - \langle w, v_0^* \rangle_{L^2} + \frac{\alpha}{2} \int_\Omega w^2 \, dx - \langle u^2, v_0^* \rangle_{L^2} \\
&\quad - \frac{K}{2} \int_\Omega u^2 \, dx - \langle u, f \rangle_{L^2} \\
&\quad - \beta \int_\Omega v_0^* \, dx \} \\
&= -\frac{1}{2} \int_\Omega \frac{(v_1^* - f)^2}{2v_0^* - K} \, dx - \frac{1}{2\alpha} \int_\Omega (v_0^*)^2 \, dx \\
&\quad - \beta \int_\Omega v_0^* \, dx,
\end{aligned}
\tag{8.9}
$$

if $-2v_0^* + K > 0$ in $\overline{\Omega}$.

Defining

$$
E = \{v_0^* \in C(\overline{\Omega}) \text{ such that } -2v_0^* + K > K/2, \text{ in } \overline{\Omega}\},
$$

where $K > 0$ is such that

$$
\frac{1}{\alpha} > \frac{32K_2^2}{K^3},
$$

for some appropriate $K_2 > 0$ to be specified, it may be proven that for $K > 0$ sufficiently large,

$$
\inf_{v_1^* \in L^2} \sup_{v_0^* \in E} \{-F^*(v_1^*) + G^*(v_1^*, v_0^*)\} \geq \inf_{u \in U} J(u).
$$

Equality concerning this last result may be obtained in a local extremal context and, under appropriate optimality conditions to be specified, also for global optimization.

At this point, we highlight that the maximization in v_0^* with the restriction $v_0^* \in E$ does not demand a Lagrange multiplier, since for the value of $K > 0$ specified the restriction is not active.

We emphasize this approach is original and substantially different from all those, of other authors, so far known.

Finally, for a more general model, in the next section we formally prove that a critical point for the primal formulation necessarily corresponds to a critical point of the dual formulation. The reciprocal may be also proven.

8.4 The duality principle for a local extremal context

In this section, we state and prove the concerning duality principle. We recall that the existence of a global minimizer for the related functional has been proven in section 8.2 of this article. At this point, it is also worth mentioning that an extensive study on duality theory and applications for such and similar models is developed in [13].

Theorem 8.4.1 *Let $\Omega, \Omega_1 \subset \mathbb{R}^3$ be open, bounded and connected sets with regular (Lipschitzian) boundaries denoted by $\partial\Omega$ and $\partial\Omega_1$, respectively.*
Assume Ω_1 is convex and $\overline{\Omega} \subset \Omega_1$. Consider the functional $J : U \to \mathbb{R}$ where

$$
\begin{aligned}
J(\phi, \mathbf{A}) \;=\; & \frac{\gamma}{2} \int_{\Omega} |\nabla\phi - i\rho\mathbf{A}\phi|^2 \, dx \\
& + \frac{\alpha}{2} \int_{\Omega} (|\phi|^2 - \beta)^2 \, dx + \frac{1}{8\pi} \| \, curl\, \mathbf{A} - \mathbf{B}_0 \|_{0,\Omega_1}^2
\end{aligned}
\tag{8.10}
$$

where $\alpha, \beta, \gamma, \rho$ are positive real constants, i is the imaginary unit and

$$
U = U_1 \times U_2,
$$

$$
U_1 = C^1(\overline{\Omega}; \mathbb{C}), \quad U_2 = C^1(\overline{\Omega_1}; \mathbb{R}^3),
$$

both with the norm $\| \cdot \|_{1,\infty}$.
Moreover,

$$
\phi : \Omega \to \mathbb{C}
$$

is the order parameter,

$$
\mathbf{A} : \Omega_1 \to \mathbb{R}^3
$$

is the magnetic potential and $\mathbf{B}_0 \in C^1(\overline{\Omega_1}, \mathbb{R}^3)$ is an external magnetic field.
Defining,

$$
B_2 = \{ \mathbf{A} \in C^1(\overline{\Omega_1}; \mathbb{R}^3) \; : \; div\, \mathbf{A} = 0 \text{ in } \Omega_1, \; \mathbf{A} \cdot \mathbf{n} = 0, \text{ on } \partial\Omega_1 \},
$$

where $\mathbf{n}$ denotes the outward normal to $\partial\Omega_1$, suppose $(\phi_0, \mathbf{A}_0) \in C^1(\overline{\Omega}; \mathbb{C}) \times B_2$ is such that

$$
\|\phi_0\|_\infty \le K_2,
$$

for some appropriate $K_2 > 0$,

$$
\delta J(\phi_0, \mathbf{A}_0) = \mathbf{0}
$$

and

$$\delta^2 J(\phi_0, \mathbf{A}_0) > 0.$$

Denoting also generically

$$(\nabla - i\rho\mathbf{A})^*(\nabla - i\rho\mathbf{A}) = |\nabla - i\rho\mathbf{A}|^2,$$

define $F : U \to \mathbb{R}$ *by*

$$F(\phi, \mathbf{A}) = \frac{\gamma}{2} \int_\Omega |\nabla\phi - i\rho\mathbf{A}\phi|^2 \, dx + \frac{K}{2} \int_\Omega |\phi|^2 \, dx,$$

$G : U \times C(\Omega) \to \mathbb{R}$ *by*

$$G(\phi, \mathbf{A}, v) = -\frac{\alpha}{2} \int_\Omega (|\phi|^2 - \beta + v)^2 \, dx - \frac{1}{8\pi} \| \operatorname{curl} \mathbf{A} - \mathbf{B}_0 \|_{0,\Omega_1}^2 + \frac{K}{2} \int_\Omega |\phi|^2 \, dx, \quad (8.11)$$

$$\begin{aligned} F^*(v_1^*, \mathbf{A}) &= \sup_{\phi \in U_1} \{ \langle \phi, v_1^* \rangle_{L^2} - F(\phi, \mathbf{A}) \} \\ &= \frac{1}{2} \int_\Omega \frac{(v_1^*)^2}{(\gamma|\nabla - i\rho\mathbf{A}|^2 + K)} \, dx, \end{aligned} \quad (8.12)$$

$$\begin{aligned} \hat{G}^*(v_1^*, v_0^*, \mathbf{A}) &= \sup_{\phi \in U_1} \inf_{v \in C(\Omega)} \{ \langle \phi, v_1^* \rangle_{L^2} - \langle v, v_0^* \rangle_{L^2} - G(\phi, \mathbf{A}, v) \} \\ &= -\frac{1}{2} \int_\Omega \frac{(v_1^*)^2}{2v_0^* - K} \, dx \\ &\quad - \frac{1}{2\alpha} \int_\Omega (v_0^*)^2 \, dx - \beta \int_\Omega v_0^* \, dx \\ &\quad + \frac{1}{8\pi} \| \operatorname{curl} \mathbf{A} - \mathbf{B}_0 \|_{0,\Omega_1}^2, \end{aligned} \quad (8.13)$$

if

$$-2v_0^* + K > 0 \text{ in } \overline{\Omega},$$

and

$$\begin{aligned} J^*(v_1^*, v_0^*, \mathbf{A}) &= -F^*(v_1^*, \mathbf{A}) + \hat{G}^*(v_1^*, v_0^*, \mathbf{A}) \\ &= -\frac{1}{2} \int_\Omega \frac{(v_1^*)^2}{(\gamma|\nabla - i\rho\mathbf{A}|^2 + K)} \, dx \\ &\quad - \frac{1}{2} \int_\Omega \frac{(v_1^*)^2}{2v_0^* - K} \, dx \\ &\quad - \frac{1}{2\alpha} \int_\Omega (v_0^*)^2 \, dx - \beta \int_\Omega v_0^* \, dx \\ &\quad + \frac{1}{8\pi} \| \operatorname{curl} \mathbf{A} - \mathbf{B}_0 \|_{0,\Omega_1}^2. \end{aligned} \quad (8.14)$$

Furthermore, define

$$\hat{v}_0^* = \alpha(|\phi_0|^2 - \beta),$$
$$\hat{v}_1^* = (2\hat{v}_0^* - K)\phi_0,$$

and

$$E = \{v_0^* \in C(\overline{\Omega}) \ : \ -2v_0^* + K > K/2, \ in \ \overline{\Omega}\}.$$

Under such hypotheses and assuming also

$$\hat{v}_0^* \in E,$$

we have

$$\delta J^*(\hat{v}_1^*, \hat{v}_0^*, \mathbf{A}_0) = \mathbf{0}.$$
$$J(\phi_0, \mathbf{A}_0) = J^*(\hat{v}_1^*, \hat{v}_0^*, \mathbf{A}_0).$$

Moreover, defining

$$J_1^*(v_1^*, \mathbf{A}) = \sup_{v_0^* \in E} J^*(v_1^*, v_0^*, \mathbf{A}),$$

for $K > 0$ such that

$$\frac{1}{\alpha} > \frac{8K_2^2}{K}$$

and sufficiently large, we have

$$\delta J_1^*(\hat{v}_1^*, \mathbf{A}_0) = \mathbf{0},$$

$$\frac{\partial^2 J_1^*(\hat{v}_1^*, \mathbf{A}_0)}{\partial(v_1^*)^2} > \mathbf{0}$$

so that there exist $r, r_1 > 0$ such that

$$
\begin{aligned}
J(\phi_0, \mathbf{A}_0) &= \min_{(\phi,\mathbf{A}) \in B_r(\phi_0, \mathbf{A}_0)} J(\phi, \mathbf{A}) \\
&= \inf_{v_1^* \in B_{r_1}(\hat{v}_1^*)} J_1^*(v_1^*, \mathbf{A}_0) \\
&= J_1^*(\hat{v}_1^*, \mathbf{A}_0) \\
&= \inf_{v_1^* \in B_{r_1}(\hat{v}_1^*)} \left\{ \sup_{v_0^* \in E} J^*(v_1^*, v_0^*, \mathbf{A}_0) \right\} \\
&= J^*(\hat{v}_1^*, \hat{v}_0^*, \mathbf{A}_0).
\end{aligned}
\tag{8.15}
$$

Proof 8.2   We start by proving that

$$\delta J^*(\hat{v}_1^*, \hat{v}_0^*, \mathbf{A}_0) = \mathbf{0}.$$

Observe that from

$$\frac{\partial J(\phi_0, \mathbf{A}_0)}{\partial \phi} = 0,$$

and

$$\frac{\partial J(\phi_0, \mathbf{A}_0)}{\partial \mathbf{A}} = 0,$$

we have

$$\begin{cases} \gamma |\nabla - i\rho \mathbf{A}_0|^2 \phi_0 + 2\alpha(|\phi_0|^2 - \beta)\phi_0 = 0, & \text{in } \Omega \\ (\nabla \phi_0 - i\rho \mathbf{A}_0 \phi_0) \cdot \mathbf{n} = 0, & \text{on } \partial\Omega, \end{cases} \quad (8.16)$$

and

$$\begin{cases} \text{curl (curl } \mathbf{A}_0) = \text{curl } \mathbf{B}_0 + 4\pi \tilde{J}_0, & \text{in } \Omega \\ \text{curl (curl } \mathbf{A}_0) = \text{curl } \mathbf{B}_0, & \text{in } \Omega_1 \setminus \overline{\Omega}, \end{cases} \quad (8.17)$$

where

$$\tilde{J}_0 = -2i\gamma\rho Im\left[(\phi_0^* \nabla \phi_0)\right] - \gamma\rho^2 |\phi_0|^2 \mathbf{A}_0.$$

Observe also that

$$\begin{aligned} \frac{\partial J^*(\hat{v}_1^*, \hat{v}_0^*, \mathbf{A}_0)}{\partial v_0^*} &= \frac{(\hat{v}_1^*)^2}{(2\hat{v}_0^* - K)^2} - \frac{\hat{v}_0^*}{\alpha} - \beta \\ &= |\phi_0|^2 - \frac{\hat{v}_0^*}{\alpha} - \beta = 0. \end{aligned} \quad (8.18)$$

Summarizing, we have got

$$\frac{\partial J^*(\hat{v}_1^*, \hat{v}_0^*, \mathbf{A}_0)}{\partial v_0^*} = 0.$$

Moreover, from the first line in equation (8.16), we obtain

$$\gamma |\nabla - i\rho \mathbf{A}_0|^2 \phi_0 + K\phi_0 + 2\alpha(|\phi_0|^2 - \beta)\phi_0 - K\phi_0 = 0, \text{ in } \Omega,$$

so that

$$\begin{aligned} \hat{v}_1^* &= (2\hat{v}_0^* - K)\phi_0 \\ &= 2\alpha(|\phi_0|^2 - \beta)\phi_0 - K\phi_0 \\ &= -\gamma |\nabla - i\rho \mathbf{A}_0|^2 \phi_0 - K\phi_0. \end{aligned} \quad (8.19)$$

Hence,

$$\phi_0 = \frac{(\hat{v}_1^*)}{2\hat{v}_0^* - K} = -\frac{\hat{v}_1^*}{\gamma |\nabla - i\rho \mathbf{A}_0|^2 + K},$$

and thus,

$$\begin{aligned} \frac{\partial J^*(\hat{v}_1^*, \hat{v}_0^*, \mathbf{A}_0)}{\partial v_1^*} &= -\frac{(\hat{v}_1^*)}{2\hat{v}_0^* - K} - \frac{\hat{v}_1^*}{\gamma |\nabla - i\rho \mathbf{A}_0|^2 + K} \\ &= -\phi_0 + \phi_0 = 0, \end{aligned} \quad (8.20)$$

Also, denoting

$$H_1 = \frac{\partial[\gamma|\nabla - i\rho\mathbf{A}_0|^2]}{\partial\mathbf{A}}\left[\frac{1}{2}\frac{(\hat{v}_1^*)^2}{(\gamma|\nabla - i\rho\mathbf{A}_0)|^2 + K)^2}\right] + \frac{1}{4\pi}\{\text{curl (curl }\mathbf{A}_0) - \text{curl }\mathbf{B}_0\}$$

$$= \frac{\partial[\gamma|\nabla - i\rho\mathbf{A}_0|^2]}{\partial\mathbf{A}}\left[\frac{|\phi_0|^2}{2}\right] + \frac{1}{4\pi}\{\text{curl (curl }\mathbf{A}_0) - \text{curl }\mathbf{B}_0\}$$

$$= \frac{1}{4\pi}\{\text{curl (curl }\mathbf{A}_0) - \text{curl }\mathbf{B}_0\} - \tilde{J}_0, \tag{8.21}$$

and

$$H_2 = \frac{1}{4\pi}\{\text{curl (curl }\mathbf{A}_0) - \text{curl }\mathbf{B}_0\},$$

we get

$$\frac{\partial J^*(\hat{v}_1^*, \hat{v}_0^*, \mathbf{A}_0)}{\partial\mathbf{A}} = \begin{cases} H_1 & \text{in } \Omega, \\ H_2, & \text{in } \Omega_1 \setminus \overline{\Omega}. \end{cases} \tag{8.22}$$

Summarizing,

$$\frac{\partial J^*(\hat{v}_1^*, \hat{v}_0^*, \mathbf{A}_0)}{\partial\mathbf{A}} = \mathbf{0}.$$

Such last results may be denoted by

$$\delta J^*(\hat{v}_1^*, \hat{v}_0^*, \mathbf{A}_0) = \mathbf{0}.$$

Recall now that

$$J_1^*(v_1^*, \mathbf{A}) = \sup_{v_0^* \in E} J^*(v_1^*, v_0^*, \mathbf{A}).$$

Thus,

$$\frac{\partial J_1^*(\hat{v}_1^*, \mathbf{A}_0)}{\partial v_1^*} = \frac{\partial J^*(\hat{v}_1^*, \hat{v}_0^*, \mathbf{A}_0)}{\partial v_1^*}$$

$$+ \frac{\partial J^*(\hat{v}_1^*, \hat{v}_0^*, \mathbf{A}_0)}{\partial v_0^*}\frac{\partial \hat{v}_0^*}{\partial v_1^*}$$

$$= \mathbf{0} \tag{8.23}$$

and

$$\frac{\partial J_1^*(\hat{v}_1^*, \mathbf{A}_0)}{\partial\mathbf{A}} = \frac{\partial J^*(\hat{v}_1^*, \hat{v}_0^*, \mathbf{A}_0)}{\partial\mathbf{A}} = \mathbf{0},$$

so that we may denote

$$\delta J_1^*(\hat{v}_1^*, \mathbf{A}_0) = \mathbf{0}.$$

Furthermore, we may easily compute,

$$J^*(\hat{v}_1^*, \hat{v}_0^*, \mathbf{A}_0) = -F^*(\hat{v}_1^*) + \hat{G}^*(\hat{v}_1^*, \hat{v}_0^*, \mathbf{A}_0)$$

$$= -\langle\phi_0, \hat{v}_1^*\rangle_{L^2} + F(\phi_0, \mathbf{A}_0) + \langle\phi_0, \hat{v}_1^*\rangle_{L^2} - \langle 0, \hat{v}_0^*\rangle_{L^2} - G(\phi_0, \mathbf{A}_0)$$

$$= J(\phi_0, \mathbf{A}_0). \tag{8.24}$$

Observe that, in particular, we have

$$J_1^*(\hat{v}_1^*, \mathbf{A}_0) = J^*(\hat{v}_1^*, \hat{v}_0^*, \mathbf{A}_0),$$

where the concerning supremum is attained through the equation

$$\frac{\partial J^*(\hat{v}_1^*, \hat{v}_0^*, \mathbf{A}_0)}{\partial v_0^*} = 0,$$

that is

$$\frac{(\hat{v}_1^*)^2}{(2\hat{v}_0^* - K)^2} - \frac{\hat{v}_0^*}{\alpha} - \beta$$

$$= |\phi_0|^2 - \frac{\hat{v}_0^*}{\alpha} - \beta = 0. \tag{8.25}$$

Taking the variation in v_1^* in such an equation, we get

$$\frac{2\hat{v}_1^*}{(2\hat{v}_0^* - K)^2} - \frac{4(\hat{v}_1^*)^2}{(2\hat{v}_0^* - K)^3} \frac{\partial \hat{v}_0^*}{\partial v_1^*} - \frac{1}{\alpha} \frac{\partial \hat{v}_0^*}{\partial v_1^*} = 0,$$

so that

$$\frac{\partial \hat{v}_0^*}{\partial v_1^*} = \frac{\frac{2\phi_0}{(2v_0^* - K)}}{\frac{1}{\alpha} + \frac{4|\phi_0|^2}{2\hat{v}_0^* - K}},$$

where, as previously indicated,

$$\phi_0 = \frac{\hat{v}_1^*}{2\hat{v}_0^* - K}.$$

At this point we observe that

$$\frac{\partial^2 J^*(\hat{v}_1^*, \mathbf{A}_0)}{\partial (v_1^*)^2}$$

$$= \frac{\partial^2 J^*(\hat{v}_1^*, \hat{v}_0^*, \mathbf{A}_0)}{\partial (v_1^*)^2}$$

$$+ \frac{\partial^2 J^*(\hat{v}_1^*, \hat{v}_0^*, \mathbf{A}_0)}{\partial v_1^* \partial v_0^*} \frac{\partial \hat{v}_0^*}{\partial v_1^*}$$

$$= -\frac{1}{\gamma |\nabla - i\rho \mathbf{A}_0|^2 + K} - \frac{1}{2\hat{v}_0^* - K}$$

$$+ \frac{\frac{4\alpha |\phi_0|^2}{(2\hat{v}_0^* - K)^2}}{\left[1 + \frac{4\alpha |\phi_0|^2}{2\hat{v}_0^* - K}\right]}$$

$$= \frac{-2\hat{v}_0^* - 4\alpha |\phi_0|^2 + K - \gamma |\nabla - i\rho \mathbf{A}_0|^2 - K}{(K + \gamma |\nabla - i\rho \mathbf{A}_0|^2)(2\hat{v}_0^* + 4\alpha |\phi_0|^2 - K)}$$

$$= \frac{-\delta_{\phi\phi}^2 J(\phi_0, \mathbf{A}_0)}{(K + \gamma |\nabla - i\rho \mathbf{A}_0|^2)(2\hat{v}_0^* + 4\alpha |\phi_0|^2 - K)}$$

$$> \mathbf{0}. \tag{8.26}$$

Summarizing,

$$\frac{\partial^2 J^*(\hat{v}_1^*, \mathbf{A}_0)}{\partial (v_1^*)^2} > \mathbf{0}.$$

From these last results, there exists $r, r_1 > 0$ such that

$$
\begin{aligned}
J(\phi_0, \mathbf{A}_0) &= \min_{(\phi, \mathbf{A}) \in B_r(\phi_0, \mathbf{A}_0)} J(\phi, \mathbf{A}) \\
&= \inf_{v_1^* \in B_{r_1}(\hat{v}_1^*)} J_1^*(v_1^*, \mathbf{A}_0) \\
&= J_1^*(\hat{v}_1^*, \mathbf{A}_0) \\
&= \inf_{v_1^* \in B_{r_1}(\hat{v}_1^*)} \left\{ \sup_{v_0^* \in E} J^*(v_1^*, v_0^*, \mathbf{A}_0) \right\} \\
&= J^*(\hat{v}_1^*, \hat{v}_0^*, \mathbf{A}_0).
\end{aligned}
\tag{8.27}
$$

The proof is complete.

Remark 8.1 At this point of our analysis and on, we consider a finite dimensional model version in a finite differences or finite elements context, even though the spaces and operators have not been relabeled. So, also in such a context, the expression

$$\int_\Omega \frac{(v_1^*)^2}{2v_0^* - K} dx,$$

indeed means

$$(v_1^*)^T (2v_0^* - K I_d)^{-1} v_1^*$$

where I_d denotes the identity matrix $n \times n$ and

$$2v_0^* - K I_d$$

denotes the diagonal matrix with the vector

$$\{2v_0^*(i) - K\}_{n \times 1}$$

as diagonal, for some appropriate $n \in \mathbb{N}$ defined in the discretization process. ■

8.5 A second duality principle

In this section, we present another duality principle, which is summarized by the next theorem.

Theorem 8.5.1 *Let $\Omega \subset \mathbb{R}^3$ be an open, bounded and connected set with a regular (Lipschitzian) boundary denoted by $\partial \Omega$.*
Let $J : U \to \mathbb{R}$ be a functional defined by

$$J(u) = G(\Lambda u) - F(\Lambda u) - \langle u, f \rangle_U, \ \forall u \in U,$$

where

$$U = W_0^{1,2}(\Omega), \ f \in L^2(\Omega), Y = Y^* = L^2,$$

$\Lambda : U \to Y$, *is a bounded linear operator which the respective adjoint is denoted by* $\Lambda^* : Y^* \to U^*$.

Suppose also $(G \circ \Lambda) : U \to \mathbb{R}$ *and* $(F \circ \Lambda) : U \to \mathbb{R}$ *are Fréchet differentiable functionals on* U *and such that* J *is bounded below.*

Define $G_K^* : Y^* \to \mathbb{R}$ *and* $F_K^* : Y^* \to \mathbb{R}$ *by*

$$G_K^*(v^*) = \sup_{v \in Y} \left\{ \langle v, v^* \rangle_{L^2} - G(v) - \frac{K}{2} \langle v, v \rangle_{L^2} \right\},$$

$$F_K^*(z^*) = \sup_{v \in Y_1} \left\{ \langle v, z^* \rangle_{L^2} - F(v) - \frac{K}{2} \langle v, v \rangle_{L^2} \right\}.$$

Assume $(\hat{v}^*, \hat{z}^*, u_0) \in Y^* \times Y^* \times U$ *is such that*

$$\delta J^*(\hat{v}^*, \hat{z}^*, u_0) = \mathbf{0},$$

where $J^* : Y^* \times Y^* \times U \to \mathbb{R}$ *is defined by*

$$J^*(v^*, z^*, u) = -G_K^*(v^*) + F_K^*(z^*) + \langle u, \Lambda^* v^* - \Lambda^* z^* - f \rangle_U.$$

Suppose also $\delta^2 J(u_0) > \mathbf{0}$ *and* $K > 0$ *is sufficiently big so that*

$$G_K^{**}(\Lambda u_0) = G_K(\Lambda u_0),$$

and

$$F_K^{**}(\Lambda u_0) = F_K(\Lambda u_0).$$

Denote also

$$A^+ = \{u \in U \ : \ u u_0 \geq 0, \ a.e. \ in \ \Omega\},$$

$$B^+ = \{u \in U \ : \ \delta^2 J(u) \geq \mathbf{0}\},$$

where we assume, there exists a linear function (in $|u|$*)* H *such that*

$$\delta^2 J(u) \geq \mathbf{0} \ if, \ and \ only \ if, \ H(|u|) \geq \mathbf{0}.$$

Moreover, defining the set

$$E = A^+ \cap B^+,$$

we have that E *is convex,*

$$\delta J(u_0) = \mathbf{0},$$

so that

$$
\begin{aligned}
J(u_0) &= \inf_{u \in E} J(u) \\
&= \inf_{u \in E}\left\{ \inf_{z^* \in Y^*}\left\{ \sup_{v^* \in Y^*} J^*(v^*, z^*, u) \right\}\right\} \\
&= \inf_{u \in E}\left\{ \sup_{v^* \in Y^*}\left\{ \inf_{z^* \in Y^*} J^*(v^*, z^*, u) \right\}\right\} \\
&= J^*(\hat{v}^*, \hat{z}^*, u_0).
\end{aligned}
\tag{8.28}
$$

Proof 8.3   Observe that

$$
\begin{aligned}
J^*(v^*, z^*, u) &= -G_K^*(v^*) + F_K^*(z^*) + \langle u, \Lambda^* v^* - \Lambda_1^* z^* - f \rangle_U \\
&\leq -\langle \Lambda u, v^* \rangle_Y + G(\Lambda u) + \frac{K}{2}\langle \Lambda u, \Lambda u \rangle_Y \\
&\quad + F_K^*(z^*) + \langle u, \Lambda^* v^* - \Lambda^* z^* - f \rangle_U \\
&= G(\Lambda u) + \frac{K}{2}\langle \Lambda u, \Lambda u \rangle_Y - \langle \Lambda u, z^* \rangle_{Y^*} + F_K^*(z^*) - \langle u, f \rangle_{L^2}, \quad (8.29)
\end{aligned}
$$

$\forall u \in U$, $v^* \in Y^*$, $z^* \in Y^*$.
   From this, we obtain

$$
\begin{aligned}
&\inf_{z^* \in Y^*}\left\{ \sup_{v^* \in Y^*} J^*(v^*, z^*, u) \right\} \\
&\leq \inf_{z^* \in Y^*}\left\{ G(\Lambda u) + \frac{K}{2}\langle \Lambda u, \Lambda u \rangle_Y - \langle \Lambda u, z^* \rangle_Y + F_K^*(z^*) - \langle u, f \rangle_U \right\} \\
&= G(\Lambda u) + \frac{K}{2}\langle \Lambda u, \Lambda u \rangle_Y \\
&\quad -F(\Lambda u) - \frac{K}{2}\langle \Lambda u, \Lambda u \rangle_Y \\
&\quad -\langle u, f \rangle_U \\
&= J(u).
\end{aligned}
\tag{8.30}
$$

Hence, we may infer that

$$
\begin{aligned}
&\inf_{u \in E}\left\{ \inf_{z^* \in Y_1^*}\left\{ \sup_{v^* \in Y^*} J^*(v^*, z^*, u) \right\}\right\} \\
&\leq \inf_{u \in E} J(u).
\end{aligned}
\tag{8.31}
$$

On the other hand, from $\delta J^*(\hat{v}^*, \hat{z}^*, u_0) = \mathbf{0}$, we have

$$
\frac{\partial G_K^*(\hat{v}^*)}{\partial v^*} - \Lambda u_0 = 0,
$$

so that from the Legendre transform properties, we obtain

$$
\hat{v}^* = \frac{\partial G_K(\Lambda u_0)}{\partial v}
$$

and

$$G_K^*(\hat{v}^*) = \langle \Lambda u_0, \hat{v}^* \rangle_{Y^*} - G_K(\Lambda u_0).$$

Similarly, from the variation in z^*, we get

$$\frac{\partial F_K^*(\hat{z}^*)}{\partial z^*} - \Lambda u_0 = 0,$$

so that from the Legendre transform properties, we obtain

$$\hat{z}^* = \frac{\partial F_K(\Lambda u_0)}{\partial v}$$

and

$$F_K^*(\hat{z}^*) = \langle \Lambda u_0, \hat{z}^* \rangle_{Y^*} - F_K(\Lambda u_0).$$

From the variation in u, we have

$$\Lambda \hat{v}^* - \Lambda^* \hat{z}^* - f = 0.$$

Joining the pieces, we have got

$$
\begin{aligned}
J^*(\hat{v}^*, \hat{z}^*, u_0) &= -G_K^*(\hat{v}^*) + F_K^*(\hat{z}^*) \\
&= -\langle \Lambda u_0, \hat{v}^* \rangle_Y + G_K(\Lambda u_0) - F_K(\Lambda u_0) + \langle \Lambda u_0, \hat{z}^* \rangle_Y \\
&= G(\Lambda u_0) - F(\Lambda u_0) - \langle u_0, f \rangle_U \\
&= J(u_0).
\end{aligned}
\tag{8.32}
$$

Moreover, from

$$\Lambda \hat{v}^* - \Lambda^* \hat{z}^* - f = 0,$$

we also have

$$\Lambda^* \left(\frac{\partial G(\Lambda u_0)}{\partial v} \right) - \Lambda^* \left(\frac{\partial F(\Lambda u_0)}{\partial v} \right) - f = 0,$$

so that

$$\delta J(u_0) = \mathbf{0}.$$

Finally, observe that if $u_1, u_2 \in A^+ \cap B^+ = E$ and $\lambda \in [0, 1]$, then

$$H(|u_1|) \geq \mathbf{0},$$

$$H(|u_2|) \geq \mathbf{0}$$

and also since

$$\text{sign } u_1 = \text{sign } u_2, \text{ in } \Omega,$$

we get

$$|\lambda u_1 + (1 - \lambda) u_2| = \lambda |u_1| + (1 - \lambda)|u_2|,$$

so that, from the hypotheses on H,

$$\lambda H(|u_1|) + (1 - \lambda) H(|u_2|) = H(|\lambda u_1 + (1 - \lambda) u_2|) \geq \mathbf{0}$$

and thus,

$$\delta^2 J(\lambda u_1 + (1-\lambda)u_2) \geq \mathbf{0}.$$

From this, we may infer that E is convex.

Moreover, since J is convex in E, and

$$\delta J(u_0) = \mathbf{0},$$

from (8.31) and (8.32), we have that

$$
\begin{aligned}
J(u_0) &= \inf_{u \in E} J(u) \\
&= \inf_{u \in E} \left\{ \inf_{z^* \in Y^*} \left\{ \sup_{v^* \in Y^*} J^*(v^*, z^*, u) \right\} \right\} \\
&= \inf_{u \in E} \left\{ \sup_{v^* \in Y^*} \left\{ \inf_{z^* \in Y^*} J^*(v^*, z^*, u) \right\} \right\} \\
&= J^*(\hat{v}^*, \hat{z}^*, u_0). \quad\quad (8.33)
\end{aligned}
$$

The proof is complete.

8.6 A third duality principle

Our third duality principle is summarized by the next theorem.

Theorem 8.6.1 *Let $\Omega \subset \mathbb{R}^3$ be an open, bounded and connected set with a regular (Lipschitzian) boundary denoted by $\partial\Omega$. Consider the functional $J : U \to \mathbb{R}$ be defined by*

$$J(u) = \frac{\gamma}{2} \int_\Omega \nabla u \cdot \nabla u \, dx + \frac{\alpha}{2} \int_\Omega (u^2 - \beta)^2 \, dx - \langle u, f \rangle_{L^2},$$

where α, β, γ are positive real constants, $U = W_0^{1,2}(\Omega)$, $f \in L^2(\Omega)$. Here, we assume

$$-\gamma\nabla^2 - 2\alpha\beta < 0$$

in an appropriate matrix sense considering, as above indicated, a finite dimensional not relabeled model approximation, in a finite differences or finite elements context.

Define $F : U \to \mathbb{R}$ and $G : U \to \mathbb{R}$, where

$$F(u) = \frac{\gamma}{2} \int_\Omega \nabla u \cdot \nabla u \, dx + \frac{K}{2} \int_\Omega u^2 \, dx,$$

and

$$G(u, v) = -\frac{\alpha}{2} \int_\Omega (u^2 - \beta + v)^2 \, dx + \frac{K}{2} \int_\Omega u^2 \, dx + \langle u, f \rangle_{L^2}$$

so that

$$J(u) = F(u) - G(u,0).$$

Define also,

$$
\begin{aligned}
F^*(v_1^*) &= \sup_{u \in U}\{\langle u, v_1^* \rangle_{L^2} - F(u)\} \\
&= \frac{1}{2}\int_\Omega \frac{(v_1^*)^2}{K - \gamma\nabla^2}\,dx
\end{aligned}
\tag{8.34}
$$

and

$$
\begin{aligned}
G^*(v_1^*, v_0^*) &= \sup_{u \in U}\inf_{v \in L^2}\{\langle u, v_1^* \rangle_{L^2} - \langle v, v_0^* \rangle_{L^2} - G(u,v)\} \\
&= -\frac{1}{2}\int_\Omega \frac{(v_1^* - f)^2}{2v_0^* - K}\,dx - \frac{1}{2\alpha}\int_\Omega (v_0^*)^2\,dx \\
&\quad -\beta\int_\Omega v_0^*\,dx,
\end{aligned}
\tag{8.35}
$$

if $-2v_0^* + K > 0$ in $\overline{\Omega}$.
Furthermore, denote

$$C = \{v_0^* \in C(\overline{\Omega}) \text{ such that } -2v_0^* + K > K/2, \text{ in } \overline{\Omega}\},$$

where $K > 0$ is such that

$$\frac{1}{\alpha} > \frac{8K_2^2}{K},$$

for some appropriate $K_2 > 0$.
   At this point suppose $u_0 \in U$ is such that $\delta J(u_0) = 0$, $\|\phi_0\|_\infty \le K_2$ and

$$\delta^2 J(u_0) > \mathbf{0}.$$

Define also,

$$A^+ = \{u \in U \ : \ uu_0 \ge 0, \text{ in } \Omega\},$$
$$B^+ = \{u \in U \ : \ \delta^2 J(u) \ge \mathbf{0}\},$$
$$E = A^+ \cap B^+,$$
$$\hat{v}_0^* = \alpha(u_0^2 - \beta),$$
$$\hat{v}_1^* = -\gamma\nabla^2 u_0 + Ku_0.$$

Under such hypothesis, assuming also $\hat{v}_0^* \in C$ and denoting

$$J^*(v_1, v_0^*) = F^*(v_1^*) - G^*(v_1^*, v_0^*),$$

$$J_1^*(v_1^*) = \sup_{v_0^* \in B} J^*(v_1^*, v_0^*),$$

we have that there exists $r > 0$ such that

$$
\begin{aligned}
J(u_0) &= \inf_{u \in E} J(u) \\
&= \inf_{v_1^* \in B_r(\hat{v}_1^*)} J_1^*(v_1^*) \\
&= J_1^*(\hat{v}_1^*) \\
&= \inf_{v_1^* \in B_r(\hat{v}_1^*)} \left\{ \sup_{v_0^* \in C} J^*(v_1^*, v_0^*) \right\} \\
&= J^*(\hat{v}_1^*, \hat{v}_0^*).
\end{aligned}
\tag{8.36}
$$

Proof 8.4   We start by proving that

$$
\delta J^*(\hat{v}_1^*, \hat{v}_0^*) = \mathbf{0}.
$$

Observe that from

$$
\frac{\partial J(u_0)}{\partial u} = 0,
$$

we have

$$
-\gamma \nabla^2 u_0 + 2\alpha(|u_0|^2 - \beta)u_0 - f = 0, \text{ in } \Omega.
$$

Observe also that

$$
\begin{aligned}
\frac{\partial J^*(\hat{v}_1^*, \hat{v}_0^*)}{\partial v_0^*} &= \frac{(\hat{v}_1^*)^2}{(2\hat{v}_0^* - K)^2} - \frac{\hat{v}_0^*}{\alpha} - \beta \\
&= |u_0|^2 - \frac{\hat{v}_0^*}{\alpha} - \beta = 0.
\end{aligned}
\tag{8.37}
$$

Summarizing, we have

$$
\frac{\partial J^*(\hat{v}_1^*, \hat{v}_0^*)}{\partial v_0^*} = 0.
$$

Moreover, from the first line in equation (8.16), we obtain

$$
-\gamma \nabla^2 u_0 + K u_0 + 2\alpha(|u_0|^2 - \beta)u_0 - K u_0 - f = 0, \text{ in } \Omega,
$$

so that

$$
\begin{aligned}
\hat{v}_1^* &= -\gamma \nabla^2 u_0 + K u_0 \\
&= -2\alpha(|u_0|^2 - \beta)u_0 + K u_0 + f.
\end{aligned}
\tag{8.38}
$$

Hence,

$$
u_0 = -\frac{(\hat{v}_1^* - f)}{2\hat{v}_0^* - K} = \frac{\hat{v}_1^*}{-\gamma \nabla^2 + K},
$$

and thus,

$$
\begin{aligned}
\frac{\partial J^*(\hat{v}_1^*, \hat{v}_0^*)}{\partial v_1^*} &= -\frac{(\hat{v}_1^* - f)}{2\hat{v}_0^* - K} - \frac{\hat{v}_1^*}{-\gamma \nabla^2 + K} \\
&= u_0 - u_0 = 0.
\end{aligned}
\tag{8.39}
$$

Such last results may be denoted by

$$\delta J^*(\hat{v}_1^*, \hat{v}_0^*) = \mathbf{0}.$$

Recall now that

$$J_1^*(v_1^*) = \sup_{v_0^* \in E} J^*(v_1^*, v_0^*).$$

Thus,

$$
\begin{aligned}
\frac{\partial J_1^*(\hat{v}_1^*)}{\partial v_1^*} &= \frac{\partial J^*(\hat{v}_1^*, \hat{v}_0^*)}{\partial v_1^*} \\
&\quad + \frac{\partial J^*(\hat{v}_1^*, \hat{v}_0^*)}{\partial v_0^*} \frac{\partial \hat{v}_0^*}{\partial v_1^*} \\
&= \mathbf{0}.
\end{aligned}
\tag{8.40}
$$

Furthermore, we may easily compute,

$$
\begin{aligned}
J^*(\hat{v}_1^*, \hat{v}_0^*) &= -F^*(\hat{v}_1^*) + \hat{G}^*(\hat{v}_1^*, \hat{v}_0^*) \\
&= -\langle u_0, \hat{v}_1^* \rangle_{L^2} + F(\phi_0) + \langle u_0, \hat{v}_1^* \rangle_{L^2} - \langle 0, \hat{v}_0^* \rangle_{L^2} - G(u_0, \mathbf{0}) \\
&= J(u_0).
\end{aligned}
\tag{8.41}
$$

Observe that, in particular, we have

$$J_1^*(\hat{v}_1^*) = J^*(\hat{v}_1^*, \hat{v}_0^*),$$

where the concerning supremum is attained through the equation

$$\frac{\partial J^*(\hat{v}_1^*, \hat{v}_0^*)}{\partial v_0^*} = 0,$$

that is

$$
\begin{aligned}
& \frac{(\hat{v}_1^* - f)^2}{(2\hat{v}_0^* - K)^2} - \frac{\hat{v}_0^*}{\alpha} - \beta \\
&= |\phi_0|^2 - \frac{\hat{v}_0^*}{\alpha} - \beta = 0.
\end{aligned}
\tag{8.42}
$$

Taking the variation in v_1^* in such an equation, we get

$$\frac{2(\hat{v}_1^* - f)}{(2\hat{v}_0^* - K)^2} - \frac{4(\hat{v}_1^*)^2}{(2\hat{v}_0^* - K)^3} \frac{\partial \hat{v}_0^*}{\partial v_1^*} - \frac{1}{\alpha} \frac{\partial \hat{v}_0^*}{\partial v_1^*} = 0,$$

so that

$$\frac{\partial \hat{v}_0^*}{\partial v_1^*} = \frac{\frac{2u_0}{(2\hat{v}_0^* - K)}}{\frac{1}{\alpha} + \frac{4|u_0|^2}{2\hat{v}_0^* - K}},$$

where, as previously indicated,

$$u_0 = \frac{\hat{v}_1^* - f}{2\hat{v}_0^* - K}.$$

At this point we observe that

$$
\begin{aligned}
\frac{\partial^2 J^*(\hat{v}_1^*)}{\partial (v_1^*)^2} &= \\
&= \frac{\partial^2 J^*(\hat{v}_1^*, \hat{v}_0^*)}{\partial (v_1^*)^2} \\
&\quad + \frac{\partial^2 J^*(\hat{v}_1^*, \hat{v}_0^*)}{\partial v_1^* \partial v_0^*} \frac{\partial \hat{v}_0^*}{\partial v_1^*} \\
&= -\frac{1}{-\gamma \nabla^2 + K} - \frac{1}{2\hat{v}_0^* - K} \\
&\quad + \frac{\frac{4\alpha |u_0|^2}{(2\hat{v}_0^* - K)^2}}{\left[1 + \frac{4\alpha |u_0|^2}{2\hat{v}_0^* - K}\right]} \\
&= \frac{-2\hat{v}_0^* - 4\alpha |u_0|^2 + K + \gamma \nabla^2 - K}{(K - \gamma \nabla^2)(2\hat{v}_0^* + 4\alpha |u_0|^2 - K)} \\
&= \frac{-\delta^2 J(u_0)}{(K - \gamma \nabla^2)(2\hat{v}_0^* + 4\alpha |u_0|^2 - K)} \\
&> 0.
\end{aligned}
$$

(8.43)

Summarizing,

$$\frac{\partial^2 J^*(\hat{v}_1^*)}{\partial (v_1^*)^2} > 0.$$

Finally, observe that

$$\delta^2 J(u) = -\gamma \nabla^2 + 6\alpha u^2 - 2\alpha \beta \geq 0,$$

if, and only if,

$$H(u) \geq 0,$$

where

$$H(u) = \sqrt{6\alpha}|u| - \sqrt{\gamma \nabla^2 + 2\alpha \beta} \geq 0.$$

Hence, if $u_1, u_2 \in A^+ \cap B^+ = E$ and $\lambda \in [0, 1]$, then

$$H(|u_1|) \geq 0,$$

$$H(|u_2|) \geq 0$$

and also since

$$\text{sign } u_1 = \text{sign } u_2, \text{ in } \Omega,$$

we get

$$|\lambda u_1 + (1-\lambda)u_2| = \lambda |u_1| + (1-\lambda)|u_2|,$$

so that,

$$H(|\lambda u_1 + (1-\lambda)u_2|) = H(\lambda |u_1| + (1-\lambda)|u_2|) = \lambda H(|u_1|) + (1-\lambda)H(|u_2|) \geq 0$$

and thus,

$$\delta^2 J(\lambda u_1 + (1-\lambda)u_2) \geq 0.$$

From this, we may infer that E is convex.

From these last results, there exists $r > 0$ such that

$$
\begin{aligned}
J(u_0) &= \inf_{u \in E} J(u) \\
&= \inf_{v_1^* \in B_r(\hat{v}_1^*)} J_1^*(v_1^*) \\
&= J_1^*(\hat{v}_1^*) \\
&= \inf_{v_1^* \in B_r(\hat{v}_1^*)} \left\{ \sup_{v_0^* \in E} J^*(v_1^*, v_0^*) \right\} \\
&= J^*(\hat{v}_1^*, \hat{v}_0^*).
\end{aligned}
\tag{8.44}
$$

The proof is complete.

8.7 A criterion for global optimality

In this section, we establish a criterion for global optimality.

Theorem 8.7.1 *Let $\Omega \subset \mathbb{R}^3$ be an open, bounded and connected set with a regular (Lipschitzian) boundary denoted by $\partial \Omega$.*

Consider the functional $J : U \to \mathbb{R}$ where

$$
\begin{aligned}
J(u) &= \frac{\gamma}{2} \int_{\Omega} \nabla u \cdot \nabla u \, dx + \frac{\alpha}{2} \int_{\Omega} (u^2 - \beta)^2 \, dx \\
&\quad - \langle u, f \rangle_{L^2}
\end{aligned}
\tag{8.45}
$$

where $\alpha > 0$, $\beta > 0$, $\gamma > 0$, $f \in C^1(\overline{\Omega})$ and $U = W_0^{1,2}(\Omega)$. Suppose also either

$$f(x) > 0, \ \forall x \in \Omega$$

or

$$f(x) < 0, \ \forall x \in \Omega.$$

Here, we assume

$$-\gamma \nabla^2 - 2\alpha \beta < 0$$

in an appropriate matrix sense considering, as above indicated, a finite dimensional not relabeled model approximation, in a finite differences or finite elements context.

Define

$$A^+ = \{u \in U \ : \ uf \geq 0, \ in \ \overline{\Omega}\},$$

$$B^+ = \{u \in U \ : \ \delta^2 J(u) \geq \mathbf{0}\}$$

and

$$E = A^+ \cap B^+.$$

Under such hypotheses, E is convex and

$$\inf_{u \in E} J(u) = \inf_{u \in U} J(u).$$

Proof 8.5   Define

$$\eta = \inf_{u \in U} J(u).$$

Let $\varepsilon > 0$

Hence, by density there exists $u_\varepsilon \in C_c^1(\Omega)$ such that

$$\delta^2 J(u_\varepsilon) \geq \mathbf{0}$$

and

$$\eta \leq J(u_\varepsilon) < \eta + \varepsilon.$$

Define

$$v_\varepsilon(x) = \begin{cases} u_\varepsilon(x), & \text{if } u_\varepsilon(x)f(x) \geq 0, \\ -u_\varepsilon(x), & \text{if } u_\varepsilon(x)f(x) < 0, \end{cases} \tag{8.46}$$

$\forall x \in \overline{\Omega}$.

Observe that

$$\delta^2 J(v_\varepsilon) = \delta^2 J(u_\varepsilon) \geq 0$$

and

$$
\begin{aligned}
J(v_\varepsilon) &= \frac{\gamma}{2} \int_\Omega \nabla v_\varepsilon \cdot \nabla v_\varepsilon \, dx + \frac{\alpha}{2} \int_\Omega (v_\varepsilon^2 - \beta)^2 \, dx \\
&\quad - \langle v_\varepsilon, f \rangle_{L^2} \\
&\leq \frac{\gamma}{2} \int_\Omega \nabla u_\varepsilon \cdot \nabla u_\varepsilon \, dx + \frac{\alpha}{2} \int_\Omega (u_\varepsilon^2 - \beta)^2 \, dx \\
&\quad - \langle u_\varepsilon, f \rangle_{L^2} \\
&= J(u_\varepsilon). \tag{8.47}
\end{aligned}
$$

Hence

$$\eta \leq J(v_\varepsilon) \leq J(u_\varepsilon) < \eta + \varepsilon.$$

From this, since $v_\varepsilon \in E$, we obtain

$$\eta \leq \inf_{u \in E} J(u) < \eta + \varepsilon.$$

Since $\varepsilon > 0$ is arbitrary, we may infer that

$$\inf_{u \in U} J(u) = \eta = \inf_{u \in E} J(u).$$

Finally, observe also that

$$\delta^2 J(u) = -\gamma \nabla^2 + 6\alpha u^2 - 2\alpha \beta \geq \mathbf{0},$$

if, and only if

$$H(u) \geq \mathbf{0},$$

where

$$H(u) = \sqrt{6\alpha}|u| - \sqrt{\gamma \nabla^2 + 2\alpha \beta} \geq \mathbf{0}.$$

Hence, if $u_1, u_2 \in A^+ \cap B^+ = E$ and $\lambda \in [0, 1]$, then

$$H(|u_1|) \geq \mathbf{0},$$

$$H(|u_2|) \geq \mathbf{0}$$

and also since

$$\operatorname{sign} u_1 = \operatorname{sign} u_2, \text{ in } \Omega,$$

we get

$$|\lambda u_1 + (1 - \lambda)u_2| = \lambda |u_1| + (1 - \lambda)|u_2|,$$

so that,

$$H(|\lambda u_1 + (1 - \lambda)u_2|) = H(\lambda |u_1| + (1 - \lambda)|u_2|) = \lambda H(|u_1|) + (1 - \lambda)H(|u_2|) \geq \mathbf{0}$$

and thus,

$$\delta^2 J(\lambda u_1 + (1 - \lambda)u_2) \geq \mathbf{0}.$$

From this, we may infer that E is convex.
The proof is complete.

8.7.1   *The concerning duality principle*

In this section, we develop a duality principle concerning the last optimality criterion established.

Theorem 8.7.2 *Let $\Omega \subset \mathbb{R}^3$ be an open, bounded and connected set with a regular (Lipschitzian) boundary denoted by $\partial \Omega$. Consider the functional $J : U \to \mathbb{R}$ be defined by*

$$J(u) = \frac{\gamma}{2} \int_{\Omega} \nabla u \cdot \nabla u \, dx + \frac{\alpha}{2} \int_{\Omega} (u^2 - \beta)^2 \, dx - \langle u, f \rangle_{L^2},$$

where α, β, γ are positive real constants, $U = W_0^{1,2}(\Omega)$, $f \in C^1(\overline{\Omega})$ and we also denote $Y = Y^* = L^2(\Omega)$.

Here we assume

$$-\gamma\nabla^2 - 2\alpha\beta < \mathbf{0}$$

in an appropriate matrix sense considering, as above indicated, a finite dimensional not relabeled model approximation, in a finite differences or finite elements context.

Suppose also either

$$f(x) > 0, \ \forall x \in \Omega$$

or

$$f(x) < 0, \ \forall x \in \Omega.$$

Given $u \in U$, define

$$L_1(u) = \sup_{v_0^* \in Y^*} \left\{ \int_\Omega v_0 u^2 \, dx - \frac{1}{2\alpha} \int_\Omega (v_0^*)^2 - \beta \int_\Omega v_0^* \, dx \right\},$$

and

$$L_2(u) = \sup_{v_0^* \in B^*} \left\{ \int_\Omega v_0 u^2 \, dx - \frac{1}{2\alpha} \int_\Omega (v_0^*)^2 - \beta \int_\Omega v_0^* \, dx \right\},$$

where

$$B^* = \{v_0^* \in Y^* \ : \ -2v_0^* + K > K/2 \ in \ \overline{\Omega}\}$$

for some $K > 0$ to be specified.

Let

$$U_1 = \{u \in U \ such \ that \ L_1(u) = L_2(u) \ and \ \|u\|_{1,\infty} \le \sqrt[4]{K}\}.$$

Moreover, define $F : U \to \mathbb{R}$ and $G : U \to \mathbb{R}$, where

$$F(u) = \frac{\gamma}{2} \int_\Omega \nabla u \cdot \nabla u \, dx + \frac{K}{2} \int_\Omega u^2 \, dx,$$

and

$$G(u,v) = -\frac{\alpha}{2} \int_\Omega (u^2 - \beta + v)^2 \, dx + \frac{K}{2} \int_\Omega u^2 \, dx + \langle u, f \rangle_{L^2}$$

so that

$$J(u) = F(u) - G(u,0).$$

Define also,

$$B_1^* = \{v_1^* \in Y^* \ : \ \|v_1^* - f\|_\infty \le K_2\},$$

where $K_2 > 0$ and $K > 0$ are such that

$$-\frac{32K_2^2}{K^3} + \frac{1}{\alpha} > 0,$$

$$C_1 = \left\{ v_1^* \in B_1^* : \text{ there exists } u \in U_1 \right.$$
$$\left. \text{such that } v_1^* = \frac{\partial F(u)}{\partial u} \right\}, \tag{8.48}$$

$$C_2 = \left\{ v_1^* \in B_1^* : \text{ there exists } u \in U_1 \right.$$
$$\left. \text{such that } v_1^* = \frac{\partial G(u,0)}{\partial u} \right\}, \tag{8.49}$$

and
$$C^* = C_1 \cap C_2.$$

Furthermore, define $F^* : C^* \to \mathbb{R}$ *by*

$$F^*(v_1^*) = \sup_{u \in U_1} \left\{ \langle u, v_1^* \rangle_{L^2} - F(u) \right\}$$

$$= \frac{1}{2} \int_\Omega \frac{(v_1^*)^2}{K - \gamma \nabla^2} \, dx \tag{8.50}$$

and $G^* : C^* \times B^* \to \mathbb{R}$ *by*

$$G^*(v_1^*, v_0^*) = \sup_{u \in U} \inf_{v \in L^2} \left\{ \langle u, v_1^* \rangle_{L^2} - \langle v, v_0^* \rangle_{L^2} - G(u,v) \right\}$$

$$= -\frac{1}{2} \int_\Omega \frac{(v_1^* - f)^2}{2v_0^* - K} \, dx - \frac{1}{2\alpha} \int_\Omega (v_0^*)^2 \, dx$$

$$- \beta \int_\Omega v_0^* \, dx. \tag{8.51}$$

Define also,
$$A^+ = \{ u \in U : uf \geq 0, \text{ in } \overline{\Omega} \},$$
$$B^+ = \{ u \in U : \delta^2 J(u) \geq \mathbf{0} \},$$
$$E = A^+ \cap B^+,$$

and
$$E_1 = E \cap U_1.$$

Moreover, define
$$\hat{v}_0^* = \alpha(u_0^2 - \beta),$$
$$\hat{v}_1^* = (2\hat{v}_0^* - K)u_0 + f$$

and assume $u_0 \in U$ *is such that* $\delta J(u_0) = \mathbf{0}$, *and*

$$u_0 \in E_1,$$

Under such hypothesis, assuming also $\hat{v}_0^* \in B^*$, $\hat{v}_1^* \in C^*$ *and denoting*

$$J^*(v_1, v_0^*) = -F^*(v_1^*) + G^*(v_1^*, v_0^*),$$

we have

$$
\begin{aligned}
J(u_0) &= \inf_{u \in E_1} J(u) \\
&= \inf_{u \in U} J(u) \\
&= \inf_{v_1^* \in C^*} \left\{ \sup_{v_0^* \in B^*} J^*(v_1^*, v_0^*) \right\} \\
&= J^*(\hat{v}_1^*, \hat{v}_0^*).
\end{aligned}
\tag{8.52}
$$

Proof 8.6   Define

$$
\eta = \inf_{u \in U} J(u).
$$

Hence

$$
\begin{aligned}
\eta &\leq J(u) \\
&\leq -\langle u, v_1^* \rangle_{L^2} + F(u) \\
&\quad + \sup_{u \in U_1} \{ \langle u, v_1^* \rangle_{L^2} - G(u, 0) \} \\
&= -\langle u, v_1^* \rangle_{L^2} + F(u) \\
&\quad + \sup_{u \in U_1} \left\{ \sup_{v_0^* \in B^*} \left\{ \langle u, v_1^* \rangle_{L^2} + \int_\Omega v_0^* u^2 \, dx - \frac{K}{2} \int_\Omega u^2 \, dx - \frac{1}{2\alpha} \int_\Omega (v_0^*)^2 \, dx - \beta \int_\Omega v_0^* \, dx \right\} \right\},
\end{aligned}
$$

$\forall u \in U_1, v_1^* \in C^*$.

Thus,

$$
\begin{aligned}
\eta &\leq J(u) \\
&\leq -\langle u, v_1^* \rangle_{L^2} + F(u) \\
&\quad + \sup_{u \in U_1} \{ \langle u, v_1^* \rangle_{L^2} - G(u, 0) \} \\
&= -\langle u, v_1^* \rangle_{L^2} + F(u) \\
&\quad + \sup_{v_0^* \in B^*} \left\{ \sup_{u \in U_1} \left\{ \langle u, v_1^* \rangle_{L^2} + \int_\Omega v_0^* u^2 \, dx - \frac{K}{2} \int_\Omega u^2 \, dx - \frac{1}{2\alpha} \int_\Omega (v_0^*)^2 \, dx - \beta \int_\Omega v_0^* \, dx \right\} \right\} \\
&= -\langle u, v_1^* \rangle_{L^2} + F(u) \\
&\quad + \sup_{v_0^* \in B^*} G^*(v_1^*, v_0^*),
\end{aligned}
$$

$\forall u \in U_1, v_1^* \in C^*$.

From this, we obtain

$$
\begin{aligned}
\eta &\leq \inf_{u \in U_1} \{ -\langle u, v_1^* \rangle_{L^2} + F(u) \} \\
&\quad + \sup_{v_0^* \in B^*} G^*(v_1^*, v_0^*) \\
&= -F^*(v_1^*) + \sup_{v_0^* \in B^*} G^*(v_1^*, v_0^*) \\
&= \sup_{v_0^* \in B^*} J^*(v_1^*, v_0^*),
\end{aligned}
\tag{8.53}
$$

$\forall v_1^* \in C^*$.

Summarizing, we have got

$$\inf_{u \in U} J(u) \leq \inf_{v_1^* \in C^*} \left\{ \sup_{v_0^* \in B^*} J^*(v_1^*, v_0^*) \right\}. \tag{8.54}$$

Similarly as in the proof of the Theorem 8.6.1, we may obtain

$$\delta J^*(\hat{v}_1^*, \hat{v}_0^*) = \mathbf{0},$$

$$J^*(\hat{v}_1^*, \hat{v}_0^*) = J(u_0),$$

and

$$J^*(\hat{v}_1^*, \hat{v}_0^*) = \sup_{v_0^* \in B^*} J^*(\hat{v}_1^*, v_0^*).$$

From the proof of Theorem 8.7.1 we may infer that E is convex.

From this, since $u_0 \in E_1 \subset E$ and $\delta J(u_0) = \mathbf{0}$ we have that, also from the Theorem 8.7.1,

$$J(u_0) = \inf_{u \in E} J(u) = \inf_{u \in U} J(u).$$

Consequently, from such a result, from $\hat{v}_1^* \in C^*$ and (8.54) we have that

$$
\begin{aligned}
J(u_0) &= \inf_{u \in E_1} J(u) \\
&= \inf_{u \in U} J(u) \\
&= \inf_{v_1^* \in C^*} \left\{ \sup_{v_0^* \in B^*} J^*(v_1^*, v_0^*) \right\} \\
&= J^*(\hat{v}_1^*, \hat{v}_0^*).
\end{aligned} \tag{8.55}
$$

The proof is complete.

8.8 Numerical results

In this section, we present some numerical results for the following sets

$$\Omega = [-1/2, 1/2] \times [-1/2, 1/2] \times [-1/2, 1/2],$$

and

$$\Omega_1 = [-3/2, 3/2] \times [-3/2, 3/2] \times [-3/2, 3/2].$$

The system of equation in question, namely, the complex Ginzburg-Landau one, is given by

$$
\begin{aligned}
&-\gamma \nabla^2 \phi - 2[i\rho\gamma((\mathbf{A} \cdot \nabla \phi) + \operatorname{div} \mathbf{A}\phi)] \\
&+ \gamma \rho^2 |\mathbf{A}|^2 \phi + \alpha |\phi|^2 \phi - \beta \phi = 0, \text{ in } \Omega,
\end{aligned} \tag{8.56}
$$

$$(\nabla\phi - i\rho\mathbf{A}\phi)\cdot\mathbf{n} = 0, \text{ on } \partial\Omega,$$

$$K_0 \text{ curl curl } \mathbf{A} = K_0 \text{ curl } \mathbf{B}_0 + \tilde{J}, \text{ in } \Omega,$$

$$\text{curl curl } \mathbf{A} = \text{ curl } \mathbf{B}_0, \text{ in } \Omega_1 \setminus \overline{\Omega}.$$

Here,

$$\tilde{J} = -2Re[i\rho\gamma\phi^*\nabla\phi] - \rho^2\gamma|\phi|^2\mathbf{A}.$$

At this point, we start to describe the process concerning the numerical method of lines.

Fixing a starting point $\{\hat{\phi}_0\}$ and $\{(\mathbf{A}_0)_n\}$ and considering the generalized method of lines, we discretize the system in partial finite differences in z, that is, we obtain $N-1$ lines which correspond to the $N-1$ partial differential equations in (x,y).

$$
\begin{aligned}
&-\gamma\frac{(\phi_{n+1} - 2\phi_n + \phi_{n-1})}{d^2} - \gamma\nabla^2\phi_n + K\phi_n - K(\hat{\phi}_0)_n \\
&-2[i\rho\gamma((\mathbf{A}_0)_n\nabla(\hat{\phi}_0)_n + div(\mathbf{A}_0)(\hat{\phi}_0)_n)] \\
&+\rho^2\gamma|(\mathbf{A}_0)_n|^2\phi_n + \alpha|(\hat{\phi}_0)_n|^2\phi_n - \beta\phi_n = 0
\end{aligned}
\tag{8.57}
$$

$\forall n \in \{1,\ldots,N-1\}$, where $d = 1/N$.

With such a equation in mind, we denote

$$\phi_{n+1} - 2\phi_n + \phi_{n-1} - K\phi_n\frac{d^2}{\gamma} + T_n(\phi_n)\frac{d^2}{\gamma} = 0, \tag{8.58}$$

where

$$
\begin{aligned}
T_n(\phi_n) &= \gamma\nabla^2\phi_n + 2[i\rho\gamma((\mathbf{A}_0)_n\nabla(\hat{\phi}_0)_n + \text{ div }(\mathbf{A}_0)_n(\phi_0)_n)] \\
&+K(\hat{\phi}_0)_n - \rho^2\gamma|(\mathbf{A}_0)_n|^2\phi_n - \alpha|(\hat{\phi}_0)_n|^2\phi_n + \beta\phi_n.
\end{aligned}
\tag{8.59}
$$

For $n = 1$, from the boundary condition

$$(\nabla\phi - i\rho\mathbf{A}\phi)\cdot\mathbf{n} = 0$$

at $x = -1/2$ we get

$$\phi_0 = H_1\phi_1,$$

for an appropriate matrix H_1.

Replacing such a relation in (8.58), we obtain

$$\phi_2 - 2\phi_1 + H_1\phi_1 - K\phi_1\frac{d^2}{\gamma} + T_1(\phi_1)\frac{d^2}{\gamma} = 0,$$

so that

$$\phi_1 = a_1 \phi_2 + b_1 T_1(\phi_1)\frac{d^2}{\gamma} + E_1, \tag{8.60}$$

where

$$a_1 = \left(2 + K\frac{d^2}{\gamma} - H_1\right)^{-1},$$

$$b_1 = a_1,$$

$$E_1 = 0.$$

For $n = 2$ replacing (8.60) into (8.58), we obtain

$$\phi_3 - 2\phi_2 + a_1 \phi_2 + b_1 T_1(\phi_1)\frac{d^2}{\gamma}$$

$$-K\phi_2\frac{d^2}{\gamma} + T_2(\phi_2)\frac{d^2}{\gamma} = 0. \tag{8.61}$$

From this, we may write

$$\phi_2 = a_2 \phi_3 + b_2 T(\phi_2)\frac{d^2}{\gamma} + E_2,$$

where

$$a_2 = \left(2 - a_1 + K\frac{d^2}{\gamma}\right)^{-1},$$

$$b_2 = a_2(b_1 + 1)$$

and

$$E_2 = a_2 b_1 (T_1(\phi_1) - T_2(\phi_2))\frac{d^2}{\gamma}.$$

At this point, we remark that the matrix concerning the operator ∇^2 must take into account the boundary conditions in (y,z) at each $n \in \{1,\dots,N-1\}$.
reasoning inductively, having

$$\phi_{n-1} = a_{n-1}\phi_n + b_{n-1}T_{n-1}(\phi_{n-1})\frac{d^2}{\gamma} + E_{n-1}$$

and replacing such an relation into (8.58) we obtain

$$\phi_{n+1} - 2\phi_n + a_{n-1}\phi_n + b_{n-1}T_{n-1}(\phi_{n-1})\frac{d^2}{\gamma} + E_{n-1}$$

$$-K\phi_n\frac{d^2}{\gamma} + T_n(\phi_n)\frac{d^2}{\gamma} = 0. \tag{8.62}$$

Hence,

$$\phi_n = a_n \phi_{n+1} + b_n T_n(\phi_n) \frac{d^2}{\gamma} + E_n$$

where

$$a_n = \left(2 - a_{n-1} + K \frac{d^2}{\gamma} \right)^{-1},$$

$$b_n = a_n(b_{n-1} + 1),$$

$$E_n = a_n b_{n-1}(T_{n-1}(\phi_{n-1}) - T_n(\phi_n)) \frac{d^2}{\gamma} + a_n E_{n-1}.$$

Thus, for $n = N - 1$ for the boundary condition

$$(\nabla \phi - i\rho \mathbf{A}\phi) \cdot \mathbf{n} = 0$$

at $x = 1/2$, we get

$$\phi_N = H_2 \phi_{N-1}$$

for an appropriate matrix H_2.

Hence, from the previous results, with $n = N - 1$, we get

$$\begin{aligned} \phi_{N-1} &= a_{N-1} \phi_N + b_{N-1} T_{N-1}(\phi_{N-1}) \frac{d^2}{\gamma} + E_{N-1} \\ &\approx a_{N-1} H_2 \phi_{N-1} + b_{N-1} T_{N-1}(\phi_{N-1}) \frac{d^2}{\gamma}. \end{aligned} \qquad (8.63)$$

Solving this last linear partial differential equation we obtain ϕ_{N-1}. Having ϕ_{N-1} we obtain ϕ_{N-2} through the equation

$$\phi_{N-2} \approx a_{N-2} \phi_{N-1} + b_{N-2} T_{N-2}(\phi_{N-2}) \frac{d^2}{\gamma}.$$

Having ϕ_{N-2} similarly we obtain ϕ_{N-3} and so on up to finding ϕ_1.

Having $\{\phi_n\}$ the next step is to calculate $\mathbf{A} = \{\mathbf{A}_n\}$ through the linear equations

$$K_0 \text{ curl curl } \mathbf{A} = K_0 \text{ curl } \mathbf{B}_0 + \tilde{J}(\{\phi_n\}, \mathbf{A}), \text{ in } \Omega,$$

$$\text{curl curl } \mathbf{A} = \text{ curl } \mathbf{B}_0, \text{ in } \Omega_1 \setminus \overline{\Omega}.$$

The idea here is to fix the Gauge of London through the equation

$$\text{div } \mathbf{A} = 0 \text{ in } \Omega_1,$$

with the boundary conditions

$$\mathbf{A} \cdot \mathbf{n} = 0, \text{ on } \partial\Omega_1.$$

Finally, we replace $\hat{\phi}_0$ and $\mathbf{A}_0$ by $\{\phi_n\}$ and $\{\mathbf{A}_n\}$ and repeat the process until an appropriate convergence criterion is satisfied.

8.8.1 A numerical example

We present numerical results for $\gamma = \alpha = \beta = K(0) = 1$. In this example,

$$\mathbf{B}_0(x,y,y) = B_0(f(x,y)\mathbf{i} + f(x,y)\mathbf{j})$$

where

$$f(x,y) = (-3/2+x)^2(-3/2+y)^2(-3/2+z)^2(x+3/2)(y+3/2)(z+3/2)/3^6$$

and

$$B_0 = 0.008.$$

For the solution $|\phi(x,y,0)|^2$ at the section $z = 0$, please see Figure 8.1.
For the solutions of $A_1(x,y,0)$ and $A_2(x,y,0)$, please see Figures 8.2 and 8.3.

For $B_0 = 0.031$, for the solution $|\phi(x,y,0)|^2$ at the section $z = 0$, please see Figure 8.4.
For such a B_0 value, for the solutions of $A_1(x,y,0)$ and $A_2(x,y,0)$ please see Figures 8.5 and 8.6.

Remark 8.2   We observe that for both values of B_0, the effect of a magnetic field on the $|\phi|^2$ distribution is more present close to the boundaries of Ω. Also, as expected, the higher value of B_0 corresponds to more decreasing in the $|\phi|^2$ distribution on its domain. We recall that $|\phi|^2$ is point-wise the proportion of electrons along the sample in the super-conducting state. It is always expected for $|\phi|^2$ point-wise, a

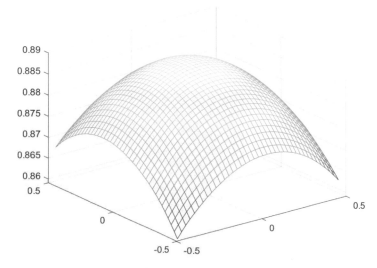

Figure 8.1: Solution $|\phi|^2(x,y,0)$ of the section $z = 0$ for $B_0 = 0.008$.

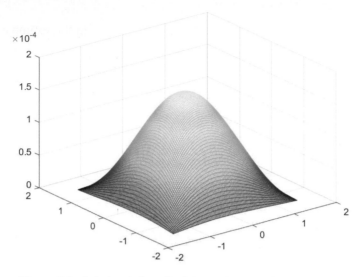

Figure 8.2: Solution $A_1(x, y, 0)$ of the section $z = 0$ for $B_0 = 0.008$.

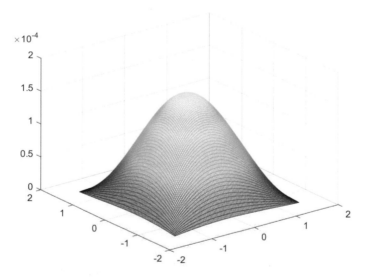

Figure 8.3: Solution $A_2(x, y, 0)$ of the section $z = 0$ for $B_0 = 0.008$.

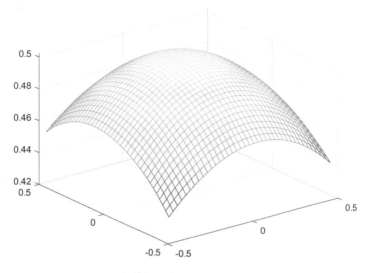

Figure 8.4: Solution $|\phi|^2(x,y,0)$ of the section $z = 0$ for $B_0 = 0.031$.

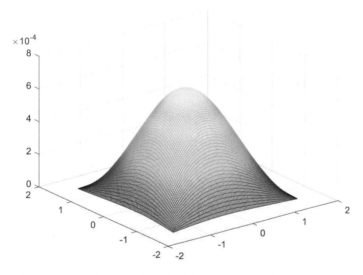

Figure 8.5: Solution $A_1(x,y,0)$ of the section $z = 0$ for $B_0 = 0.031$.

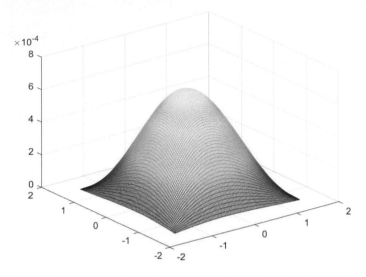

Figure 8.6: Solution $A_2(x, y, 0)$ of the section $z = 0$ for $B_0 = 0.031$.

value between 0 and 1, with $\phi = 0$ corresponding to the normal state and $\phi = 1$ corresponding to the super-conducting state. ∎

Chapter 9

A Classical Description of the Radiating Cavity Model in Quantum Mechanics

9.1 Introduction

In this final chapter, we address a solution for a problem presented in the first chapter of the David Bohm book on quantum theory [12] (our main reference in this chapter), namely, the problem of obtaining the equilibrium distribution of electromagnetic radiation in a hollow cavity.

We present a classical description of such a result, originally established through the Planck hypothesis of radiation quantization in order to fit the theoretical results with the experimental data.

Consider a cavity of dimensions $L \times L \times L$ wherein the walls are radiating electromagnetic energy so that the amount of energy between the frequencies v and $v + \Delta v$ is approximately $U(v)\Delta v$.

In [12], Chapter 1, it is indicated the intensity of radiation per unit of solid angle, coming through the hole, is given by

$$I(v) = \frac{c}{4\pi}U(v),$$

where c is the speed of light in a vacuum. Observe that, at low frequencies, we have experimentally the validity of the Rayleigh-Jeans law

$$U(v)dv \propto KTv^2 dv,$$

as predicted by classical physics.

For higher frequencies, we have,

$$U(v)dv \propto v^3 e^{\frac{-\hbar v}{KT}} dv$$

where $K > 0$ is the Boltzmann constant and T is the temperature. This is the Wien Law, which is in accordance with experimental data for higher frequencies. Planck succeeded in unifying two results such as those given above through his hypothesis of quantization of the radiation energy, assuming

$$E_n = n\hbar v, \ \forall n \in \mathbb{N}.$$

From such an assumption, he got the following expression

$$U(v) = \frac{8\pi V}{c^3} \hbar v^3 \left(\frac{e^{-\frac{\hbar v}{KT}}}{1 - e^{-\frac{\hbar v}{KT}}} \right), \tag{9.1}$$

where $V = L^3$

Such an expression for $U(v)$ is in good agreement with experimental data for all ranges of frequencies.

Remark 9.1 The main reference for this work is [12]. Another important reference on statistical physics is [40]. Regarding function spaces and Lagrange multipliers theory, we would cite [1, 13]. Finally, reference [37] addresses the mathematical foundations of quantum mechanics. ∎

9.2 Regarding an approximate classical description of the last section previous result

The idea here is to re-obtain an approximation of the formula (9.1), through a classical physics argumentation, which leads to a natural quantization process. Denoting $\Omega = [0, L] \times [0, L] \times [0, L]$, $x = (x_1, x_2, x_3)$ and $dx = dx_1 dx_2 dx_3$, the concerning electromagnetic energy is defined by $J : U \to \mathbb{R}$,

$$
\begin{aligned}
J(\mathbf{A}) \ = \ & -\frac{1}{8\pi c^2} \int_0^{T_f} \int_\Omega \left(\frac{\partial \mathbf{A}}{\partial t} \right)^2 dx dt \\
& + \frac{1}{8\pi} \int_0^{T_f} \int_\Omega |\operatorname{curl} \mathbf{A}|^2 dx dt
\end{aligned} \tag{9.2}
$$

where $[0, t_f]$ is a time interval, the magnetic potential $\mathbf{A} \in U$, where

$$U = \{\mathbf{A} \in C^1(\Omega \times [0, T_f]; \mathbb{R}^3)$$
$$\text{such that } \mathbf{A}(x, 0) = \mathbf{A}(x, T_f) = 0, \text{ in } \Omega\}. \tag{9.3}$$

We consider the problem of finding a critical point of J subject to

$$\int_0^{T_f} \int_\Omega |\mathbf{A}|^2 \, dxdt = 1,$$

which corresponds to normalize and restrict the amount of energy. Hence, already including a Lagrange multiplier, the final Lagrangian functional $J(\mathbf{A}, \lambda)$ is expressed by

$$\begin{aligned} J(\mathbf{A}, \lambda) = &-\frac{1}{8\pi c^2} \int_0^{T_f} \int_\Omega \left(\frac{\partial \mathbf{A}}{\partial t}\right)^2 \, dxdt \\ &+ \frac{1}{8\pi} \int_0^{T_f} \int_\Omega |\operatorname{curl} \mathbf{A}|^2 \, dxdt \\ &+ \frac{\lambda}{2}\left(\int_0^{T_f} \int_\Omega |\mathbf{A}|^2 dxdt - 1\right) \end{aligned} \tag{9.4}$$

The corresponding Euler Lagrange equations for such a function are given by

$$\frac{1}{4\pi c^2}\frac{\partial^2 \mathbf{A}}{\partial t^2} + \frac{1}{4\pi}\operatorname{curl} \operatorname{curl} \mathbf{A} + \lambda \mathbf{A} = \mathbf{0},$$

in $\Omega \times [0, T]$. with the boundary conditions

$$\operatorname{curl} \mathbf{A} \cdot \mathbf{n} = 0, \text{ on } \partial\Omega \times [0, T_f],$$

where $\mathbf{n}$ denotes the outward normal to the box surface in question. We shall look for a concerning solution which satisfies

$$\operatorname{div} \mathbf{A} = 0, \text{ in } \Omega \times [0, T_f],$$

corresponding to fix the gauge. So, more specifically, recalling that in such a case

$$\operatorname{curl} \operatorname{curl} \mathbf{A} = -\nabla^2 \mathbf{A},$$

we shall look for a solution

$$A_k = T_n(t)X_k(x_1)Y_k(x_2)Z_k(x_3), \ \forall k \in \{1, 2, 3\}.$$

Through such a separation of variables procedure, we may find a solution such that

$$\frac{T_n''(t)}{4\pi c^2} + \lambda_n T_n(t) = 0,$$

$$T(0) = T(T_f) = 0,$$

$$X_k(x_1) = c_k, \; Y_k(x_2) = d_k, \; Z_k(x_3) = e_k, \; \forall k \in \{1,2,3\},$$

In particular

$$\operatorname{div} \mathbf{A} = 0,$$

and

$$T_n(t) = a_n \sin\left(\frac{2\pi nt}{T_f}\right),$$

where

$$4\pi c^2 \lambda_n = \frac{4\pi^2 n^2}{T_f^2},$$

so that

$$\lambda_n = \frac{\pi n^2 v^2}{c^2},$$

where for a period T_f the corresponding frequency is

$$v = 1/T_f.$$

Observe that the constants are such that

$$
\begin{aligned}
E_n &= \frac{1}{8\pi c^2} \int_0^{T_f} \int_\Omega \left(\frac{\partial \mathbf{A}}{\partial t}\right)^2 dxdt \\
&\quad - \frac{1}{8\pi} \int_0^{T_f} \int_\Omega |\operatorname{curl} \mathbf{A}|^2 \, dxdt \\
&= \frac{\lambda_n}{2} \int_0^{T_f} \int_\Omega |\mathbf{A}|^2 \, dxdt \\
&= \frac{\lambda_n}{2} \\
&= \frac{\pi n^2 v^2}{2c^2}.
\end{aligned}
\tag{9.5}
$$

Denote now

$$B_n = \frac{\sqrt{E_n}}{\alpha F(T)},$$

with $\alpha \in \mathbb{R}$ and $F(T)$ to be specified in the next lines. For N possible degrees of freedom, consider the statistics where defining for $n \in \mathbb{N} \cup \{0\}$, fixed,

$$W_n^N = \frac{0! B_n^0 + 1! B_n^1 + 2! B_n^2 + \cdots + N! B_n^N}{N! B_n^N},$$

which takes into account two identical degrees of freedom. Letting $N \to \infty$, we obtain

$$
\begin{aligned}
W_n &= \lim_{N \to \infty} W_n^N \\
&= e^{-B_n} \\
&= e^{-\frac{\sqrt{E_n}}{\alpha F(T)}},
\end{aligned}
\tag{9.6}
$$

Normalizing such a density, we get

$$
\hat{W}_n = \frac{W_n}{\sum_{n=0}^{\infty} W_n} = \frac{e^{-B_n}}{\sum_{n=0}^{\infty} e^{-B_n}},
$$

so that

$$
\hat{W}_n = \frac{e^{-\frac{\sqrt{E_n}}{\alpha F(T)}}}{\sum_{n=0}^{\infty} e^{-\frac{\sqrt{E_n}}{\alpha F(T)}}},
$$

that is

$$
\hat{W}_n = \frac{e^{-n\nu H(\alpha,T)}}{\sum_{n=0}^{\infty} e^{-n\nu H(\alpha,T)}} = e^{-n\nu H(\alpha,T)} \left(1 - e^{-\nu H(\alpha,T)} \right).
$$

where

$$
H(\alpha,T) = \frac{\sqrt{\pi}}{\alpha F(T) c \sqrt{2}}.
$$

Hence, the average

$$
\begin{aligned}
\overline{B} &= \overline{\left(\frac{\sqrt{E}}{\alpha F(T)} \right)} \\
&= \sum_{n=0}^{\infty} \frac{\sqrt{E_n}}{\alpha F(T)} \hat{W}_n \\
&= \left(1 - e^{-\nu H(\alpha,T)} \right) \left(\sum_{n=0}^{\infty} \left[e^{-n\nu H(\alpha,T)} n\nu H(\alpha,T) \right] \right).
\end{aligned}
\tag{9.7}
$$

Recalling that the calculation in the next lines may be found in Chapter 1 of the book [12], at this point we observe that denoting

$$
S = \sum_{n=0}^{\infty} n e^{-n\gamma},
$$

we have

$$
\begin{aligned}
S &= \sum_{n=0}^{\infty} n e^{-n\gamma} \\
&= -\sum_{n=0}^{\infty} \left(\frac{d e^{-\gamma n}}{d\gamma} \right) \\
&= -\frac{d}{d\gamma} \left(\sum_{n=0}^{\infty} e^{-n\gamma} \right) \\
&= -\frac{d}{d\gamma} \left(\frac{1}{1 - e^{-\gamma}} \right) \\
&= \frac{e^{-\gamma}}{(1 - e^{-\gamma})^2}.
\end{aligned}
\tag{9.8}
$$

From such a result and (9.7), we obtain

$$
\overline{\left(\frac{\sqrt{E}}{\alpha F(T)} \right)} = \frac{e^{-\nu H(\alpha,T)}}{\left(1 - e^{-\nu H(\alpha,T)} \right)} H(\alpha,T)\nu.
$$

It follows, from the expected expression

$$
U(\nu) \approx \frac{8\pi V}{c^3} \nu^2 E,
$$

we obtain

$$
U(\nu) \approx \frac{8\pi V}{c^3} \nu^2 \beta G(T) \left(\overline{\left(\frac{\sqrt{E}}{\alpha F(T)} \right)} \right)^2
$$

for an appropriate function $G(T)$ and appropriate $\beta \in \mathbb{R}$, so that

$$
U(\nu) \approx \frac{8\pi V}{c^3} \nu^4 H(\alpha,T)^2 \beta G(T) \left(\frac{e^{-\nu H(\alpha,T)}}{1 - e^{-\nu H(\alpha,T)}} \right)^2.
$$

In such a case, the functions $F(T), G(T)$ and the constants $\alpha, \beta \in \mathbb{R}$, must be such that this last expression better fits the experimental data, that is, α, β and $F(T)$, $G(T)$ must be obtained through the solution of an appropriate inverse optimization problem.

9.3 Conclusion

In this chapter, we have obtained a classical description of the problem of radiating cavity. Such a model was one of the first examples of quantum mechanics nature. As mentioned above, we re-obtain an approximation for such a result through a classical approach.

References

[1] R.A. Adams and J.F. Fournier, Sobolev Spaces, 2nd edn. (Elsevier, Oxford, UK, 2003).

[2] R.A. Adams, *Sobolev Spaces,* Academic Press, New York, 1975.

[3] J.F. Annet, Superconductivity, Superfluids and Condensates, 2nd edn. (Oxford Master Series in Condensed Matter Physics, Oxford University Press, New York, Reprint, 2010).

[4] T.M. Apostol, Calculus, Vol. 2, 2nd edition, Wiley India, 2007.

[5] R. Abraham, J.E. Marsden and T. Ratiu, Manifolds, Tensor Analysis and Applications, Applied Mathematical Sciences, Vol. 75, 2nd edition, Springer, New York, 1988.

[6] G. Bachman and L. Narici, *Functional Analysis*, Dover Publications, Reprint, 2000.

[7] W.R. Bielski, A. Galka and J.J. Telega, The Complementary Energy Principle and Duality for Geometrically Nonlinear Elastic Shells. I. Simple case of moderate rotations around a tangent to the middle surface. Bulletin of the Polish Academy of Sciences, Technical Sciences, 38(7-9), 1988.

[8] W.R. Bielski and J.J. Telega, A Contribution to Contact Problems for a Class of Solids and Structures, Arch. Mech., 37(4-5): 303–320, Warszawa 1985.

[9] F.S. Botelho, Functional Analysis and Applied Optimization in Banach Spaces, Springer Switzerland, 2014.

[10] F.S. Botelho, Functional Analysis, Calculus of Variations and Numerical Methods in Physics and Engineering, Taylor and Francis, Florida, 2020.

[11] D. Bohm, A Suggested Interpretation of the Quantum Theory in Terms of Hidden Variables I, Phys. Rev. 85(2), 1952.

[12] D. Bohm, Quantum Theory (Dover Publications INC., New York, 1989).

[13] F.S. Botelho, Functional Analysis and Applied Optimization in Banach Spaces, Springer Switzerland, 2014.

[14] F.S. Botelho, Functional Analysis, Calculus of Variations and Numerical Methods in Physics and Engineering, CRC Taylor and Francis, Florida, 2020.

[15] F.S. Botelho, *Variational Convex Analysis*, Ph.D. thesis, Virginia Tech, Blacksburg, VA-USA, 2009.

[16] F.S. Botelho, Variational Convex Analysis, Applications to Non-Convex Models, Lambert Academic Publishing, Berlin, June, 2010.

[17] F.S. Botelho, *Dual Variational Formulations for a Non-linear Model of Plates*, Journal of Convex Analysis, 17(1): 131–158, 2010.

[18] F.S. Botelho, *Existence of solution for the Ginzburg-Landau system, a related optimal control problem and its computation by the generalized method of lines*, Applied Mathematics and Computation, 218: 11976–11989, 2012.

[19] F.S. Botelho, *Topics on Functional Analysis, Calculus of Variations and Duality*, Academic Publications, Sofia, 2011.

[20] F.S. Botelho, *On Duality Principles for Scalar and Vectorial Multi-Well Variational Problems*, Nonlinear Analysis 75: 1904–1918, 2012.

[21] F.S. Botelho *On the Lagrange multiplier theorem in Banach spaces*, Computational and Applied Mathematics, 32: 135–144, 2013.

[22] F.S. Botelho, A Classical Description of Variational Quantum Mechanics and Related Models, Nova Science Publishers, New York, 2017.

[23] F.S. Botelho, Real Analysis and Applications (Springer Switzerland, 2018).

[24] F.S. Botelho, A variational formulation for the relativistic Klein-Gordon equation, Ciência e Natura, V. 40: e57, 2018.

[25] F.S. Botelho, A variational formulation for relativistic mechanics based on Riemannian geometry and its application to the quantum mechanics context, arXiv:1812.04097v2[math.AP], 2018.

[26] H. Brezis, Functional Analysis, Sobolev Spaces and Partial Differential Equations, Springer New York, 2010.

[27] P. Ciarlet, *Mathematical Elasticity*, Vol. I—Three Dimensional Elasticity, North Holland, Elsevier, 1988.

[28] P. Ciarlet, *Mathematical Elasticity*, Vol. II—Theory of Plates, North Holland Elsevier, 1997.

[29] P. Ciarlet, *Mathematical Elasticity*, Vol. III—Theory of Shells, North Holland Elsevier, 2000.

[30] F. Clarke, Functional Analysis, Calculus of Variations and Optimal Control, Springer New York, 2013.

[31] I. Ekeland and R. Temam, *Convex Analysis and Variational Problems*. North Holland, Elsevier, 1976.

[32] L.C. Evans, *Partial Differential Equations*, Graduate Studies in Mathematics, 19, AMS, 1998.

[33] M. Giaquinta and S. Hildebrandt, Calculus of Variations I, a series of comprehensive studies in mathematics, Vol. 310, Springer, Berlin, 1996.

[34] M. Giaquinta and S. Hildebrandt, Calculus of Variations II, a series of comprehensive studies in mathematics, Vol. 311, Springer, Berli, 1996.

[35] E. Giusti, *Direct Methods in the Calculus of Variations*, World Scientific, Singapore, Reprint, 2005.

[36] L. Godinho and J. Natario, An Introduction to Riemannian Geometry, with Applications to Mechanics and Relativity, Springer Switzerland, 2014.

[37] B. Hall, Quantum Theory for Mathematicians (Springer, New York, 2013).

[38] K. Ito and K. Kunisch, Lagrange Multiplier Approach to Variational Problems and Applications, Advances in Design and Control, SIAM, Philadelphia, 2008.

[39] E.L. Lima, Variedades Diferenciáveis, Impa publicações, Rio de Janeiro, 2011.

[40] L.D. Landau and E.M. Lifschits, Course of Theoretical Physics, Vol. 5—Statistical Physics, Part 1. (Butterworth-Heinemann, Elsevier, Reprint, 2008).

[41] D.G. Luenberger, *Optimization by Vector Space Methods*, John Wiley and Sons, Inc., New York, 1969.

[42] M. Reed and B. Simon, *Methods of Modern Mathematical Physics, Volume I, Functional Analysis*, Reprint Elsevier (Singapore, 2003).

[43] S.M. Robinson, *Strongly regular generalized equations*, Math. of Oper. Res., 5: 43–62, 1980.

[44] R.T. Rockafellar, *Convex Analysis*, Princeton Univ. Press, New Jersey, 1970.

[45] W. Rudin, *Functional Analysis*, second edition, McGraw-Hill, New York, 1991.

[46] W. Rudin, *Real and Complex Analysis*, third edition, McGraw-Hill, New York, 1987.

[47] H. Royden, *Real Analysis*, third edition, Prentice Hall, India, 2006.

[48] E.M. Stein and R. Shakarchi, *Real Analysis*, Princeton Lectures in Analysis III, Princeton University Press, New Jersey, 2005.

[49] J.C. Strikwerda, *Finite Difference Schemes and Partial Differential Equations*, SIAM, second edition, Philadelphia, 2004.

[50] J.F. Toland, *A duality principle for non-convex optimisation and the calculus of variations*, Arch. Rath. Mech. Anal., 71(1): 41–61, 1979.

[51] J.L. Troutman, *Variational Calculus and Optimal Control*, second edition, Springer, New York, 1996.

[52] R.M. Wald, General Relativity, University of Chicago Press, Chicago, 1984.

[53] S. Weinberg, Gravitation and Cosmology, Principles and Applications of the General Theory of Relativity, Wiley and Sons (Cambridge, Massachusetts, 1972).

Index

A

action 190, 191, 196, 201, 202, 204, 206
advanced calculus 1
angular momentum 148–151, 272–274
angular momentum operator 272, 274
atomic model 234, 252, 253
atoms 148, 164, 170, 176, 182, 183, 188, 206, 252, 253, 257, 265

B

base 102, 103
basic 61
bilinear form 89
Bohmian mechanics 129, 234
Bohr atomic model 234
Boltzmann constant 312
boundary conditions 138, 203, 241, 278, 304–306, 313
bounded operator 289
bounded sequence 218
bounded set 5, 38, 52, 98, 120, 129, 139, 144, 192, 196, 204, 207, 235, 257, 277, 279, 282, 288, 292, 297, 299

C

calculus of variations 280
Cauchy sequence 17, 95, 219
causal structure 214, 233
cavity model 311
closed set 37, 41
closure 120
compact set 98, 223
cone 215

conjugate 213
connected set 43, 92, 129, 139, 144, 192, 196, 204, 207, 216, 235, 257, 277, 279, 282, 288, 292, 297, 299
connection 140, 189, 194–196, 212
constraints 32, 142, 145, 146, 155, 158, 161, 167, 168, 171, 179, 201, 252, 255
continuity 6, 11, 18, 20, 23, 69, 215, 225
continuous function 11, 41, 92, 95, 96, 111
convergence 219, 220, 230, 306
convergent 162
convex function 280
convex set 18, 44
curvature 50, 52, 53, 56, 59, 60, 128, 130, 141, 145, 190, 192, 199, 201, 204, 212, 234, 235, 242, 254
curve 46, 48–53, 55, 70, 89–93, 97, 118, 122, 215–218, 220–226

D

David Bohm 129, 234, 311
density field 254, 271
density functional theory 160, 161
derivative 4, 7, 11, 13, 15, 19, 42, 43, 47, 79, 80, 89–91, 118, 129, 160, 194, 235, 242
differentiability 24, 42, 43
differentiable function 16, 44, 45, 48, 61, 66, 72, 151, 289
differential 46, 49, 61, 92, 98, 100–102, 105, 107, 111, 123, 126, 133, 135, 197, 198, 207, 209, 214, 233, 238, 245, 248, 304, 306

differential forms 49, 61, 98, 102, 105, 107, 111, 123, 126
differential geometry 46, 207, 214, 233
divergence theorem 116, 117, 119, 120
domain 11, 42, 43, 79, 117, 178, 185, 223, 224, 280, 307
dual space 81, 193
duality 253, 275, 279, 282, 288, 292, 299

E

Einstein equations 190, 192, 202, 204, 247
electric field 157, 248–250
electron 145, 148, 164, 166, 171, 176, 177, 182–185, 253, 254, 257, 268
electronic field 144, 248
energy 128–133, 135, 138, 140, 141, 145, 147, 156, 157, 167, 170, 176, 178, 185, 188–190, 198, 199, 201, 203, 206, 209, 212, 213, 234–238, 240–245, 248, 249, 251–255, 262, 263, 275, 276, 311–313
entropy 261–263
equivalent 28, 38, 65, 93, 261
exterior differential 101, 123, 126
exterior product 86

F

field 68, 85–87, 91, 101, 102, 120, 121, 124, 125, 128–130, 139–141, 144, 145, 156, 157, 160, 177, 188–191, 194, 196, 197, 204–210, 214, 226, 233–235, 241, 242, 248–250, 253–255, 257, 259, 262, 263, 268–272, 276, 277, 282, 307
frequency 314
function 2–5, 7–9, 11–13, 15–19, 25, 26, 29, 32, 37, 41, 43–52, 61–64, 66, 67, 70–74, 83, 92, 96, 101, 103, 111, 118, 119, 125, 128, 139, 140, 144, 155, 166, 167, 170, 176, 191, 192, 194, 197, 199, 208, 214, 233, 236, 253, 262, 263, 271, 272, 276, 289, 312, 313, 316
functional 81, 83, 133, 140–142, 146, 147, 151, 157, 158, 160, 161, 167, 168, 170, 171, 180, 186–188, 205, 214, 226, 231, 245, 249, 252, 255, 265, 276, 277, 279, 280, 282, 288, 292, 297, 299, 313

G

general 28, 81, 92, 102, 103, 105, 106, 111, 113, 114, 128, 139, 140, 161, 188, 206, 207, 226, 233, 234, 282
generically 53, 80, 85, 99, 102, 131, 140, 145, 146, 149, 179, 198, 199, 208, 213, 267, 271, 276, 283
global maximum 45
global minimum 45
Green theorem 119

H

Hilbert space 81
hydrogen atom 148, 164, 170, 176, 182

I

imaginary unit 145, 253, 277, 282
immersion 61, 63, 65, 66, 74
implicit function 2, 3, 8, 17, 19, 25, 29, 46, 71
implicit function theorem 2, 3, 8, 17, 19, 25, 29, 46, 71
implicit function theorem vectorial case 19, 29
inequality 15, 16, 19, 21, 33, 95, 97, 278
infinite limit 38
infinite set 38
infinity 167
inner product 45, 140
integral 89–92, 97, 102, 105, 111, 122
integral curve 89–92, 97, 122
integration 102, 105, 107, 250
interior 259
intermediate value theorem 4
interval 92, 129, 141, 167, 170, 178, 191, 197, 204, 207, 235, 241, 252, 263, 313
inverse function 32, 37, 62, 64, 73
inverse function theorem 32, 37, 62, 64, 73

J

Jacobian 13, 65, 68, 69, 197
Jacobian matrix 13, 65, 69, 197

K

Klein-Gordon Equation 128, 132, 135, 139, 188, 202, 204, 206, 238, 241

L

Lagrange multipliers 25, 28, 32, 45, 131,
  142, 146, 160, 168, 171, 179, 201, 205,
  213, 214, 237, 243, 252, 255, 264, 281,
  312, 313
Lie bracket 87, 89, 189, 196, 212
Lie derivative 89, 91
limit 9–12, 38, 39, 46, 47, 232, 259
limit point 38
linear 61, 71, 74–76, 78, 79, 81, 83, 84,
  193, 212, 275, 289, 306
linear function 71, 76, 289
local minimum 25, 26, 28, 29
Lorentz transformation 247, 250
lower bounded 229

M

magnetic field 157, 160, 276, 277, 282, 307
magnetic potential 157, 250, 276, 282, 313
manifold 61, 102, 110, 126, 132, 139, 140,
  191, 207, 208, 212, 214–217, 220, 221,
  223–226, 233, 245
mass 129, 133, 135, 141, 144–146, 158,
  168, 171, 176, 178, 183, 190, 191, 197,
  201, 206, 208, 209, 213, 235, 238, 245,
  248, 254, 264, 268
maximum 45
mean value inequality 15, 16, 19, 21, 33
mean value theorem 5, 6, 16
measure 49
metric 205
minimizer 226, 280, 282
minimum 25, 26, 28, 29, 45

N

necessary conditions 25–27, 109
neighborhood 4, 7, 8, 37, 46, 67, 71, 215,
  223, 224
neutrons 265
Newtonian mechanics 234
norm 215, 252, 282
normal field 121, 128, 130, 139, 140, 145,
  156, 157, 188, 209, 210, 214, 233, 242,
  255, 268
null space 225
numerical 154, 155, 162, 167, 170, 176,
  182, 183, 303, 304, 307

numerical methods 304
numerical results 155, 162, 176, 303, 307

O

one-sided limit 47
open ball 252
open set 3, 16, 19, 25, 28, 32, 35, 37, 38,
  40, 41, 44, 61–67, 70, 72–75, 77, 79, 92,
  96, 98, 103, 120, 125, 217, 223
optimality conditions 29, 32, 281
optimization 26, 29, 109, 141, 252, 280,
  281, 316

P

parametrization 49, 65–67, 69–71, 74, 75,
  77–80, 191
path connected 140
period 314
Planck 311, 312
Planck constant 312
positive definite matrix 45
protons 164, 166, 167, 170, 177, 182, 183,
  185, 252, 257, 259, 265, 268

Q

quantum mechanics 127–129, 139, 140,
  144, 145, 148, 150, 170, 189, 192, 196,
  204, 206, 234, 241, 252, 311, 312, 316

R

radiating cavity 311, 316
range 69, 74, 75, 192, 233, 312
real sequence 222
real set 40, 81, 86, 277, 279, 282, 292, 299
relativistic Klein-Gordon equation 128,
  188, 202, 206
relativistic mechanics 189, 207, 233
relativity 127, 247, 272
Riemannian geometry 189, 192
rotation 150, 151, 272

S

scalar function 10, 114, 116
Schrödinger equation 132, 139, 144, 155,
  204, 241
sequence 17, 18, 93, 95, 162, 165, 217–219,
  222, 227–229, 232, 277, 278

set 3, 5, 8–13, 16, 18, 19, 22, 25, 27, 28, 32, 35, 37, 38, 40, 41, 43, 44, 52, 61–68, 70, 72–75, 77–79, 81, 82, 85–87, 92, 96, 98, 100, 103, 111, 120, 121, 125, 129, 132, 137, 139, 144, 154, 167, 182, 192–194, 196, 204, 207, 216, 217, 223, 224, 235, 239, 244, 248, 257, 261, 269, 270, 277, 279, 282, 288, 289, 292, 297, 299, 303

Sobolev spaces 128, 139, 192, 234

space 63, 71–81, 86, 89, 122, 128, 139, 192, 193, 203, 208, 214–217, 220, 221, 223–226, 233, 234, 276, 288, 312

speed of light 132, 196, 208, 244, 312

spherical coordinates 148, 149, 172, 180, 186, 187, 253

spin 155, 271, 274

spin operator 271

Stokes theorem 61, 102, 113–115, 117, 124, 125

sufficient conditions 205

surface 45, 52, 58, 60, 61, 65–71, 73–75, 77, 78, 80, 81, 83, 85–92, 101, 102, 105, 107, 110, 113, 114, 119–126, 192–196, 224, 226, 313

T

time interval 129, 141, 167, 170, 178, 191, 197, 204, 207, 235, 241, 252, 263, 313

U

uniform convergence 219, 220

uniformly continuous 95, 218

unique 4, 5, 8, 18–20, 22, 23, 29, 34, 71, 78, 81, 82, 97

V

vacuum 204, 208, 312

variational quantum mechanics 127, 128

vector 50–52, 59, 63, 65, 68, 69, 76–78, 81, 85–87, 89, 91, 105, 107, 108, 111, 113, 114, 121, 189, 192, 194, 196, 215, 218, 259, 270, 272, 288

vector analysis 192

vector calculus 114

vector field 68, 85–87, 91, 121, 189, 194, 196, 259

velocity field 269, 270

Volume form 107

W

wave function 128, 139, 140, 144, 155, 166, 167, 170, 182, 183, 199, 208, 262, 263, 271

Z

Zero 43, 56, 59, 62, 65, 108, 167, 170